AF675313

Intelligent Technologies for Vehicular Networks

Intelligent Technologies for Vehicular Networks

Guest Editor

Yolanda Blanco Fernández

MDPI

Basel • Beijing • Wuhan • Barcelona • Belgrade • Novi Sad • Cluj • Manchester

Guest Editor
Yolanda Blanco Fernández
atlanTTic research Center for
Telecommunication
Technologies
University of Vigo
Vigo
Spain

Editorial Office
MDPI AG
Grosspeteranlage 5
4052 Basel, Switzerland

This is a reprint of the Special Issue, published open access by the journal *Electronics* (ISSN 2079-9292), freely accessible at: https://www.mdpi.com/journal/electronics/special_issues/6BQ3FQI446.

For citation purposes, cite each article independently as indicated on the article page online and as indicated below:

Lastname, A.A.; Lastname, B.B. Article Title. *Journal Name* **Year**, *Volume Number*, Page Range.

ISBN 978-3-7258-7384-5 (Hbk)
ISBN 978-3-7258-7385-2 (PDF)
https://doi.org/10.3390/books978-3-7258-7385-2

© 2026 by the authors. Articles in this reprint are Open Access and distributed under the Creative Commons Attribution (CC BY) license. The reprint as a whole is distributed by MDPI under the terms and conditions of the Creative Commons Attribution-NonCommercial-NoDerivs (CC BY-NC-ND) license (https://creativecommons.org/licenses/by-nc-nd/4.0/).

Contents

About the Editor . . . . **vii**

Preface . . . . **ix**

Yolanda Blanco Fernández
Intelligent Technologies for Vehicular Networks
Reprinted from: *Electronics* **2024**, *13*, 3990, https://doi.org/10.3390/electronics13203990 . . . . . **1**

Haochong Chen, Xincheng Cao, Levent Guvenc and Bilin Aksun-Guvenc
Deep-Reinforcement-Learning-Based Collision Avoidance of Autonomous Driving System for Vulnerable Road User Safety
Reprinted from: *Electronics* **2024**, *13*, 1952, https://doi.org/10.3390/electronics13101952 . . . . . **4**

Inseok Song, Prohim Tam, Seungwoo Kang, Seyha Ros and Seokhoon Kim
DRL-Based Backbone SDN Control Methods in UAV-Assisted Networks for Computational Resource Efficiency
Reprinted from: *Electronics* **2023**, *12*, 2984, https://doi.org/10.3390/electronics12132984 . . . . . **20**

Christopher Conrad, Saba Al-Rubaye and Antonios Tsourdos
Intelligent Embedded Systems Platform for Vehicular Cyber-Physical Systems
Reprinted from: *Electronics* **2023**, *12*, 2908, https://doi.org/10.3390/electronics12132908 . . . . . **36**

Kasun Moolikagedara, Minh Nguyen, Wei Qi Yan and Xue Jun Li
Video Blockchain: A Decentralized Approach for Secure and Sustainable Networks with Distributed Video Footage from Vehicle-Mounted Cameras in Smart Cities
Reprinted from: *Electronics* **2023**, *12*, 3621, https://doi.org/10.3390/electronics12173621 . . . . . **64**

Prohim Tam, Seyha Ros, Inseok Song and Seokhoon Kim
QoS-Driven Slicing Management for Vehicular Communications
Reprinted from: *Electronics* **2024**, *13*, 314, https://doi.org/10.3390/electronics13020314 . . . . . . **75**

Sukru Yaren Gelbal, Bilin Aksun-Guvenc and Levent Guvenc
Vulnerable Road User Safety Using Mobile Phones with Vehicle-to-VRU Communication
Reprinted from: *Electronics* **2024**, *13*, 331, https://doi.org/10.3390/electronics13020331 . . . . . . **93**

Ritesh Yaduwanshi, Sushil Kumar, Arvind Kumar, Omprakash Kaiwartya, Deepti, Mohammad Aljaidi and Jaime Lloret
Efficient Route Planning Using Temporal Reliance of Link Quality for Highway IoV Traffic Environment
Reprinted from: *Electronics* **2023**, *12*, 130, https://doi.org/10.3390/electronics12010130 . . . . . . **121**

J. de Curtò, I. de Zarzà, Juan Carlos Cano, Pietro Manzoni and Carlos T. Calafate
Adaptive Truck Platooning with Drones: A Decentralized Approach for Highway Monitoring
Reprinted from: *Electronics* **2023**, *12*, 4913, https://doi.org/10.3390/electronics12244913 . . . . . **143**

Kiseok Kim, Sangmin Lee, Taehoon Yoo and Hwangnam Kim
Vehicular Localization Framework with UWB and DAG-Based Distributed Ledger for Ensuring Positioning Accuracy and Security
Reprinted from: *Electronics* **2023**, *12*, 4756, https://doi.org/10.3390/electronics12234756 . . . . . **167**

Irene Kilanioti, Nikolaos Astrinakis and Symeon Papavassiliou
Content Caching and Distribution Policies for Vehicular Ad-Hoc Networks (VANETs): Modeling and Simulation
Reprinted from: *Electronics* **2023**, *12*, 2901, https://doi.org/10.3390/electronics12132901 . . . . . **188**

About the Editor

Yolanda Blanco Fernández

Yolanda Blanco Fernández is a senior researcher at atlanTTic (Research Center for Telecommunication Technologies at the University of Vigo) and an associate professor specializing in intelligent personalization systems across domains such as audiovisual media and cultural heritage. Her recent work focuses on Deep Learning and NLP, particularly Transformer-based models.

She has published 57 JCR articles, 11 book chapters, and 90+ conference papers, and has participated in multiple national and European research projects. Yolanda also drives technology transfer, with experience in industry collaborations and proof-of-concept initiatives leveraging advanced AI models.

Preface

This Reprint focuses on the latest research and developments in intelligent vehicular networks and intelligent transport systems. Its scope covers critical aspects of data management, security, privacy, and communication in high mobility environments. The purpose of this Reprint is to present innovative approaches and practical solutions that enhance connectivity, safety, and efficiency in modern transportation infrastructures. It is motivated by the rapid evolution of connected and autonomous vehicles and the increasing need for secure, reliable, and scalable vehicular communication networks.

This Reprint is intended for researchers, engineers, and practitioners working in intelligent transportation, vehicular communications, autonomous driving, and related fields, providing them with current insights and methodologies to address emerging challenges in next-generation vehicular systems.

Yolanda Blanco Fernández
Guest Editor

electronics MDPI

Editorial

Intelligent Technologies for Vehicular Networks

Yolanda Blanco Fernández

atlanTTic Research Center for Telecommunication Technologies, University of Vigo, 36310 Vigo, Spain; yolanda@det.uvigo.es

Intelligent transport systems (ITS) [1] have made remarkable strides in several critical areas, including the Internet of Vehicles (IoV) [2,3], vehicular network security [4], vehicular clouds [5–7], and advanced routing protocols [8]. The IoV facilitates real-time data exchange between vehicles and infrastructure, significantly enhancing traffic management and safety [9]. Meanwhile, advanced encryption and intrusion detection systems are pivotal for securing vehicular communication networks [10,11]. Vehicular clouds offer scalable data storage and processing solutions, supporting various applications such as traffic monitoring [12,13]. Enhanced routing protocols optimize travel routes using real-time traffic data. Collectively, these advancements have fostered the development of smart vehicular systems, paving the way for innovations in autonomous driving and advanced driver assistance systems (ADAS) [14,15]. These innovations have markedly improved safety, efficiency, and connectivity within transportation networks [16,17].

This Special Issue addresses crucial gaps in data management, security, privacy, and communication within high-mobility environments through ten groundbreaking papers. Yaduwanshi et al. (Contributor 1) propose a novel geocast routing approach that addresses challenges in highway environments by improving throughput and message delivery despite intermittent connectivity and GPS outages. Enhancements to the veins tool are discussed by Kilanioti et al. (Contributor 2), incorporating new caching and content distribution features along with machine learning algorithms for more effective VANET simulations. In another paper, Conrad et al. (Contributor 3) explore a customizable cyber-physical system (CPS) designed for the development and testing of intelligent vehicular systems, bridging theoretical concepts with practical applications. Song et al. (Contributor 4) use UAVs to extend the coverage of mobile edge computing (MEC) and propose a deep reinforcement learning-based SDN controller (DRL-SDNC) to optimize resource allocation, such as computational resources, bandwidth, and storage. The DRL-SDNC adjusts these resources based on task requirements and network conditions, enhancing efficiency and quality of service in UAV-assisted 5G networks. The approach shows improved resource efficiency compared to traditional methods. Moolikagedara et al. (Contributor 5) introduce a video blockchain framework that enhances security, privacy, and scalability in smart city vehicular networks by safeguarding video data storage and retrieval. This advancement not only improves situational awareness and data integrity but also removes the need for third-party intermediaries. Kim et al. (Contributor 6) present advances in secure positioning for vehicle swarms through a DAG-based distributed ledger combined with ultra-wideband (UWB) positioning, ensuring data integrity and security against tampering. In another paper, de Curtò et al. (Contributor 7) propose a drone-based decentralized framework for truck platooning, utilizing drones for real-time communication to enhance platoon management, collision avoidance, and traffic safety. Tam et al. (Contributor 8) explore network slicing techniques for vehicular communications, leveraging deep learning and reinforcement learning to optimize resource allocation and scheduling. Gelbal et al. (Contributor 9) introduce an Android app utilizing low-energy Bluetooth (BLE) to enhance safety for vulnerable road users (VRUs) by smoothing motion data, predicting movement, and issuing driver warnings. Chen et al. (Contributor 10) contribute with a hierarchical deep reinforcement learning framework to bolster collision avoidance with pedestrians in autonomous driving systems by integrating adaptive control mechanisms.

Future research in intelligent vehicular networks should focus on addressing key challenges related to security and privacy by developing advanced cybersecurity methods and data privacy techniques. Integrating 5G and 6G technologies for enhanced vehicle-to-everything (V2X) communication, creating new communication protocols, and leveraging artificial intelligence for route optimization and anomaly detection are essential. Additionally, investigating IoT interoperability and efficient data management, enhancing vehicle cooperation for autonomous driving, and developing robust testing methods are crucial. Emphasis should also be placed on energy management for electric vehicles, smart charging infrastructure, and the integration of heterogeneous vehicular networks. Edge computing for real-time processing, advanced traffic simulation tools, and the development of global standards and regulatory frameworks are vital for the ongoing advancement and practical implementation of intelligent vehicular networks.

As the Guest Editor, I could not finish without extending my heartfelt gratitude to the authors, reviewers, and the editorial team for their invaluable contributions and support. I trust this Special Issue will inspire continued research and collaboration in the pursuit of intelligent and sustainable vehicular networks.

Conflicts of Interest: The authors declare no conflicts of interest.

List of Contributions:

1. Efficient Route Planning Using Temporal Reliance of Link Quality for Highway IoV Traffic Environment. Available online: https://www.mdpi.com/2079-9292/12/1/130 (accessed on 9 October 2024).
2. Content Caching and Distribution Policies for Vehicular Ad-Hoc Networks (VANETs): Modeling and Simulation. Available online: https://www.mdpi.com/2079-9292/12/13/2901 (accessed on 9 October 2024).
3. Intelligent Embedded Systems Platform for Vehicular Cyber-Physical Systems. Available online: https://www.mdpi.com/2079-9292/12/13/2908 (accessed on 9 October 2024).
4. DRL-Based Backbone SDN Control Methods in UAV-Assisted Networks for Computational Resource Efficiency. Available online: https://www.mdpi.com/2079-9292/12/13/2984 (accessed on 9 October 2024).
5. Video Blockchain: A Decentralized Approach for Secure and Sustainable Networks with Distributed Video Footage from Vehicle-Mounted Cameras in Smart Cities. Available online: https://www.mdpi.com/2079-9292/12/17/3621 (accessed on 9 October 2024).
6. Vehicular Localization Framework with UWB and DAG-Based Distributed Ledger for Ensuring Positioning Accuracy and Security. Available online: https://www.mdpi.com/2079-9292/12/23/4756 (accessed on 9 October 2024).
7. Adaptive Truck Platooning with Drones: A Decentralized Approach for Highway Monitoring. Available online: https://www.mdpi.com/2079-9292/12/24/4913 (accessed on 9 October 2024).
8. QoS-Driven Slicing Management for Vehicular Communications. Available online: https://www.mdpi.com/2079-9292/13/2/314 (accessed on 9 October 2024).
9. Vulnerable Road User Safety Using Mobile Phones with Vehicle-to-VRU Communication. Available online: https://www.mdpi.com/2079-9292/13/2/331 (accessed on 9 October 2024).
10. Deep-Reinforcement-Learning-Based Collision Avoidance of Autonomous Driving System for Vulnerable Road User Safety. Available online: https://www.mdpi.com/2079-9292/13/10/1952 (accessed on 9 October 2024).

References

1. Vijayakumar, S.; Singaravelu, R. Time Aware Long Short-Term Memory and Kronecker Gated Intelligent Transportation for Smart Car Parking. *J. Adv. Res. Appl. Sci. Eng. Technol.* **2024**, *44*, 134–150. [CrossRef]
2. Liang, C. Intelligent monitoring methodology for large-scale logistics transport vehicles based on parallel Internet of Vehicles. *EURASIP J. Wirel. Commun. Netw.* **2023**, *2023*, 75. [CrossRef]
3. Omar, N.; Yaakob, N.; Ahmed, M.E.; Husin, Z.; Ismail, I.; Dafhalla, A.K.Y. Velocity Based Performance Analysis of GreedLea Routing Protocol in Internet of Vehicle (IoV). *J. Adv. Res. Appl. Sci. Eng. Technol.* **2024**, *42*, 199–214. [CrossRef]

4. Kushardianto, N.C.; Ribouh, S.; El Hillali, Y.; Tatkeu, C. Vehicular network anomaly detection based on 2-step deep learning framework. *Veh. Commun.* **2024**, *49*, 100802. [CrossRef]
5. Mistareeh, H.; Tennyson, M.; Salameh, H.B. Enhancing privacy-preserving in vehicular cloud through an incentive-based strategy. *Int. J. Electr. Comput. Eng.* **2024**, *14*, 4078–4087. [CrossRef]
6. Zhu, L.; Tan, L. Task offloading scheme of vehicular cloud edge computing based on Digital Twin and improved A3C. *Internet Things* **2024**, *26*, 101192. [CrossRef]
7. Ajaz, F.; Naseem, M.; Shabaz, M.; Khan, M.A. An architectural view of VANETs cloud: Its models, services, applications and challenges. *Int. J. Web Grid Serv.* **2024**, *20*, 292–341. [CrossRef]
8. Kait, R.; Kaur, S.; Sharma, P.; Ankita, C.; Kumar, T.; Cheng, X. Fuzzy logic-based trusted routing protocol using vehicular cloud networks for smart cities. *Expert Syst.* **2024**, e13561. [CrossRef]
9. Brik, B.; Lagraa, N.; Tamani, N.; Lakas, A.; Ghamri-Doudane, Y. Renting out cloud services in mobile vehicular cloud. *IEEE Trans. Veh. Technol.* **2018**, *67*, 9882–9895. [CrossRef]
10. Deng, J.; Deng, J.; Liu, P.; Wang, H.; Yan, J.; Pan, D.; Liu, J. A Survey on Vehicular Cloud Network Security. *IEEE Access* **2023**, *11*, 136741–136757. [CrossRef]
11. Aloqaily, M.; Otoum, S.; Ridhawi, I.A.; Jararweh, Y. An intrusion detection system for connected vehicles in smart cities. *Ad Hoc Netw.* **2019**, *90*, 101842. [CrossRef]
12. Ranjan Senapati, B.; Mohan Khilar, P.; Ranjan Swain, R. Environmental monitoring through Vehicular Ad Hoc Network: A productive application for smart cities. *Int. J. Commun. Syst.* **2021**, *34*, e4988. [CrossRef]
13. Yuan, H.; Guo, D.; Liao, H.; Wu, R.; Li, J. Enabling Road Detection Tasks via Collaborative Smart Vehicles. *IEEE Internet Things J.* **2022**, *9*, 4269–4279. [CrossRef]
14. Gaouar, N.; Lehsaini, M.; Nebbou, T. CCITL: A cloud-based smart traffic management protocol using intelligent traffic light system in VANETs. *Concurr. Comput. Pract. Exp.* **2023**, *35*, e7686. [CrossRef]
15. Lee, E.K.; Gerla, M.; Pau, G.; Lee, U.; Lim, J.H. Internet of Vehicles: From intelligent grid to autonomous cars and vehicular fogs. *Int. J. Distrib. Sens. Netw.* **2016**, *12*, 1–14. [CrossRef]
16. Sharma, K.; Butler, B.; Jennings, B. Scaling and Placing Distributed Services on Vehicle Clusters in Urban Environments. *IEEE Trans. Serv. Comput.* **2023**, *16*, 1402–1416. [CrossRef]
17. Nkenyereye, L.; Park, Y.; Rhee, K.H. Secure vehicle traffic data dissemination and analysis protocol in vehicular cloud computing. *J. Supercomput.* **2018**, *74*, 1024–1044. [CrossRef]

Disclaimer/Publisher's Note: The statements, opinions and data contained in all publications are solely those of the individual author(s) and contributor(s) and not of MDPI and/or the editor(s). MDPI and/or the editor(s) disclaim responsibility for any injury to people or property resulting from any ideas, methods, instructions or products referred to in the content.

electronics MDPI

Article

Deep-Reinforcement-Learning-Based Collision Avoidance of Autonomous Driving System for Vulnerable Road User Safety

Haochong Chen, Xincheng Cao, Levent Guvenc and Bilin Aksun-Guvenc *

Automated Driving Laboratory, Ohio State University, Columbus, OH 43212, USA; chen.9286@osu.edu (H.C.); cao.1375@osu.edu (X.C.); guvenc.1@osu.edu (L.G.)
* Correspondence: aksunguvenc.1@osu.edu

Abstract: The application of autonomous driving system (ADS) technology can significantly reduce potential accidents involving vulnerable road users (VRUs) due to driver error. This paper proposes a novel hierarchical deep reinforcement learning (DRL) framework for high-performance collision avoidance, which enables the automated driving agent to perform collision avoidance maneuvers while maintaining appropriate speeds and acceptable social distancing. The novelty of the DRL method proposed here is its ability to accommodate dynamic obstacle avoidance, which is necessary as pedestrians are moving dynamically in their interactions with nearby ADSs. This is an improvement over existing DRL frameworks that have only been developed and demonstrated for stationary obstacle avoidance problems. The hybrid A* path searching algorithm is first applied to calculate a pre-defined path marked by waypoints, and a low-level path-following controller is used under cases where no VRUs are detected. Upon detection of any VRUs, however, a high-level DRL collision avoidance controller is activated to prompt the vehicle to either decelerate or change its trajectory to prevent potential collisions. The CARLA simulator is used to train the proposed DRL collision avoidance controller, and virtual raw sensor data are utilized to enhance the realism of the simulations. The model-in-the-loop (MIL) methodology is utilized to assess the efficacy of the proposed DRL ADS routine. In comparison to the traditional DRL end-to-end approach, which combines high-level decision making with low-level control, the proposed hierarchical DRL agents demonstrate superior performance.

Keywords: autonomous driving system; deep reinforcement learning; collision avoidance

1. Introduction

At present, rapid urbanization and technological development have led to a steady rise in privately owned vehicles. This widespread adoption of private cars makes our lives more convenient by letting us travel easily between our workplaces and homes. However, this convenience is shadowed by growing concerns that are mostly due to the accompanying rapid rise in vehicular accidents. According to the World Health Organization's Global Status Report on Road Safety, over 50 million individuals get injured and approximately 1.3 million lives are lost in car accidents globally each year [1]. A significant portion of these accidents, estimated at around 75%, are directly attributable to driver errors, including drowsy driving, driving under the influence (DUI), and distracted driving [2]. This statistic highlights the critical need for interventions aimed at mitigating human error on the roads. How to effectively reduce traffic accidents has become a challenge that modern cities must confront.

The ADS benefits from powerful and robust autonomous driving algorithms, which can significantly reduce car accidents caused by human mistakes. The Society of Automotive Engineers (SAE) categorized autonomous driving into six levels, ranging from level 0, no automation, to level 5, full automation [3]. The high levels of autonomous driving, especially levels 4 and 5, have the potential to significantly reduce accidents caused by human error. This includes algorithms to improve the safety of VRUs in their interactions with autonomous vehicles (AVs). Therefore, the development of AV collision avoidance algorithms for VRU safety has become a popular and promising area of research.

Extensive research has already been conducted in this field [4–7]. Currently, most path planning and collision avoidance research follows two major directions. The first is the optimization-based approach, which conceptualizes path planning and collision avoidance as an optimization problem with specific constraints and aims to plan an optimal collision-free trajectory. Guvenc et al. considered the collision avoidance problem as one of waypoint optimization and applied the Elastic Band algorithm to iteratively generate a new trajectory for autonomous vehicles [8]. Wang et al. further applied this algorithm in a low-speed electric shuttle and modified the parameter to make the ego-vehicle maintain a socially acceptable distance with VRUs [9]. Morsali et al. proposed a Support Vector Machine (SVM)-based spatial–temporal planning method, which, combined with the A* path search algorithm and SVM-based heuristics, can efficiently identify optimal collision-free paths in complex traffic situations [10]. Zhu treated collision avoidance path planning as a quintic splines optimization problem and applied look-up tables to enhance computational efficiency [11]. Chen et al. introduced an enhanced spatio-temporal obstacle avoidance algorithm that benefited from a 3D spatio-temporal grid map to achieve better efficiency and performance compared to the traditional hybrid A* algorithm [12]. Meanwhile, a large number of similar studies have been performed to tackle the same problem [13–16]. However, the optimization-based approach often cannot guarantee satisfactory real-time performance due to computational complexity, and, sometimes, it does not take control feasibility into consideration. Flow-based path planning and collision avoidance has also been the focus of recent research, like the unmanned air mobility study in reference [17], with a simulated annealing and parallel computing solution, and path planning for shared autonomous vehicles in a dynamic traffic network in reference [18].

The second approach is the machine learning method, which considers autonomous driving and collision avoidance tasks as Markov Decision Processes (MDPs) and utilizes the reinforcement learning method to find optimal solutions. Kendall et al. pioneered the application of the DRL framework in autonomous driving, an innovation proposing an end-to-end model structure for autonomous driving [19]. Yurtsever et al. proposed an innovative hybrid DRL framework to develop ADS [20]. Peng et al. developed an end-to-end ADS utilizing a Dueling Double Deep Q-Network (DDDQN) framework. The Open Racing Car Simulator (TORCS) is utilized to validate the method's efficiency and effectiveness [21]. Jaritz et al. introduced an Asynchronous-Actor-Critic-based method for autonomous driving that maps the RGB image from the front camera to driving actions and using a realistic rally game environment for training. The approach demonstrates faster convergence and robust performance compared to other DRL-based end-to-end methods, indicating its potential for practical applications in autonomous vehicles [22]. In order to handle critical pre-accident scenarios in emergency situations, Merola et al. proposed a Deep Q-Network (DQN)-based approach to design an ADS and to train the system to execute emergency maneuvers to minimize or avoid damage [23]. Cao et al. introduced a hierarchical reinforcement and imitation learning (H-REIL) approach for autonomous driving to handle near-accident scenarios. By integrating a low-level imitation learning controller with a high-level reinforcement learning controller, their approach demonstrated the capability of balancing safety and efficiency [24]. Additionally, a substantial amount of research is currently ongoing within machine learning approaches [25–29]. Many studies combine DRL with traditional approaches, intending to leverage DRL to enhance the performance of conventional methods [30,31]. However, a notable disadvantage of the machine-learning-based approach is the instability in model performance under normal traffic conditions due to the absence of hard-coded safety protocols. Moreover, the overall performance of the machine-learning-based approach heavily relies on the quality of the training data.

This paper introduces a novel, hybrid, hierarchical DRL framework aimed at developing high-efficiency collision avoidance controllers for autonomous driving. This innovative controller integrates a PID-based pure pursuit path-tracking controller with a DRL-based collision avoidance agent. In normal, collision-free situations, a low-level path-following controller is employed for precise navigation. Conversely, when vulnerable road users are detected near the vehicle, a high-level DRL collision avoidance controller is activated to help the vehicle avoid the potential collision by decelerating or changing the trajectory. The major contribution of the research is mainly in the integration of traditional optimization-based path-following methods with machine-learning-based collision-avoidance algorithms. This proposed hybrid approach aims to enhance both path tracking and collision avoidance performance, thereby establishing solid groundwork for future research in hybrid control strategies for autonomous vehicles.

2. Methodology

2.1. Vehicle Model

A simplified vehicle dynamic model is used in this paper to guarantee meaningful training results. This model is a longitudinal dynamic augmented linear single-track lateral model. The geometry of the longitudinal vehicle model is illustrated in Figure 1. The symbols in Figure 1 mean (1) F_a = aerodynamic drag caused by headwind with velocity V_{wind}; (2) F_r = rolling resistance; (3) θ = road grade; (4) Mg = vehicle gravitational force; (5) F_g = force component of vehicle gravitational force along the road grade; and (6) F_x = longitudinal tire force. The model's input–output structure is illustrated in Figure 2. It uses throttle and brake pedal actions together with headwind velocity and road grade as inputs and longitudinal velocity as the output. It is worth noting that for the sake of simplicity, the headwind and road grade are assumed to be non-existent/zero. The detailed structure of this longitudinal model is illustrated in Figure 3, where Figure 3a demonstrates the calculation procedure from longitudinal forces to the longitudinal vehicle velocity, and Figure 3b demonstrates the calculation procedure from the input engine and brake torques to the longitudinal tire force. Table 1 lists the parameters that appear in Figure 3.

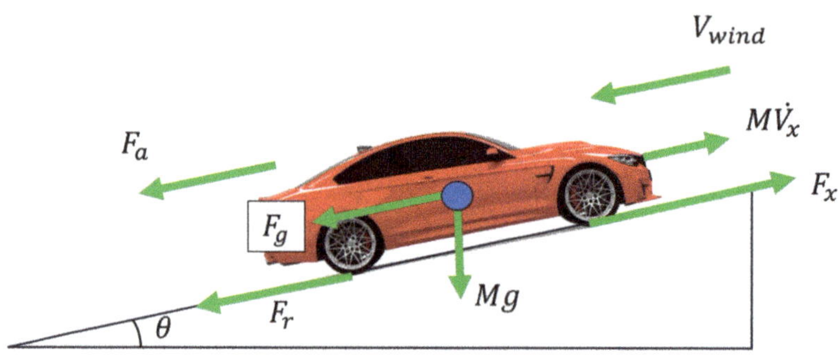

Figure 1. Longitudinal vehicle model geometry.

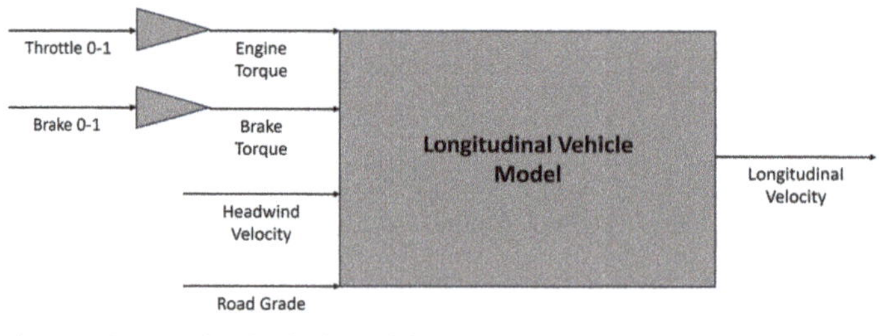

Figure 2. Longitudinal vehicle model input–output structure.

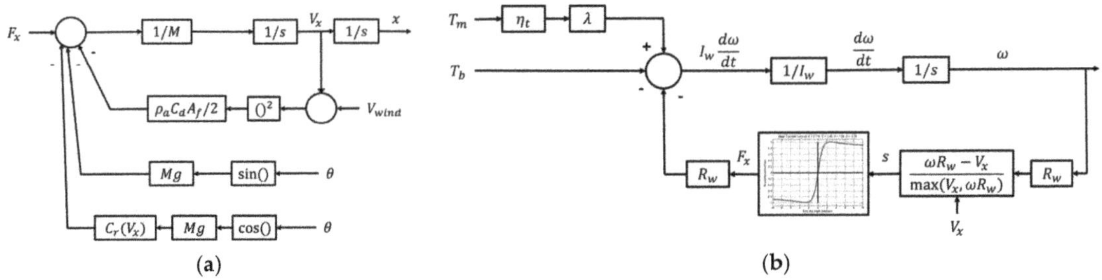

Figure 3. Detailed components in the longitudinal vehicle model. (**a**) Longitudinal forces to longitudinal vehicle velocity calculation. (**b**) Input torques to longitudinal tire force calculation.

Table 1. Longitudinal model parameters.

Symbol	Parameter
F_x	Longitudinal tire force
M	Vehicle mass
V_x	Vehicle longitudinal velocity
X	Vehicle longitudinal position
ρ_a	Air density
C_d	Air drag coefficient
A_f	Vehicle cross-sectional area
V_{wind}	Headwind velocity
θ	Road grade
C_r	Rolling resistance coefficient
T_m	Motor torque
T_b	Brake torque
η_t	Transmission efficiency
λ	Gear ratio
I_w	Wheel moment of inertia
ω	Wheel angular velocity
R_ω	Wheel radius
s	Longitudinal tire slip

The lateral single-track model geometry is displayed in Figure 4, and the state-space form representation of the simplified linear single-track model is illustrated in Equation (1) [8]. The parameters used in this lateral model are listed in Table 2. The inputs are the front and rear wheel steer angle as well as the vehicle yaw disturbance, and the outputs are the vehicle side-slip angle and the vehicle yaw rate. In this case, we assumed a front-wheel-steer vehicle and no external yaw disturbance; thus, the input is simply the front wheel steer angle. It is also worth noting that in this simplified linear model, vehicle's longitudinal speed is treated as being constant.

$$\begin{bmatrix} \dot{\beta} \\ \dot{r} \end{bmatrix} = \begin{bmatrix} \frac{-C_f - C_r}{MV} & -1 + \frac{C_r l_r - C_f l_f}{MV^2} \\ \frac{C_r l_r - C_f l_f}{I_z} & \frac{-C_f {l_f}^2 - C_r {l_r}^2}{I_z V} \end{bmatrix} \begin{bmatrix} \beta \\ r \end{bmatrix} + \begin{bmatrix} \frac{C_f}{MV} & \frac{C_r}{MV} \\ \frac{C_f l_f}{I_z} & \frac{C_r l_r}{I_z} \end{bmatrix} \begin{bmatrix} \delta_f \\ \delta_r \end{bmatrix} + \begin{bmatrix} 0 \\ \frac{1}{I_z} \end{bmatrix} M_{zd} \quad (1)$$

The structure of the overall full vehicle model is displayed in Figure 5. This model uses throttle, brake, and steering actions as its inputs and outputs vehicle X and Y positions as well as the vehicle yaw angle. As mentioned in the previous paragraph, the lateral portion of the model assumes constant speed. As a result, the lateral component must be implemented as a time-varying model, where its parameters can change at each time step.

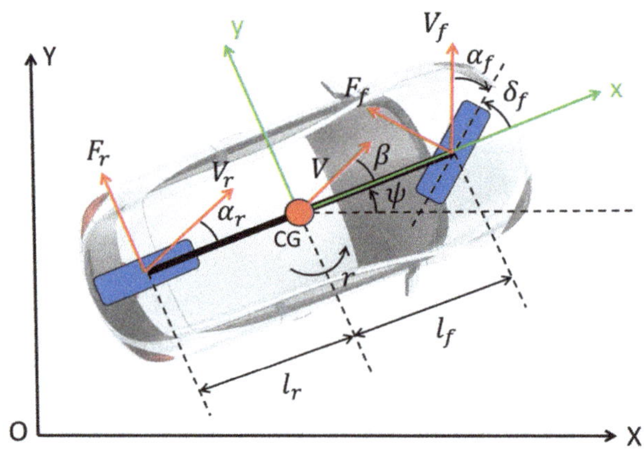

Figure 4. Lateral single-track vehicle model geometry.

Table 2. Lateral model parameters.

Symbol	Parameter
X, Y	Earth-fixed frame coordinate
x, y	Vehicle-fixed frame coordinate
V	Vehicle center-of-gravity (CG) velocity
M	Vehicle mass
I_z	Vehicle yaw moment of inertia
β	Vehicle side-slip angle
ψ	Vehicle yaw angle
r	Vehicle yaw rate
M_{zd}	Yaw disturbance moment
δ_f, δ_r	Front and rear wheel steer angle
F_f, F_r	Front and rear lateral tire force
V_f, V_r	Front and rear axle velocity
α_f, α_r	Front and rear tire slip angle
l_f, l_r	Distance between vehicle CG and front and rear axle
C_f, C_r	Front and rear tire cornering stiffness

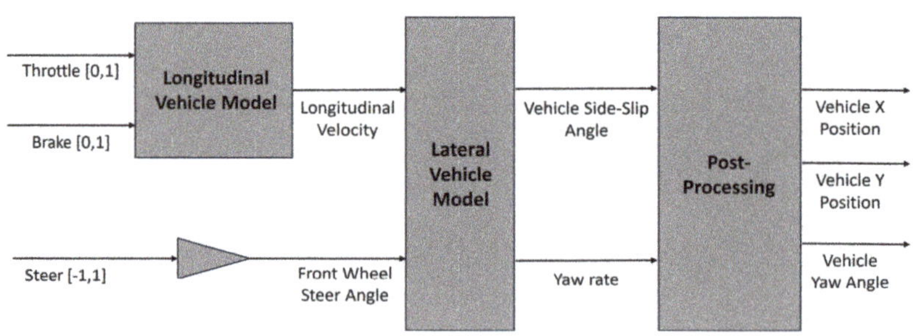

Figure 5. Full vehicle model structure.

2.2. Autonomous Driving System Design

In this paper, an innovative, hybrid, hierarchical, deep-reinforcement-learning-based controller is proposed for autonomous vehicles for tracking a pre-determined path and performing an evasive collision avoidance maneuver when necessary. The architecture of this hybrid controller is designed to leverage the strengths of both the PID pure pursuit controller for precise path-following and the DRL-based collision avoidance controller for safe and efficient collision avoidance. The detailed design of this hybrid control system is presented in the following.

2.2.1. PID Pure Pursuit Controller Design

A PID pure pursuit path-tracking controller is utilized to perform precise path-following under normal traffic conditions. The proposed PID controller consists of a longitudinal PID controller and a lateral PID controller. The longitudinal PID controller is primarily responsible for controlling the vehicle's speed. The difference in speed between the current speed of the vehicle and the desired speed, demonstrated in Equation (2), is used as the input to generate a Speed command. Meanwhile, the lateral PID controller is mainly used for controlling the vehicle's steering. The difference in the angle between the vehicle's current moving direction and the desired path direction, given in Equation (3), is taken as the input to generate a vehicle steering command. Figure 6 demonstrates the method for calculating the angle difference input to the lateral PID controller. The controller gains (K_p, K_i, K_d) of both the longitudinal and lateral PID controllers have been manually turned to obtain reasonable values according to the vehicle model.

$$\varepsilon_{longitudinal} = v_{vehicle} - v_{desired} \tag{2}$$

$$\varepsilon_{lateral} = \theta_{vehicle} - \theta_{path} \tag{3}$$

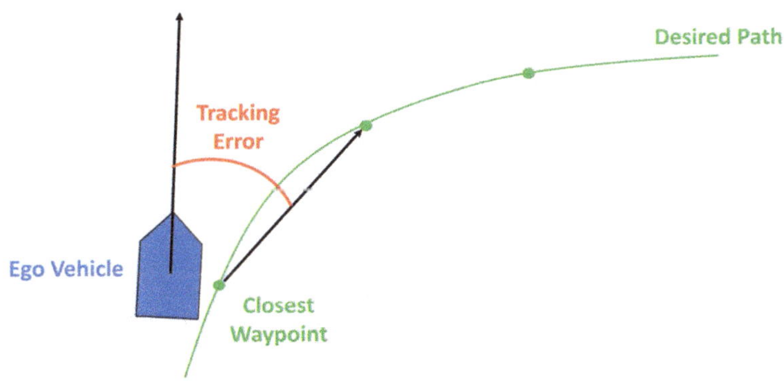

Figure 6. Lateral PID controller path-tracking error.

2.2.2. Markov Decision Process and Deep Reinforcement Learning

The collision avoidance of an autonomous vehicle is a dynamic and continuous decision-making process. Initially, the AV must identify the positions, velocities, and trajectories of nearby road users. Subsequently, based on the vehicle's own location and velocity, the AV must make prompt and accurate decisions to control the vehicle and dodge other road users safely. This indicates that all decisions made by the AV are correlated with the ego-vehicle's current state as well as the states of nearby road users. This process is very similar to an MDP. Given this analogy, the task of collision avoidance can be treated as a classic MDP, wherein MDPs are employed to model and address the uncertainties inherent in traffic environments. To tackle this complex MDP, the deep reinforcement learning method is utilized to design an autonomous driving system, which can make optimal decisions that aim to maximize expected rewards and enhance the safety and efficiency of the collision avoidance procedure. The fundamental elements of an MDP consist of states (S), actions (A), transition probabilities (P), and rewards (r). The setup of the MDP used in this paper is demonstrated as follows.

- State space (S): The state space contains a collection of states that represent the current traffic environment's information. Each state within the state space consists of four essential components. Firstly, an occupancy grid represented by a 2D array is utilized to map the surrounding obstacles relative to the vehicle. The occupancy grid can identify other road users as obstacles within a specified range and assign a weight to each grid based on the importance of nearby road users. A relatively higher weight will be assigned to vulnerable road users (VRUs). In this paper, the detection ranges are defined as 20 m forward, 5 m backward, and 7 m to the sides. Figures 7 and 8 demonstrate the occupancy grid for a vehicle with zero yaw angle orientation and non-zero yaw angle orientation, respectively. In the figure, the cross symbols are used to indicate the proposed occupancy grid: white crosses for collision-free areas, black for potential collisions with the pedestrian, red for the vehicle's geometric center, and blue for the vehicle's vertical and horizontal coordinates. In practical implementations, the occupancy grid is typically derived through the fusion of data from multiple sensors, such as the lidar, camera, and radar. However, the purpose of this study is to validate the proposed hybrid controller. Thus, to simplify the data collection procedure, the occupancy grid array data are extracted and processed from the simulation environment through frame transformation techniques, enabling precise collision detection for each array point with obstacles. The second component contains the ego-vehicle's data, such as its location, orientation, and velocity. The third component contains information regarding the pre-calculated path, including target-tracking waypoints. Lastly, the fourth component contains obstacle information, including the vehicle time-to-collision-zone (TTZ_v), the pedestrian time-to-collision-zone (TTZ_p), and the difference between vehicle and pedestrian time-to-collision-zones (TTZ_diff).

- Action space (*A*): The action space contains a collection of discrete actions available to the ego-vehicle in response to varying traffic scenarios. The action is defined as a control command tuple consisting of steering, throttle, and brake commands. The configuration of the action space is designed according to the specific requirements of different test cases.
- Transition model (*P*): The transition model is the key component of traffic simulation designed to simulate next states based on the execution of a given action at the current state and the transition probability. This paper divides the transition model into two major parts. Firstly, a SIMULINK vehicle model simulates the motion of the ego-vehicle. At the same time, the CARLA simulator is used to simulate the traffic environment and the motion of other road users. The details of the vehicle model have been discussed in previous sections, while the details of the traffic simulator are elaborated within the case study section.
- Reward (*r*): A reward function is used to calculate the immediate reward for each step based on the transition from the current state to next state after executing specific actions. The details of the reward function's design are elaborated within the case study section.

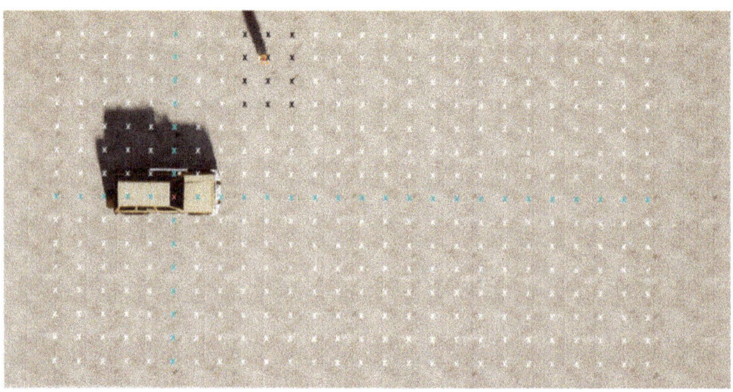

Figure 7. Occupancy grid 2D array with zero yaw angle.

Figure 8. Occupancy grid 2D array with 30° yaw angle.

To design a high-performance collision avoidance system, the Double Deep Q-Network (DDQN) approach is utilized to address the previously outlined MDP and seek an optimal policy that maximizes cumulative rewards. The DDQN is an enhanced version of the DQN, which itself is a breakthrough in the field of reinforcement learning. DQN, developed by DeepMind Technologies in 2013, successfully combined Q-learning, a popular reinforcement learning algorithm, with deep neural networks to handle high-dimensional state spaces [32–34]. DQN's architecture contains many important innovations, such as the replay buffer and two distinct neural networks, the Q-network and the target-network. However, one of the limitations of the original DQN model is its tendency to overestimate action values due to the maximization step in the Q-learning update. This overestimation can lead to suboptimal policy and unstable learning. Hasselt et al. proposed DDQN in 2015, which successfully addressed this issue by decoupling the action selection from the target Q-value generation during the learning process [34]. Like its predecessor, DDQN leverages the strength of both the neural network and Q-learning, and it is designed to address complex MDP problems. DDQN stands out from conventional collision avoidance algorithms, such as the optimization-based approach or the supervised learning method. Unlike the optimization-based approach, DDQN can handle high-dimensional inputs, like the aforementioned occupancy grid array. Moreover, in contrast to supervised learning methods, which depend on pre-labeled training datasets, DDQN learns directly from interactions with the environment, making it exceptionally suitable for training in autonomous driving tasks. Thus, DDQN is expected to be a superior alternative for collision avoidance compared to traditional approaches. Table 3 presents the comparison between traditional optimization-based approaches and the proposed DDQN framework to demonstrate the benefits of choosing DDQN.

Table 3. Comparison to traditional optimization-based approaches.

Approaches	Pros	Cons
Elastic Band [9]	1. Easy to implement. 2. Can avoid getting stuck at local minimum. 3. Flexibility in local path modification.	1. Path shape may be irregular, especially in complex environment. 2. Computational complexity may increase with number of obstacles. 3. Path may be control-infeasible.
Potential-Field-related	1. Easy to implement. 2. Can achieve real-time performance. 3. Path easy to visualize and understand.	1. Sometimes stuck at local minimum, especially in complex environment. 2. Oscillations may occur around obstacles. 3. Path may be control-infeasible.
SVM-based optimization [10]	1. Path planning in spatial–temporal region. 2. Can find optimal, efficient, and control-feasible path.	1. Sometimes stuck at local minimum, especially in complex environment. 2. Oscillations may occur around obstacles. 3. Path may be control-infeasible.
Other optimization-based method [11,35]	1. Can generate control-feasible and optimal (either time- or fuel-efficient) path. 2. Can adapt to different traffic scenarios.	1. Computationally inefficient and may not achieve real-time performance. 2. Performance of the optimization might be sensitive to the tuning of parameters.
Proposed DDQN-based ADS	1. Learning ability. 2. Model can achieve real-time performance. 3. Can adapt to different traffic scenarios.	1. Training requires good computational resources. 2. Performance of the model depends on training data quality.

Compared to other DRL algorithms, DDQN has its unique advantages. Currently, there are two primary approaches in the DRL field: policy-based and value-based methods. Policy-based methods, such as Proximal Policy Optimization (PPO), focus on directly optimizing the policy that dictates the agent's actions, aiming to improve the expected long-term rewards. These methods are particularly effective in handling high-dimensional or continuous action spaces, and they are known for their robustness and stability during training [36]. However, such an approach can be less training-efficient than value-based methods like DQN and DDQN, requiring more interactions with the environment to achieve similar performance because it must discard old data after updating the policy. On the other hand, value-based methods, such as DDQN, focus on estimating the values of actions from each state, thereby optimizing a policy indirectly by selecting actions that maximize these estimated values. Because value-based methods are usually off-policy, they can reuse the previously generated data to enhance the training efficiency, and they have been used in collision avoidance research [37–40]. Moreover, another popular approach is the Deep Deterministic Policy Gradient (DDPG). DDPG blends both policy-based and value-based elements. It uses a policy network to determine actions and a value network to evaluate them, and it is suitable for environments with continuous action spaces. However, DDPG is very complex to implement in the CARLA setting, and the performance of DDPG is very sensitive to hyperparameter settings, which require extensive computational resources. Thus, in this paper, DDQN is utilized as the DRL framework to develop ADS collision avoidance.

The purpose of DDQN is to learn from interactions with the environmental and to let the neural network model approximate the optimal action–value function. Unlike DQN, which directly employs the Q function to determine the maximum future rewards, DDQN introduces a crucial modification to mitigate the overestimation of action values. In DDQN, the loss function is:

$$L_i(\theta) = \mathbb{E}_{(s,a,r)}\left[\left(r + \gamma \max_{a_{t+1}} Q_{\theta_i^-}\left(s_{t+1}, \underset{a_{t+1}}{argmax} Q_{\theta_i}(s_{t+1}, a_{t+1})\right) - Q_{\theta_i}(s_t, a_t)\right)^2\right] \tag{4}$$

where i represents the iteration number, s represents the state, a represents the action, r represents the immediate reward, θ_i represents the Q-network, and θ_i^- represents the target-network. In DDQN, $\max_{a_{t+1}} Q_{\theta_i^-}\left(s_{t+1}, \underset{a_{t+1}}{argmax} Q_{\theta_i}(s_{t+1}, a_{t+1})\right)$ is used to replace the $\max_{a_{t+1}} Q_{\theta_i^-}(s_{t+1}, a_{t+1})$, which decouples the selection of the action from the evaluation of that action's value and can effectively reduce the overestimation bias of the Q-value.

The architecture of the neural network employed in the DDQN is shown in Figure 9. This network consists of four fully connected hidden layers: three layers each containing 128 units, followed by a final layer with 32 units. The process begins with the occupancy grid being flattened and fed into the first, second, and third 128-unit hidden layers in sequence. Following the third layer, additional sensor data, including information about the ego-vehicle, path-tracking waypoints, and other road users, are combined with the output from the third layer and fed into the fourth hidden layer, which contains 32 units. This design strategy is intentional, aiming to preserve important information that might otherwise be lost in the initial processing layers, thereby ensuring that important details are considered in the network's final output. In addition, the detailed flowchart of the DDQN algorithm is shown in Algorithm 1. The DDQN algorithm proposed in this study employs a dual neural network architecture to optimize decision making. The online neural network Q is tasked with generating optimal actions for the given state. Meanwhile, the target neural network $\hat{Q}$ is utilized for performing gradient descent, with its parameters being updated only after a predefined number of steps. This staged update procedure is designed to stabilize the training process by mitigating rapid fluctuations in learning targets. The simulation steps generated by the SIMULINK vehicle model and the CARLA environment are stored in a replay buffer, which enables the breaking of correlations between consecutive training samples. This approach significantly reduces the variance in model training and enhances the efficiency of data utilization by ensuring a more uniform and comprehensive sampling from the replay buffer during training sessions. In this paper, the implementation details for the training of the DDQN are as follows. 1. The learning rate is set at 0.01. 2. The initial probability of selecting an action randomly is 1, indicating that exploration starts at its maximum. 3. The decay period for the probability of randomly selecting an action is 200,000 steps. 4. The final probability of selecting an action randomly is reduced to 0.05, balancing exploration with exploitation. 5. The reward discount factor is set at 0.9, influencing how future rewards are valued relative to immediate rewards. 6. Each episode is limited to a maximum of 4000 steps to constrain the training duration per episode.

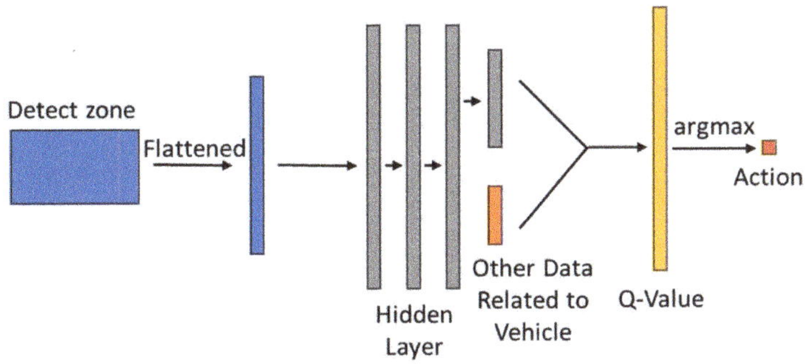

Figure 9. DDQN framework neural network structure.

Algorithm 1. DDQN

Initialize replay memory D
Initialize target network $\hat{Q}$ and Online Network Q with random weights θ
for each episode **do**
  Initialize traffic environment
  for t = 1 to T **do**
    With probability ϵ select a random action a_t
    Otherwise select $a_t = \max_a Q^*(s_t, a; \theta)$
    Execute a_t in CARLA and extract reward r_t and next state s_{t+1}
    Store transition $(s_t,\ a_t,\ r_t,\ s_{t+1})$ in D
    if t mod training frequency == 0 **then**
      Sample random minibatch of transitions $(s_j,\ a_j,\ r_j,\ s_{j+1})$) from D
      Set $y_j = r_j + \gamma \max_{a_{j+1}} \hat{Q}\left(s_{j+1}, \text{argmax}_{a_{j+1}} Q(s_j, a_{j+1}; \theta); \theta\right)$
      for non-terminal s_{j+1}
      or $y_j = r_j$ for terminal s_{j+1}
      Perform a gradient descent step to update θ
      Every N steps reset $\hat{Q} = Q$
    end if
    Set $s_{t+1} = s_t$
  end for
end for

2.3. *Vehicle-in-Virtual-Environment (VVE)*

ADAS testing involving pedestrians poses a significant safety concern. As a result, a safer and more efficient approach is required to reliably implement the proposed pedestrian collision avoidance experiment. Vehicle-in-virtual-environment (VVE) is an ideal alternative for this purpose. VVE allows for the motion synchronization between the real test vehicle operating in a safe, open space and a virtual vehicle operating in a highly detailed and realistic 3D virtual environment. It also allows for the motion synchronization between real and virtual pedestrians, where the virtual pedestrian operates in the same environment as the virtual vehicle. This enables the pedestrian collision avoidance experiment to be carried out as depicted in Figure 10. Preliminary results of vehicle motion synchronization and vehicle-to-pedestrian (V2P) connectivity using the VVE method are discussed in detail in [41].

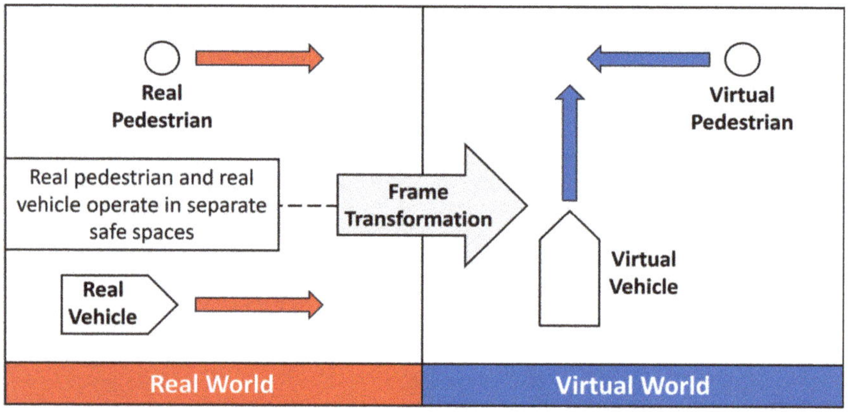

Figure 10. One possible pedestrian collision avoidance test setup using the VVE method.

3. Case Study

To demonstrate the capabilities of the proposed routine, two traffic scenarios of pedestrian collision avoidance are introduced for model-in-the-loop evaluation implementation. The test is conducted in the CARLA realistic animation environment with the SIMULINK vehicle dynamic model being used to facilitate the model-in-the-loop test configuration.

3.1. Scenario 1

The first traffic scenario is illustrated in Figure 11. In this scenario, the pedestrian enters the crosswalk as the vehicle approaches. The crosswalk hence becomes the potential collision zone, and the vehicle must execute a gradual slow-down and stop before it reaches the edge of the zone. It should be noted that the DRL module action space will only include throttle and brake actions, as only a longitudinal motion maneuver needs to be performed. It should be additionally noted that for this scenario, several exemplary slowing-down speed profiles are generated for the DRL module to learn from so that the routine does not rely on random action selection as much. This significantly reduces the time needed to train the model.

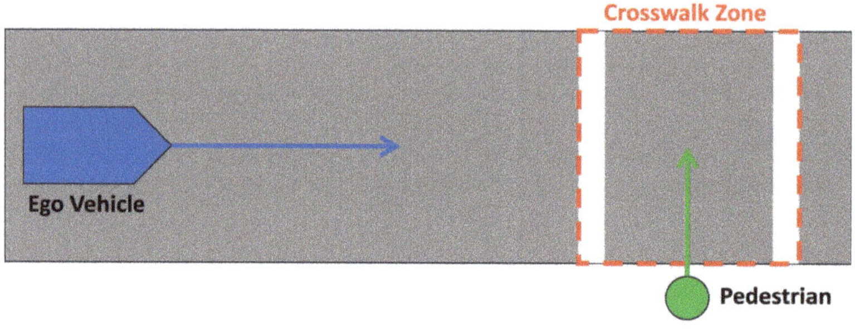

Figure 11. Traffic Scenario 1 setup.

Figure 12 shows the progression of the step rewards as the training episodes increase. It can be observed that the rewards increase with episodes and tend to plateau after 1000 episodes, indicating convergence and hence a successful search for the optimal policy. It is notable that this particular test case is relatively straightforward, allowing for a scenario where even randomly generated actions have a reasonable probability of completing the driving task successfully. Consequently, the reward progression does not show significant improvement across training episodes after the initial period, reflecting the simplicity of the task. However, the proposed deep reinforcement learning (DRL) model demonstrates its effectiveness by enabling the vehicle to successfully complete test cases all of the time, showcasing its robust capability in typical traffic scenarios.

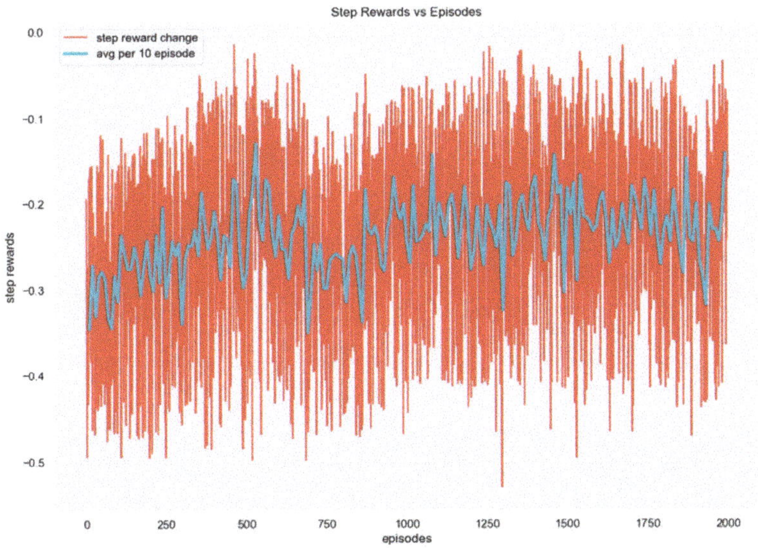

Figure 12. Scenario 1 step rewards vs. training episodes.

Figure 13 illustrates the evolution of TTZ for both the pedestrian and the vehicle. The three colored arcs denote the boundaries of the severity levels, with level one being the most critical, requiring emergent actions to avoid likely collisions, and level three being the least urgent, where the vehicle has plenty of time and space to react to possible collision risks. It can be observed that the TTZ for both the pedestrian and the vehicle is consistently larger than four seconds, indicating the low likelihood of collision, proving the efficacy of the optimal policy. Figure 14 displays the progression of vehicle speed versus time, where it can be observed that the DRL agent is capable of reasonable speed following performance. It is worth noting that the vehicle's speed does not track the desired slow-down profile initially, and this is most likely due to actuator saturation. The link to the Scenario 1 demo video is attached here: https://youtu.be/fgXN1hhA6qk (accessed on 14 May 2024).

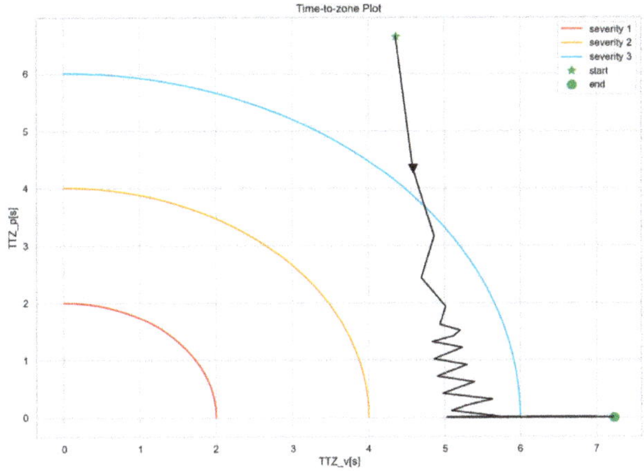

Figure 13. Scenario 1 TTZ progression.

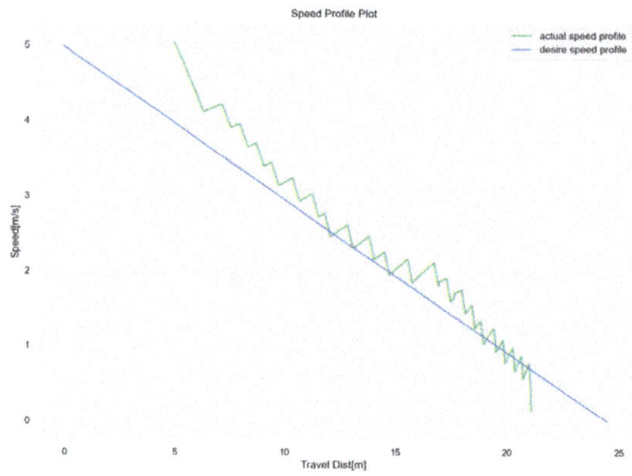

Figure 14. Scenario 1 speed following performance.

3.2. *Scenario 2*

The second traffic scenario is displayed in Figure 15. In this scenario, the pedestrian enters the road at a random point as the vehicle approaches, establishing a potential collision risk. In this case, the pedestrian is treated as a moving obstacle for the vehicle, and the vehicle must go around the pedestrian by steering clear to avoid possible collisions. With this setup, the action space of the DRL module will include not only longitudinal controls of throttle and brake inputs but also steering actions.

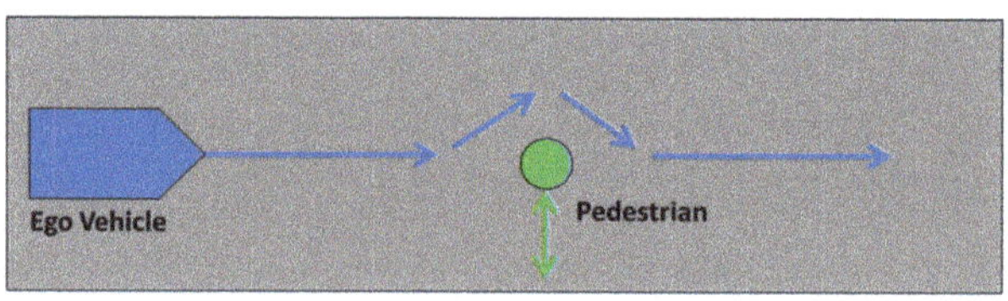

Figure 15. Traffic Scenario 2 setup.

Figure 16 illustrates the reward progression of Scenario 2 training. It can again be observed that the rewards increase as more episodes are completed, and they eventually level off at a high level, indicating a successful optimal policy search. In particular, the convergence of the TTZ reward suggests that the agent learns to avoid collision with the pedestrian, a point that is proven by Figure 17, where collision risk is low. The convergence of the steering reward indicates that the agent also tends to select a reasonable amount of steering during the maneuver. It is also worth noting that the agent tends to slow down as it approaches the pedestrian and then speeds up when the collision risk is clear, as shown in Figure 18, and this is the desirable reaction during this kind of maneuver. The link to the Scenario 2 demo video is attached here: https://youtu.be/CmGtaAjZ_x4 (accessed on 14 May 2024).

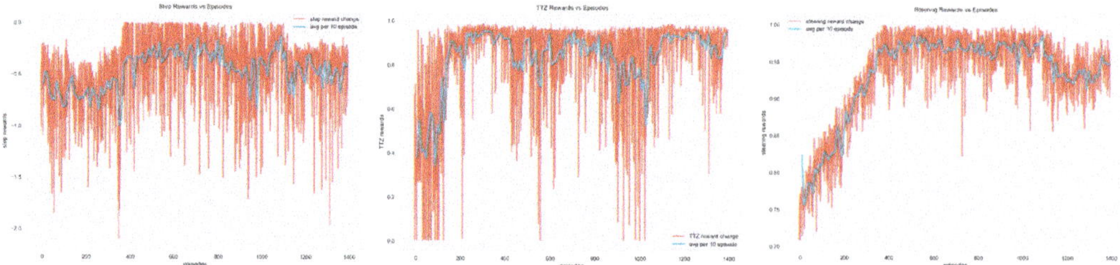

(a) (b) (c)

Figure 16. Training rewards vs. episodes. (**a**) Step reward. (**b**) TTZ reward. (**c**) Steering reward.

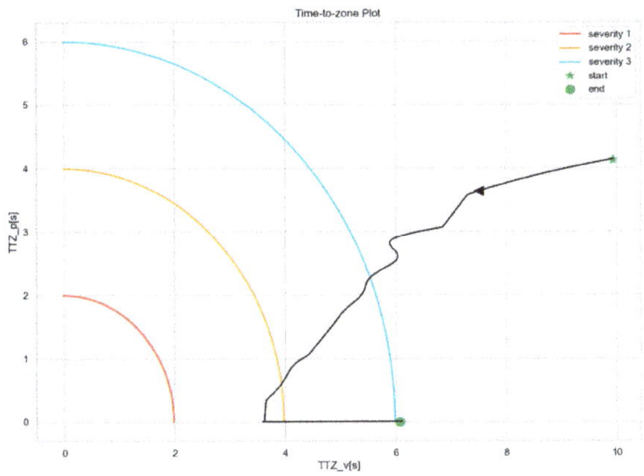

Figure 17. Scenario 2 TTZ progression.

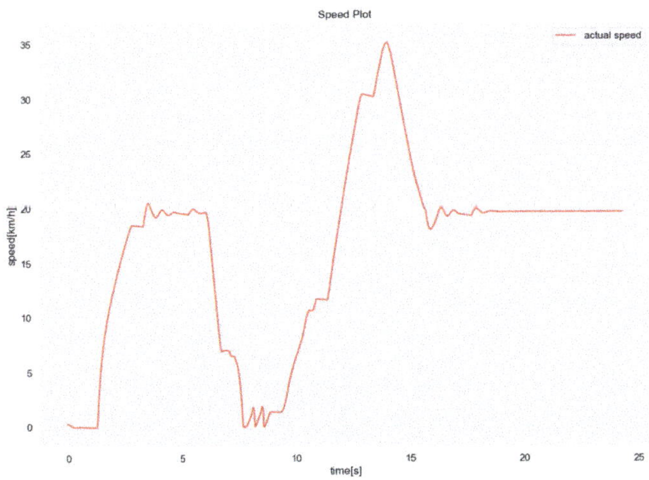

Figure 18. Scenario 2 vehicle speed progression.

4. Conclusions and Future Work

This paper addressed vulnerable road user (VRU) safety by integrating a novel hierarchical deep reinforcement learning (DRL) framework into the autonomous driving system (ADS) to enable automated collision avoidance. A traditional PID controller was used for path-following and speed-following when no VRU was in the vicinity of the vehicle, and the DRL module was activated to start collision avoidance when VRU(s) (pedestrians) enter(s) the vehicle detection zone. A model-in-the-loop (MIL) case study was conducted, and the proposed approach demonstrated satisfactory performance, which demonstrates the capability of the proposed DRL framework to handle moving obstacle problems, instead of the traditional static obstacle navigation problem. For future work, this approach can be further implemented using the vehicle-in-virtual-environment (VVE) method. Other DRL frameworks and neural network structures can also be applied. CARLA is also not mandatory for testing. Instead, other virtual environments, possibly combined with more powerful and realistic vehicle models, such as high-fidelity CarSim models, can be utilized to show its multi-actor capability. In addition, mixed-reality (XR) goggles can be integrated into the test, which, together with the VVE method, can facilitate both safe testing and enhanced realism for the participants.

Author Contributions: Conceptualization, all authors; methodology, H.C. and X.C.; validation, H.C. and X.C.; formal analysis, all authors; investigation, all authors; resources, B.A.-G. and L.G.; data curation, H.C. and X.C.; writing—original draft preparation, H.C. and X.C.; writing—review and editing, B.A.-G. and L.G.; visualization, all authors; supervision, B.A.-G. and L.G.; project administration, B.A.-G. and L.G.; funding acquisition, B.A.-G. and L.G. All authors have read and agreed to the published version of the manuscript.

Funding: This project is funded in part by Carnegie Mellon University's Safety21 National University Transportation Center, which is sponsored by the US Department of Transportation.

Data Availability Statement: Data are contained within the article.

Acknowledgments: The authors thank NVIDIA for its GPU donations. The authors thank the Automated Driving Lab at Ohio State University for its support.

Conflicts of Interest: The authors declare no conflicts of interest.

References

1. World Health Organization. *Global Status Report on Road Safety 2015*; World Health Organization: Geneva, Switzerland, 2015; Available online: https://iris.who.int/handle/10665/189242 (accessed on 24 October 2023).
2. Medina, A.; Lee, S.; Wierwille, W.; Hanowski, R. Relationship between Infrastructure, Driver Error, and Critical Incidents. *Proc. Hum. Factors Ergon. Soc. Annu. Meet.* **2004**, *48*, 2075–2079. [CrossRef]
3. J3016_202104: Taxonomy and Definitions for Terms Related to Driving Automation Systems for On-Road Motor Vehicles—SAE International. Available online: https://www.sae.org/standards/content/j3016_202104/ (accessed on 24 October 2023).
4. Ye, F.; Zhang, S.; Wang, P.; Chan, C.-Y. A Survey of Deep Reinforcement Learning Algorithms for Motion Planning and Control of Autonomous Vehicles. In Proceedings of the 2021 IEEE Intelligent Vehicles Symposium (IV), Nagoya, Japan, 11–17 July 2021; pp. 1073–1080. [CrossRef]
5. Kiran, B.R.; Sobh, I.; Talpaert, V.; Mannion, P.; Al Sallab, A.A.; Yogamani, S.; Pérez, P. Deep Reinforcement Learning for Autonomous Driving: A Survey. *IEEE Trans. Intell. Transp. Syst.* **2022**, *23*, 4909–4926. [CrossRef]
6. Zhu, Z.; Zhao, H. A Survey of Deep RL and IL for Autonomous Driving Policy Learning. *IEEE Trans. Intell. Transp. Syst.* **2022**, *23*, 14043–14065. [CrossRef]
7. Zha, Y.; Deng, J.; Qiu, Y.; Zhang, K.; Wang, Y. A Survey of Intelligent Driving Vehicle Trajectory Tracking Based on Vehicle Dynamics. *SAE Int. J. Veh. Dyn. Stab. NVH* **2023**, *7*, 221–248. [CrossRef]
8. Autonomous Road Vehicle Path Planning and Tracking Control | IEEE eBooks | IEEE Xplore. Available online: https://ieeexplore.ieee.org/book/9645932 (accessed on 24 October 2023).
9. Wang, H.; Tota, A.; Aksun-Guvenc, B.; Guvenc, L. Real time implementation of socially acceptable collision avoidance of a low speed autonomous shuttle using the elastic band method. *Mechatronics* **2018**, *50*, 341–355. [CrossRef]
10. Morsali, M.; Frisk, E.; Åslund, J. Spatio-Temporal Planning in Multi-Vehicle Scenarios for Autonomous Vehicle Using Support Vector Machines. *IEEE Trans. Intell. Veh.* **2021**, *6*, 611–621. [CrossRef]
11. Zhu, S. Path Planning and Robust Control of Autonomous Vehicles. Ph.D. Thesis, The Ohio State University, Columbus, OH, USA, 2020. Available online: https://www.proquest.com/docview/2612075055/abstract/73982D6BAE3D419APQ/1 (accessed on 24 October 2023).
12. Chen, G.; Yao, J.; Gao, Z.; Gao, Z.; Zhao, X.; Xu, N.; Hua, M. Emergency Obstacle Avoidance Trajectory Planning Method of Intelligent Vehicles Based on Improved Hybrid A*. *SAE Int. J. Veh. Dyn. Stab. NVH* **2023**, *8*, 3–19. [CrossRef]
13. Ararat, Ö.; Güvenç, B.A. Development of a Collision Avoidance Algorithm Using Elastic Band Theory. *IFAC Proc. Vol.* **2008**, *41*, 8520–8525. [CrossRef]
14. Emirler, M.T.; Wang, H.; Güvenç, B. Socially Acceptable Collision Avoidance System for Vulnerable Road Users. *IFAC Pap.* **2016**, *49*, 436–441. [CrossRef]
15. Gelbal, S.Y.; Guvenc, B.A.; Guvenc, L. SmartShuttle: A Unified, Scalable and Replicable Approach to Connected and Automated Driving in A Smart City. In *Proceedings of the 2nd International Workshop on Science of Smart City Operations and Platforms Engineering*; In SCOPE '17; Association for Computing Machinery: New York, NY, USA, 2017; pp. 57–62. [CrossRef]
16. Guvenc, L.; Guvenc, B.A.; Emirler, M.T. Connected and Autonomous Vehicles. In *Internet of Things and Data Analytics Handbook*; John Wiley & Sons, Ltd.: Hoboken, NJ, USA, 2017; pp. 581–595. [CrossRef]
17. Wang, Z.; Delahaye, D.; Farges, J.L.; Alam, S.S. A quasi-dynamic air traffic assignment model for mitigating air traffic complexity and congestion for high-density UAM operations. *Transp. Res. Part C Emerg. Technol.* **2023**, *154*, 104279. [CrossRef]
18. Maruyama, R.; Seo, T.T. Integrated public transportation system with shared autonomous vehicles and fixed-route transits: Dynamic traffic assignment-based model with multi-objective optimization. *Int. J. Intell. Transp. Syst. Res.* **2023**, *21*, 99–114. [CrossRef]
19. Kendall, A.; Kendall, A.; Hawke, J.; Janz, D.; Mazur, P.; Reda, D.; Allen, J.M.; Lam, V.-D.; Bewley, A.; Shah, A. Learning to Drive in a Day. In Proceedings of the 2019 International Conference on Robotics and Automation (ICRA), Montreal, QC, Canada, 20–24 May 2019; pp. 8248–8254. [CrossRef]

20. Yurtsever, E.; Capito, L.; Redmill, K.; Ozgune, U. Integrating Deep Reinforcement Learning with Model-based Path Planners for Automated Driving. In Proceedings of the 2020 IEEE Intelligent Vehicles Symposium (IV), Las Vegas, NV, USA, 19 October–13 November 2020; pp. 1311–1316. [CrossRef]
21. Peng, B.; Sun, Q.; Li, S.E.; Kum, D.; Yin, Y.; Wei, J.; Gu, T. End-to-End Autonomous Driving Through Dueling Double Deep Q-Network. *Automot. Innov.* **2021**, *4*, 328–337. [CrossRef]
22. Jaritz, M.; de Charette, R.; Toromanoff, M.; Perot, E.; Nashashibi, F. End-to-End Race Driving with Deep Reinforcement Learning. In Proceedings of the 2018 IEEE International Conference on Robotics and Automation (ICRA), Brisbane, QLD, Australia, 21–25 May 2018. [CrossRef]
23. Merola, F.; Falchi, F.; Gennaro, C.; Di Benedetto, M. Reinforced Damage Minimization in Critical Events for Self-driving Vehicles. In *Proceedings of the 17th International Joint Conference on Computer Vision, Imaging and Computer Graphics Theory and Applications*; Online Streaming, —Select a Country—: SCITEPRESS—Science and Technology Publications; SciTePress: Setúbal, Portugal, 2022; pp. 258–266. [CrossRef]
24. Cao, Z.; Bıyık, E.; Wang, W.Z.; Raventos, A.; Gaidon, A.; Rosman, G.; Sadigh, D. Reinforcement Learning based Control of Imitative Policies for Near-Accident Driving. *arXiv* **2020**, arXiv:2007.00178. [CrossRef]
25. Nageshrao, S.; Tseng, H.E.; Filev, D. Autonomous Highway Driving using Deep Reinforcement Learning. In Proceedings of the 2019 IEEE International Conference on Systems, Man and Cybernetics (SMC), Bari, Italy, 6–9 October 2019; pp. 2326–2331. [CrossRef]
26. Deep Reinforcement-Learning-Based Driving Policy for Autonomous Road Vehicles—Makantasis—2020—IET Intelligent Transport Systems—Wiley Online Library. Available online: https://ietresearch.onlinelibrary.wiley.com/doi/full/10.1049/iet-its.2019.0249 (accessed on 24 October 2023).
27. Aksjonov, A.; Kyrki, V. A Safety-Critical Decision-Making and Control Framework Combining Machine-Learning-Based and Rule-Based Algorithms. *SAE Int. J. Veh. Dyn. Stab. NVH* **2023**, *7*, 287–299. [CrossRef]
28. Knox, W.B.; Allievi, A.; Banzhaf, H.; Schmitt, F.; Stone, P. Reward (Mis)design for autonomous driving. *Artif. Intell.* **2023**, *316*, 103829. [CrossRef]
29. Wang, Y.; Wei, H.; Yang, L.; Hu, B.; Lv, C. A Review of Dynamic State Estimation for the Neighborhood System of Connected Vehicles. *SAE Int. J. Veh. Dyn. Stab. NVH* **2023**, *7*, 367–385. [CrossRef]
30. Lu, S.; Xu, R.; Li, Z.; Wang, B.; Zhao, Z. Lunar Rover Collaborated Path Planning with Artificial Potential Field-Based Heuristic on Deep Reinforcement Learning. *Aerospace* **2024**, *11*, 253. [CrossRef]
31. Xi, Z.; Han, H.; Cheng, J.; Lv, M. Reducing Oscillations for Obstacle Avoidance in a Dense Environment Using Deep Reinforcement Learning and Time-Derivative of an Artificial Potential Field. *Drones* **2024**, *8*, 85. [CrossRef]
32. Mnih, V.; Kavukcuoglu, K.; Silver, D.; Graves, A.; Antonoglou, I.; Wierstra, D.; Riedmiller, M. Playing Atari with Deep Reinforcement Learning. *arXiv* **2013**, arXiv:1312.5602. [CrossRef]
33. Mnih, V.; Kavukcuoglu, K.; Silver, D.; Rusu, A.A.; Veness, J.; Bellemare, M.G.; Graves, A.; Riedmiller, M.; Fidjeland, A.K.; Ostrovski, G.; et al. Human-level control through deep reinforcement learning. *Nature* **2015**, *518*, 7540. [CrossRef]
34. van Hasselt, H.; Guez, A.; Silver, D. Deep Reinforcement Learning with Double Q-Learning. *Proc. AAAI Conf. Artif. Intell.* **2016**, *30*, 1. [CrossRef]
35. Zhang, X.; Liniger, A.; Borrelli, F. Optimization-Based Collision Avoidance. *IEEE Trans. Control Syst. Technol.* **2021**, *29*, 972–983. [CrossRef]
36. Mu, C.; Liu, S.; Lu, M.; Liu, Z.; Cui, L.; Wang, K. Autonomous spacecraft collision avoidance with a variable number of space debris based on safe reinforcement learning. *Aerosp. Sci. Technol.* **2024**, *149*, 109131. [CrossRef]
37. Feng, S.; Sebastian, B.; Ben-Tzvi, P. A Collision Avoidance Method Based on Deep Reinforcement Learning. *Robotics* **2021**, *10*, 73. [CrossRef]
38. Wang, C.; Zhang, X.; Yang, Z.; Bashir, M.; Lee, K. Collision avoidance for autonomous ship using deep reinforcement learning and prior-knowledge-based approximate representation. *Front. Mar. Sci.* **2023**, *9*, 1084763. [CrossRef]
39. Sun, Z.; Fan, Y.; Wang, G. An Intelligent Algorithm for USVs Collision Avoidance Based on Deep Reinforcement Learning Approach with Navigation Characteristics. *J. Mar. Sci. Eng.* **2023**, *11*, 812. [CrossRef]
40. de Curtò, J.; de Zarzà, I. Analysis of Transportation Systems for Colonies on Mars. *Sustainability* **2024**, *16*, 3041. [CrossRef]
41. Cao, X.; Chen, H.; Gelbal, S.Y.; Guvenc, B.A.; Guvenc, L. *Vehicle-in-Virtual-Environment Method for ADAS and Connected and Automated Driving Function Development, Demonstration and Evaluation*; SAE Technical Paper 2024-01–1967; SAE International: Warrendale, PA, USA, 2024; Available online: https://www.sae.org/publications/technical-papers/content/2024-01-1967/ (accessed on 7 March 2024).

Disclaimer/Publisher's Note: The statements, opinions and data contained in all publications are solely those of the individual author(s) and contributor(s) and not of MDPI and/or the editor(s). MDPI and/or the editor(s) disclaim responsibility for any injury to people or property resulting from any ideas, methods, instructions or products referred to in the content.

electronics

MDPI

Article

DRL-Based Backbone SDN Control Methods in UAV-Assisted Networks for Computational Resource Efficiency

Inseok Song [1], Prohim Tam [1], Seungwoo Kang [1], Seyha Ros [1] and Seokhoon Kim [1,2,*]

[1] Department of Software Convergence, Soonchunhyang University, Asan 31538, Republic of Korea; sis5041@sch.ac.kr (I.S.); prohimtam@gmail.com (P.T.); oooksw12@sch.ac.kr (S.K.); seyha@sch.ac.kr (S.R.)

[2] Department of Computer Software Engineering, Soonchunhyang University, Asan 31538, Republic of Korea

* Correspondence: seokhoon@sch.ac.kr

Abstract: The limited coverage extension of mobile edge computing (MEC) necessitates exploring cooperation with unmanned aerial vehicles (UAV) to leverage advanced features for future computation-intensive and mission-critical applications. Moreover, the workflow for task offloading in software-defined networking (SDN)-enabled 5G is significant to tackle in UAV-MEC networks. In this paper, deep reinforcement learning (DRL) SDN control methods for improving computing resources are proposed. DRL-based SDN controller, termed DRL-SDNC, allocates computational resources, bandwidth, and storage based on task requirements, upper-bound tolerable delays, and network conditions, using the UAV system architecture for task exchange between MECs. DRL-SDNC configures rule installation based on state observations and agent evaluation indicators, such as network congestion, user equipment computational capabilities, and energy efficiency. This paper also proposes the training deep network architecture for the DRL-SDNC, enabling interactive and autonomous policy enforcement. The agent learns from the UAV-MEC environment through experience gathering and updates its parameters using optimization methods. DRL-SDNC collaboratively adjusts hyperparameters and network architecture to enhance learning efficiency. Compared with baseline schemes, simulation results demonstrate the effectiveness of the proposed approach in optimizing resource efficiency and achieving satisfied quality of service for efficient utilization of computing and communication resources in UAV-assisted networking environments.

Keywords: computational resource efficiency; deep reinforcement learning; mobile edge computing; software-defined networking; unmanned aerial vehicles

1. Introduction

Mobile cloud computing (MCC) is a highly efficient system located in central cloud; however, the end user faces several computing inadequacies due to backbone congestion and offloading delays. Therefore, mobile edge computing (MEC) method is proposed which offers extraction by leveraging storage, computing, communication, and network capacities [1–3]. The integration of networking system and cloud/edge computing brings network cloudification which leverages existing cloud and edge computing infrastructure to host virtual network functions. The convergence of networking and cloud/edge computing necessitates a comprehensive perspective on the advancements and the delivery of composite network-cloud/edge services, which offers advantages, such as enhanced resource utilization, cost reduction, and emerging new prospects for stakeholders [4].

However, without mobility, the coverage extension is deficient, which further leads numerous researchers/organizations to opt for the cooperation between unmanned aerial vehicles (UAV) and MEC that would bring a variety of benefits. With the growing popularity and deployment of UAV in various applications, there is an increasing demand for efficient and reliable communication networks to support UAV-assisted operations [5,6]. UAV-assisted networks have the potential to enable a wide range of applications, including aerial surveillance, disaster management, etc. [7–9]. Nevertheless, UAVs have several limitations in terms of energy, computation resources, etc. Moreover, the dynamic and unpredictable nature of UAV mobility poses significant challenges in ensuring seamless connectivity and resource efficiency in control entities, which requires a thorough proactive and autonomous model.

Software-defined networking (SDN) has emerged as a promising paradigm for managing and controlling network resources in a flexible and centralized manner, which can be abstracted from the UAV states for intelligent flow management [10–13]. The need for wireless communication infrastructure in critical applications is assisted by utilizing UAVs equipped with NFV/SDN capabilities, which are crucial technologies for enabling flexible control of UAV networks and computing [14]. By decoupling the control plane from the data plane, SDN allows for dynamic resource allocation and efficient network management [15]. However, the conventional SDN architectures are primarily designed for terrestrial networks and may not be well-suited for UAV-assisted networks due to the unique characteristics, such as mobility, limited energy, and limited processing capabilities. Therefore, intelligent agent modeling using deep learning or deep reinforcement learning (DRL) is suggested for integration [16,17].

In [18], authors present a state-specific environment on UAV observations, including SINR measurement, height of UAVs, and spectral efficiency, that offer insightful context into the agent on applied actions on (1) base station selection with transmission power settings, and (2) UAV elevation. The proposed system evaluates the action efficiency by rewarding joint maximization of energy and throughput until the converged policy is obtained. Furthermore, in [19], an analytical framework is proposed to assess the coverage probability in mobile networks assisted by UAVs, considering clustered users, varying UAV heights, and imperfect beam alignment. Figure 1 presents the overview interactions of unmanned aerial resource system (UARS) states gathering for agent training, action configuration, and policy installation in SDN controller (SDNC). With the advancement of intelligent modeling, the controlling agent can be generated in a flexible way using complete network state observation, action-based autonomous configuration, and prediction models [20–22]. By leveraging the capabilities of DRL, the proposed approach aims to enhance the resource efficiency and adaptability of SDN-based control mechanisms in UAV-assisted networks.

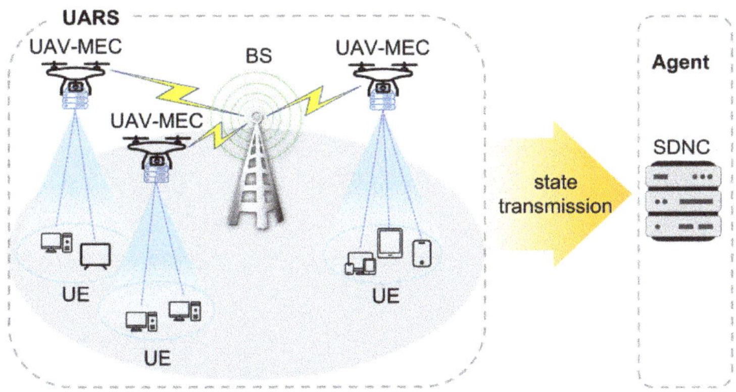

Figure 1. UARS environment for state gathering to interact with the agent controller.

The main contribution of this paper is the development of an agent with computational resource efficiency (CRE) objective function to optimize performance metrics, such as throughput, latency, and energy consumption in UAV-assisted networks. The proposed system utilizes DRL agents to learn the optimal control policies for handling computing resource, appropriate node selection, and network management tasks based on various state observation features. Overall, we can highlight the following contributions:

- The working flow of the system architecture are given in this work by mentioning the main interfaces, such as (1) UE-to-UAV, (2) UAV-to-MEC, (3) UAV/MEC-to-SDN, (4) SDN-to-DRL, and (5) SDN, as a central controller to the whole network architecture. We provide insights into the feasibility and effectiveness of DRL in optimizing UAV states in dynamic and resource-constrained environments.
- The reward evaluation metrics are proposed to formulate the measurement of resource efficiency and adaptability in UAV-assisted networks through intelligent network management and SDN flow controls.
- We offer a comprehensive evaluation of the proposed system using OpenAI and network simulation software to highlight its advantages over existing solutions in several terms of resource utilization, latency, and energy scoring metrics.

2. Related Works

2.1. UAV-MEC for Resource Efficiency

The issue of energy-efficient resource allocation has been of note in modern networks, and has motivated researchers towards the implementation of UAV-enabled MEC systems. In [23], the authors aim to minimize the combined power consumption of UE and UAV. The minimization problem is well-formulated on total power consumption while considering latency and coverage constraints. The proposed algorithm addresses nonconvex problems through iterative optimization of (1) user association, (2) power control, (3) computation capacity allocation, and (4) location planning. In [24], a novel approach utilizing UAV in a MEC-enabled VANET is proposed to deliver low latency and reliable computing services to vehicles. The optimization objective focuses on minimizing comprehensive task processing delay by considering transmission models, security assurance, and task computation models. To optimize VANETs, the authors propose a network scheme that jointly considers these 3-tuple policies, and by leveraging the MEC-enabled UAV-assisted VANET architecture, the methods harness the communication capabilities of UAVs to enhance VANETs' computational abilities. An iterative algorithm based on the relax-and-rounding method and the Lagrangian method is used, and the simulation results capture the key performance metrics on successful task processing ratio and task processing delay. Furthermore, in [25], authors investigated the interaction between multiple users, drones, edge servers, and the cloud, aiming to optimize resource allocation and offloading decisions in a multi-user scenario. The minimization of energy consumption and delay is formulated to consider UAV offloading, edge server computing cost, and communication processing delay, in terms of local device computing power. The authors proposed MEC architecture for IoT devices, a drone-based edge computing layer, and edge computing server layer. The offloading strategy to minimize energy costs and maximum delay is formulated as a non-convex quadratic constrained quadratic program and a heuristic algorithm based on semi-definite relaxation and adaptive adjustment is used as a solver.

2.2. DRL for UAV-MEC Control

The autonomy of DRL can be appended as a solving agent in the UAV-MEC systems in terms of resource allocation, offloading decisions, and other optimization approaches. In [26], authors focus on a UAV-assisted wireless IoT network and introduces a multi-agent DRL framework combined with a round-robin resource-scheduling algorithm to optimize joint resource management. A system model is designed to capture the dynamic and heterogeneous nature of the environment, considering various constraints, such as user count, channel gains, noises, and power consumption. K-means and round-robin algorithm are used to efficiently handle service requests from IoT users in urban and sub-urban clusters. The proposed DRL framework is applied to optimize resource management for UAV-assisted IoT devices in the primary system model. In [27], an innovative application scenario where a UAV operates in a complex urban environment is considered. The UAV receives computing tasks from clients within a specific flight range, leverages mobile edge computing servers for task processing, and optimizes its trajectory to enhance communication quality and improve the computing rate of the MEC network. By employing DRL and experience replay techniques to generate optimal offloading decisions and introducing a time frame allocation algorithm for resource allocation, the problem becomes predictive and is effectively solved [28,29]. The proposed scheme, combining UAV-assisted MEC networks, DRL, and radio map optimization, represents a novel and systematic approach to resource optimization. By integrating UAV and MEC, the accessibility and cost-effectiveness are reached, and by utilizing DRL, the computation offloading in a UAV-assisted MEC network can be optimized. The integrated approach can simplify the complexity of the original problem, resulting in improved computational and energy efficiencies with maximized computing rates [30].

2.3. DRL-Based SDN for Efficient Network Management

SDN enhances the scalability, flexibility, and control efficiencies of UAV-MEC networks by simplifying the addition or removal of UAVs and MEC servers, allowing traffic optimization/service provisioning, and enabling automated network orchestration, policy enforcement, and real-time monitoring, thereby enhancing operational efficiency in UAV-MEC networks [31]. To further advance SDNC in complex network environments, DRL has been integrated in several studies. In [32], SDN-based dynamic task scheduling and resource management approach using DRL for IoT traffic scheduling is proposed. The objective is to achieve high network performance by minimizing latency and ensuring energy efficiency. The proposed solution introduces an architectural design and formulates a task assignment, also called a scheduling problem. The approach offers effective trade-offs between response time constraints, model fidelity, inference accuracy, and task schedule-ability. Furthermore, in [33], an integrated intelligent algorithm in SDN-based QoS-routing optimization is proposed to enable more intelligent dynamic routing. Offline training methods are used for supervised learning-based models, while DRL-based models can be trained both online and offline. The study achieved intelligent online QoS-routing optimization solution using SDN and asynchronous advantage actor-critic, which introduces the algorithm for dynamic routing decisions and the DRL-enabled framework for real-time data collection and strategy learning.

3. DRL-Based Backbone SDN Control Methods in UAV-Assisted Networks

This section covers the system architecture, environment modeling, and algorithm designs with specified interfaces of the proposed framework. Throughout these subsections, the proposed framework is described overall by how the agent connects and operates for policy installation to achieve the objectives of computational resource efficiency.

3.1. System Architecture

The system architecture, as illustrated in Figure 2, emphasizes the relations of UAV-MEC in UARS with the base stations (BS), 5G service-based architectures (SBA), UAV access and mobility management function (*U-AMF*), and UAV session management function (*U-SMF*), and the interactions with the agent in SDNC through user plane function (UPF).

The workflow of task offloading in SDN-enabled 5G UAV-MEC networks starts from the task identification, where UE identifies computation-intensive tasks from resource-intensive applications that can be offloaded to the selected MEC for processing. Task offloading request is sent by UE to the *DRL-SDNC*, which includes the states on computational requirements, data size, and QoS constraints. *DRL-SDNC* receives the task offloading request and analyzes it for determining the appropriate network resources and MEC allocation for a complete/partial task processing. Comprehensively, *DRL-SDNC* allocates the necessary resources, including computational resources in MEC, bandwidth, and storage, based on the task requirements and network conditions. UPF is used to exchange tasks between MEC. *DRL-SDNC* configures the rule installation based on state observations and the proposed agent evaluation indicators, such as network congestion, computational capabilities of the UE, and energy efficiency factors. Furthermore, *DRL-SDNC* provides the

UE with the necessary information and interfaces to intelligentlytransmit the task data to the designated MEC for processing. Once the task processing is complete, SDNC notifies the UE about the completion status and provides the necessary interfaces for retrieving the processed outputs.

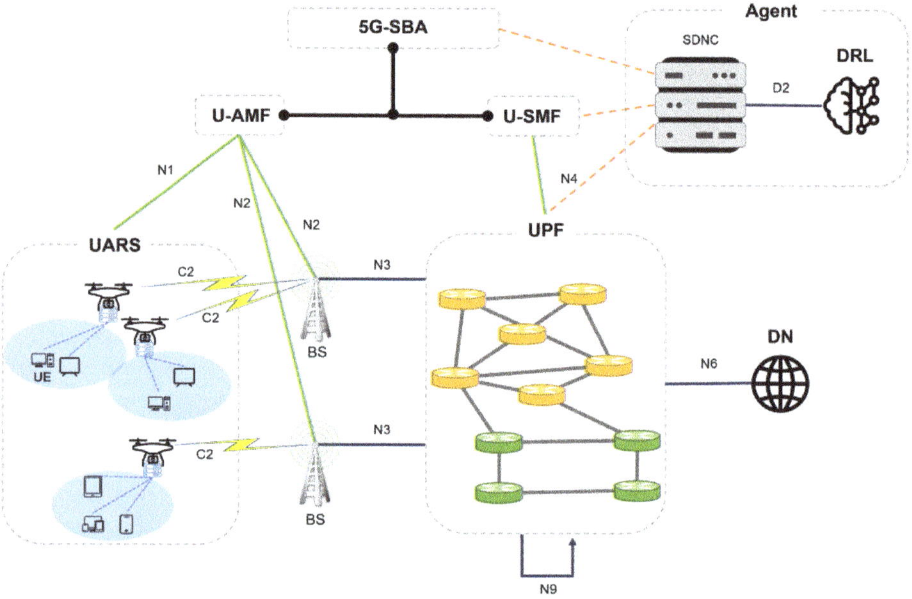

Figure 2. System architecture of the UAV-MEC environment in SDN-enabled 5G networks.

Through the access part of 5G UAV interfaces, *UARS* includes the radio interface between the UAV and the ground network, which enables wireless communication between the UAV and the 5G network infrastructure. *N*1 interface covers the interaction between UE or UAV and the 5G access network by handling the wireless communication protocols and connectivity between the UE and the access network. *N*2 interfaces cover the interactions between different 5G access network nodes by handling communication and coordination, such as base stations, in order to provide coverage and handover support for our proposed UAV-MEC. And for *C*2 interface, it handles the UAV and the 5G control plane by allowing the control and management of the UAV's movements, configurations, and interactions with the 5G network.

*N*3 handles signaling and control messages related to mobility management, session establishment, and authentication. *N*9 is the interface within UPF (e.g., SGW and PGW) which carries user data and controls messages for packet forwarding and routing purposes, which is later used for abstracting information and state observation to *DRL-SDNC*. *N*6 and *N*4 offer the interactions between UPF (e.g., PGW) and the data network (DN) or external control entities by handling the control plane signaling for mobility management and policy installation between the 5G core network and the proposed *DRL-SDNC*.

In our studies, the state gathering process and action configuration are important to collaborate with UPF, which includes the user plane data forwarding, routing, and optimization information, by performing tasks, such as packet filtering, forwarding, and traffic management. Moreover, 5G-SBA is also an entity to observe the states and enable the action deployment of flexible and scalable services in UAV-MEC environment. 5G-SBA provides a service-based interface model for efficient communication and interaction between network functions and services.

For *U-AMF*, the agent leveraged the capability for UAV to manage access and mobility-related functions. And for *U-SMF*, the proposed system aims to handle the UAV session establishment, policy enforcement, and management of data flows between the UE, UAV, and the network infrastructure. These interfaces and entities work collaboratively for global UAV-assisted network state abstractions, environment connectivity, agent's action-based control, and data management in our proposed system architecture for ensuring reliable communication, and adequate computational resources between UE tasks and UAV-MEC selection/placement.

3.2. Environment Modeling

From the above-mentioned system architecture, the primary entities that generate states and configure the action rules must be modeled into softwarization for interacting and ordering the offloading policies into uses by the agent controller. Table 1 presents the important notations and its description used in this proposed DRL-based system.

Table 1. Key notations used in the environment initialization and the proposed agent interactions.

Notation	Description
$T = \{1, 2, \ldots, t\}$	Set of timeslots
$I = \{1, 2, \ldots, i\}$	Set of UAVs
$M = \{1, 2, \ldots, m\}$	Set of MECs
$N = \{1, 2, \ldots, n\}$	Set of UEs
$J = \{1, 2, \ldots, j\}$	Set of offloading tasks
$S = \{1, 2, \ldots, s\}$	Set of small base stations
$P_i^t(x_t,\ y_t,\ z_t)$	Position coordination (x-axis, y-axis, height) of UAV-i at timeslot-t
$D_i^t(d_{n,i},\ d_{s,i})$	Distance metrics of UAV-i between UE-n and BS-s
$(e_i^t,\ e_n^t)$	States of remaining energy metrics of UAV-i and UE-n
$(C_{\max.m}^t,\ C_{\text{re}.m}^t)$	Maximum computing and remaining resources of MEC-m at timeslot-t
T_{bw}	Total system bandwidth
$(bw_{n\to i}^t,\ g_{n\to i}^t)$	States of bandwidth allocation and channel gain between UE-to-UAV
$(bw_{i\to s}^t,\ g_{i\to s}^t)$	States of bandwidth allocation and channel gain between UAV-to-BS
$(p_n^t,\ \forall\delta)$	Transmission power of the devices and overall noises
$\tau_n^t(\rho_n^t, \gamma_n^t,\ C_n^t)$	States of tasks from UE-n at timeslot-t with 3-tuple information of task sizes, upper-bound tolerable delays, and computation workload
$R_{env}^t(r_d^t, r_e^t, r_{cr}^t)$	Primary reward evaluation metrics of the environment per episode, which formulating by 3 sub-rewards on joint delay, energy, and computational resource efficiency
$e_t(s_t,\ a_t,\ r_t,\ s_{t+1})$	Experience replay batches at t including the state, action, reward, and transited next-state.

The task process can be described in 4 primary phases as follows:

1. ***Task offloading from UE to the selected UAV-MEC***: In each timeslot-t, UE-n can offload task-j to the integrated paring UAV-MEC $(i - m)$ node. The task offloading decision is given by action output from the proposed agent, which can be presented as $a_i^{off}(t) = \{n_j, i, m\}$, indicating the index of UE-n which offloads a task-j to UAV-MEC system composed of UAV-i and MEC-m. The communication model, denoted as data rate U_{n-i}^t, between UE-to-UAV is presented in Equation (1) which is associated with the states of allocating bandwidth $bw_{n\to i}^t$, channel gain $g_{n\to i}^t$, transmission power p_n^t,

and overall noises $\forall\delta$, including the interference between UAVs and BSs and also additive white Gaussian noise.

$$U^t_{n-i} = bw^t_{n\rightarrow i}\log_2(1 + \frac{p^t_n g^t_{n\rightarrow i}}{\forall\delta}) \quad (1)$$

2. ***UAV-MEC task processing***: The integrated UAV-MEC system receives the offloaded tasks from UEs and performs the computing to get the expected task output. The task processing can be represented as $j^{comp}_{i-m}(t)$ indicating the processed task, consumed energy/resource, and executed times.
3. ***Task offloading from UAV-MEC to BS***: The UAV-MEC system can further offload processed tasks to BS for transmission or communication purposes. In our environment, within small BS-*s*, there is a single UAV and single MEC which assists with the task computation. The communication model of UAV-to-BS is presented in Equation (2) by formulating uplink data rates associated with the states of allocating bandwidth $bw^t_{i\rightarrow s}$, channel gain $g^t_{i\rightarrow s}$, transmission power p^t_i, and overall noises/interference.

$$U^t_{i-s} = bw^t_{i\rightarrow s}\log_2(1 + \frac{p^t_i g^t_{i\rightarrow s}}{\forall\delta}) \quad (2)$$

4. ***Base Station task processing***: Base stations receive tasks from the UAV-MEC system and perform task processing or transmission. The state information from the tasks is gathered for the *DRL-SDNC* agent.

For DRL environment and agent initialization, the components can be described following Markov decision process, which primarily consists of states, actions, and reward, which are expressed in the following sub-sections, and in Equations (3)–(5), respectively.

3.2.1. States of UAV-MEC and Tasks from UE

- $P^t_i(x_t,\ y_t,\ z_t)$ represents the coordination of geographical UAV positions, and its trajectory P^{t+1}_i is set by the agent output in terms of distance metrics and offloading task patterns.
- $D^t_i(d_{n,i},\ d_{s,i})$ represents the distance between UAV-*i* to local task owner, UE-*n*, and its pairing small BS-*s*.
- e^t_i and e^t_n represent the states of energy of UAV-*i* and UE-*n* at timeslot-*t*, respectively.
- Within paired MEC-*m*, the states of total computation workload and remaining resources at timeslot-*t* are denoted as $(C^t_{\max.m},\ C^t_{\mathrm{re}.m})$, respectively, for further re-allocations and enhancing decision-making policies.
- In communication model, the states of total bandwidth T_{bw} is observed from the environment. And the bandwidth allocation and channel gain between UE-to-UAV and UAV-to-BS, $(bw^t_{n\rightarrow i},\ g^t_{n\rightarrow i})$ and $(bw^t_{i\rightarrow s},\ g^t_{i\rightarrow s})$, are gathered, respectively.
- From local UE, the states of tasks and local computational resources are observed for formulating the offloading decision of local computation, and complete, or partial task offloading. τ^t_n represents the tasks from UE-*n* at timeslot-*t*, which consists of a 3-tuple information of task sizes, upper-bound tolerable delays, and computation workload, denoted as ρ^t_n, γ^t_n, and $\mathbb{C}^t_n$, respectively.
- An experience tuple $j^{comp}_{i-m}(t)$ from processed tasks including the consumed energy/resource, and time spent from experiences.

$$s_t = \{P^t_i,\ D^t_i,\ e^t_i, e^t_n, C^t_{\max.m},\ C^t_{\mathrm{re}.m}, T_{bw}, bw^t_{n\rightarrow i},\ g^t_{n\rightarrow i} bw^t_{i\rightarrow s},\ g^t_{i\rightarrow s},\ \tau^t_n,\ j^{comp}_{i-m}(t)\} \quad (3)$$

3.2.2. Actions of DRL-SDNC

The offloading decision-maker and flow scheduling, represented by $a^{off}_i(t)$, from the backbone SDN control methods are synchronized as an action from the agent, which have the global views of the UAV data plane and abstraction programmability. The mechanism of the DRL-SDNC interacts with the controllers for understanding the experience batches of resource allocation and performance patterns in each observed state iteration. The actions also cover (1) the optimal MEC selection $a^{sel}_{m|c}(t)$ based on computational resource and (2) load balancing $a^{bal}_{\forall m}(t)$ over all the edge servers. SDN actions collaborates with the proposed agent by optimizing flow rule installations and alleviating heavy congestions for efficient task completion times within upper-bound tolerable delays.

$$a_t = \{a^{off}_i(t),\ a^{sel}_{m|c}(t),\ a^{bal}_{\forall m}(t)\} \tag{4}$$

3.2.3. Rewards and Evaluation on Optimal Policy Selection

The evaluation of the proposed agent and SDN control methods measures the efficiency of applying action a_t into UAV-MEC environment state s_t by obtaining reward r_t before transiting to the next state s_{t+1}. In our proposed method, the primary reward, denoted as R^t_{env}, is a complete formulation from sub-rewards on delay, energy, and computational resources, denoted as r^t_d, r^t_e, and r^t_{cr}, respectively. The critical weights ω on computational resources r^t_{cr} is adjusted for adapting partial of exceptional mandatory delay and energy requirements to primarily serve the efficiency of MEC computational resource placement and offloading decisions.

$$R^t_{env} = r^t_d + r^t_e + \omega r^t_{cr} \tag{5}$$

The optimal policy $\pi(t)^*$ is the SDN backbone control policy from the overall batches that maximize the long-term reward expectation by experiencing each rule condition, which is presented in (6). The value function formulates the expectation of increasing reward summation following the selected policy π from state s. Equation (7) presents the standard optimal q-value function following the bellman equation.

$$\pi(t)* = \underset{\pi}{\text{argmax}}\mathbb{E}_\pi\left[\sum_{t\in T}\gamma^t R^t_{env}\right] \tag{6}$$

$$Q*(s,a) = \mathbb{E}_{\sim s'}\left[R^t_{env} + \gamma\underset{a'}{\text{argmax}}Q*(s',a')\right] \tag{7}$$

3.3. Algorithm Flows

The execution phases of this proposed control method can be trained through these four following phases:

1. ***Hyperparameter/parameter initialization***: (1) Define the number of episodes, representing the iterations of the DRL training process. (2) Initialize the value function, which estimates the expected return for each state–action pair in the environment. (3) Initialize the policy function, which maps states to actions, and set the value of epsilon for epsilon-greedy exploration, with a decaying value over each iteration.
2. ***State initialization***: This phase includes the capturing process from data plane context, such as the geographical positions of the UAVs, calculating the distance between UAV and the local task owner, recording the energy states of the UAVs, tracking the total computation workload and remaining resources, and evaluating the bandwidth allocation and channel gain between UE-to-UAV and UAV-to-BS. This information is stored as an experience tuple, as expressed in Equation (3).
3. ***Iterative learning***: Within each episode, the system performs the eight following steps: (1) Select an action based on the current state using the policy function. Actions can include optimal MEC selection, load balancing across edge servers, and offloading decision-making and flow scheduling as $\{a^{off}_i(t),\ a^{sel}_{m|c}(t),\ a^{bal}_{\forall m}(t)\}$ tuples. (2) Execute the selected action in the environment and observe the resulting state. (3) Calculate the reward based on a complete formulation in Equation (5) that includes sub-rewards related to delay r^t_d, energy consumption r^t_e, and computational resources ωr^t_{cr}. (4) Up-

date the value function and policy function using the observed state, action, reward, and next state. (5) Store the experience tuple $e_t(s_t, a_t, r_t, s_{t+1})$ in a memory buffer for training the online and target networks (using per batch sampling). (6) Perform gradient descent optimization on the loss function to update the model's parameters. (7) Decay the value of epsilon to reduce exploration over each iteration. The above steps are repeated for the specified number of setting number of episodes from phase 1.

4. ***Post-training completion***: The DRL model can be used for decision-making in real-time scenarios, considering the optimal learned policy for selecting optimal actions based on observed states. The context and improved accuracy of policy selection will be input into SDNC for flow installation and rule settings.

Figure 3 presents a single-looped schematic flows starting from (1) initializing the states from Equation (3), (2) selecting action (whether explored or exploited) from Equation (4),
(3) calculating the total rewards from Equation (5), (4) formulating the transition probability to next state, (5) storing the experience batches which includes $e_t(s_t, a_t, r_t, s_{t+1})$, (6) feeding (s_t, a_t, r_t) to train the online network, and (s_{t+1}) to the target network, also exchanging weights for network improvement, and finally (7) obtaining the output as the recommended next action a_{t+1}.

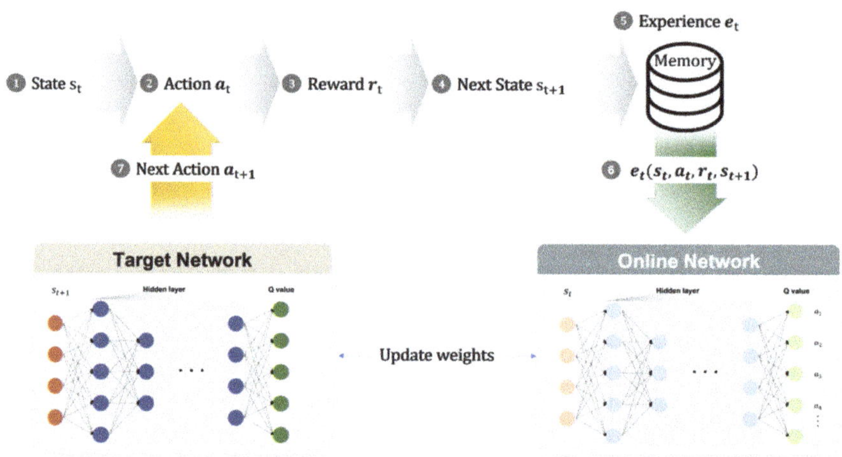

Figure 3. Schematic flows (states input and configured actions) for efficient computational rewards.

Building the neural network architecture that will serve as the online network in the DRL-SDNC will keep the action configuration or policy enforcement interactive and autonomous for the proposed UAV-MEC systems. Depending on the characteristics of the state representation and the temporal dynamics of the UAV-MEC networking environment, the weights are exchanged between online and target networks for iterative improvement. Experience gathering enables the agent to interact with the UAV-MEC networking environment and obtain the hidden patterns. The proposed agent acts as double Q-networks and updates the parameters of the online network using optimization methods like stochastic gradient descent and perform the iterative learning. The UAV-MEC environment states are gathered by the agent to gain new experiences, update the online network, and improve its offloading decision-making capabilities. SDNC monitors the performance of the agent during training and DRL-SDNC collaboratively adjusts hyperparameters or network architecture to enhance learning efficiency and achieve the long-term computational efficiency of the offloading tasks. Therefore, the proposed DRL-SDNC can learn to make intelligent decisions in a UAV-assisted networking environment, optimizing resource efficiency based on the evaluated reward function and achieving efficient utilization of computing/communicating resources.

In the process, first, the proposed DRL agent receives input about the network state and flow statuses from the SDNC. Neural networks are used to estimate the value or policy function based on s_t. Next, the DRL-SDNC agent configures action a_t, which represents a flow scheduling decision, and communicates it to the SDNC. The SDNC applies the action by updating the flow rules through UPF. The updated UAV network state s_{t+1} is then fed back to the DRL-SDNC, allowing it to observe the outcome and update its neural network parameters. This iterative process continues to refine the DRL-SDNC agent's flow scheduling strategies to the optima.

4. Performance Evaluation

4.1. Simulation Setup

Table 2 presents the primary parameter configuration used in this system. In this system flows, there are five primary phases:

1. ***Network topology and environment settings***: A simulated SDN environment is created with a network topology consisting of switches, controllers, and hosts. Hosts can be customized as UEs and MECs with specific requirements of resource settings. In OpenAI, the replication of network topology is carried out as an environment initialization (with randomness and reset every episode) following the state tuple information as listed in Equation (3). The state presentation of UAV network is added using OpenAI to complete the missing topology in SDN, and later collected and represented in a suitable format for the DRL-SDNC agent. The state representation serves as input to the agent's neural network, which constructs using TensorFlow.
2. ***Task offloading***: A set of tasks with random requirements and capacities, such as application complexity, latency constraints, and energy/resource consumption/requirement are generated. Each UAV-MEC pair represents a potential offloading option for task execution, and the flow rule is synced to central SDNC.
3. ***DRL agent***: An instance of the DRL ***agent function*** using OpenAI libraries with TensorFlow is initialized. The ***environment function*** samples the state, and ***action function*** alters the state information into new space after every iteration action is applied. **Reset function** is executed when the episode ends, or optima is reached. ***Immediate reward*** and ***Q-value function*** is co-located with the ***agent function*** to evaluate the state–action pair performance.
4. ***Training and iterative learning***: The proposed agent is trained using a sampling state matrix generated from the UAV-MEC environment. The agent's online/target network parameters are updated through backpropagation and gradient descent to optimize its policy/value function. The hyperparameters and network configurations are iteratively adjusted to optimize the ordering flow for task offloading in the UAV-MEC environment.
5. ***Evaluation***: The reward per episode can be captured within OpenAI-TensorFlow simulation, and network configuration of sending payload task sizes based on agent action identifies the efficient/inefficient computational resources which possibly leads to high drop ratios.

Table 2. Primary parameter configuration for DRL-based SDN backbone environment.

Purpose/Platform	Specification
Hosting infrastructure	Intel(R) Xeon(R) Silver 4280 CPU @ 2.10 GHz, 128 GB, NVIDIA Quadro RTX 4000 GPU
Number of UAVs, MECs, and BS	5
Task generation rate, constraints, bandwidth, resource scales, channel gain, power, UAV coordination/distances, and speed	Ratio scale (0 to 1) for agent sampling and adjusting based on state-action forward-backward calculation
Task complexity and sizes	Random set (intensive, normal, non-intensive) ranging (256 Kbits, 512 Kbits, 1024 Kbits)
Learning rate	0.001
Discount factor	0.95
Batch size	Random set (32, 64, 128, 256) by congestion states
Exploration	0.5
Number of episodes	500
Simulation times	5000 s
SDN-UE, control, and interfaces	Mininet

With the integration of DRL and SDN [34,35], this simulation setup enables the DRL agent to learn and adapt its decision-making process to schedule for computational resource efficiency in UAV-MEC task offloading, leading to improved energy efficiency, reduced overhead latency, and enhanced resource utilization.

4.2. Proposed and Reference Schemes

This sub-section presents the proposed and reference schemes applied in the experiments to illustrate the performance differences in terms of different congestion conditions, task complexities, and intensity level of the heavy tasks compared with the simulation setting and episode numbers of the learning iteration.

- ***PDRL-SDNC-UAV*** indicates the proposed DRL-based SDN backbone control utilizing double Q-network to approximate the Q-value function and collaboratively configures the SDN flow rules for efficient scheduling and resource placement. This approach intelligently train the function approximator to handle high-dimensional state space observation and representation. The experience replay, network training, and SDNC synchronization in UAV-MEC networks follow the expression in Section 3 to learn complex policies for emphasizing the critical weights on sub-reward r^t_{cr}, and handles different congestion states from multi-diversity complex application with high computation-intensive tasks.
- ***T-SDNC-UAV*** represents the traditional SDNC for UAV-MEC-assisted network environment with task offloading problem handling. This baseline involves a centralized controller that manages the network's resources and controls the behavior of the UAVs, which obeys the default SDN rules. The controller communicates with the UAVs and makes decisions based on network-wide information, such as topology, traffic conditions, and default offloading policies. *T-SDNC-UAV* approach focuses on network-level optimization, ensuring efficient routing, resource allocation, and QoS provisioning for UAV-assisted networks.
- ***SRL-SDNC-UAV*** indicates the single reinforcement learning-based SDNC approach, in which a single Q-learning is employed to control the UAV network. The agent learns through trial and error, optimizing its policy by iterative querying through Q-learning processing (e.g., using Q-table).

4.3. Results and Discussion

In this subsection, we present the results of the proposed *PDRL-SDNC-UAV* and reference schemes, namely *SRL-SDNC-UAV* and *T-SDNC-UAV*, in terms of rewards and task delivery/drop ratios, primarily based on the efficient/inefficient computational resource conditions. The simulation setup conducts on the same topology; only the controller and agent are different for the performance metrics.

Figure 4 demonstrates the total rewards R^t_{env} throughout each scheme's exploration and exploitation within 500 episodes. In the exploration phase, the proposed scheme, *PDRL-SDNC-UAV*, achieves a significantly higher R^t_{env} compared to other schemes, with a difference of 11 and 32 positive scores when compared to *SRL-SDNC-UAV* and *T-SDNC-UAV*, respectively. In the exploitation phase, deep (RL)-based approaches achieve a completely stable reward compared to traditional control, which is 60% difference. At the ending episode, the proposed scheme reaches 34 positive scores which is 11.76% and 47.05% higher efficiency points compared to *SRL-SDNC-UAV* and *T-SDNC-UAV*, respectively. Figure 5 illustrates the main scoring metrics, which is r^t_{cr}. The proposed scheme configured the emphasized weight ω for efficient computational resources, which achieves 50% of the total rewards for this particular sub-reward, compared to 30% and 25% from *SRL-SDNC-UAV* and *T-SDNC-UAV*, respectively. *PDRL-SDNC-UAV*, *SRL-SDNC-UAV* and *T-SDNC-UAV* reached 17, 9, and 4.5, respectively, at the ending episode, which shows that the proposed scheme highly outperforms both baseline approaches.

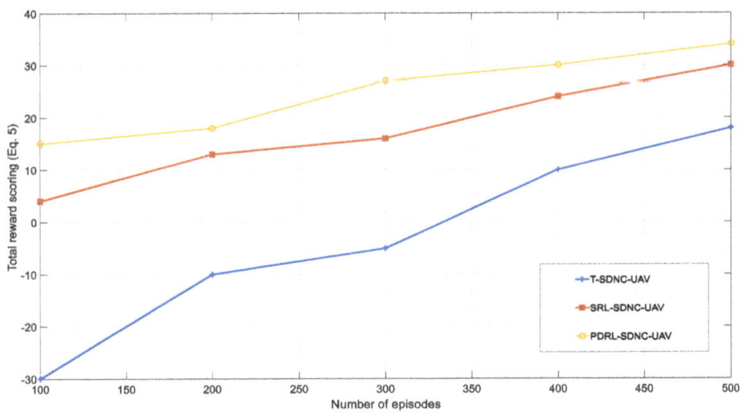

Figure 4. Performance metrics on total reward scoring.

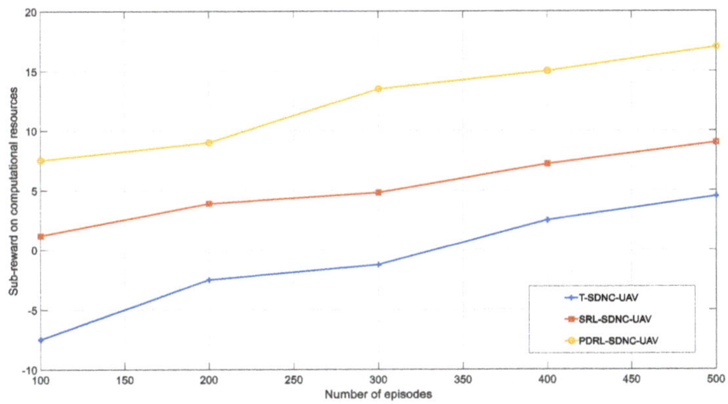

Figure 5. Performance metrics on sub-reward of computational resources.

The sub-rewards on energy r_e^t and latency r_d^t scoring points, presented in Figures 6 and 7, are primarily used to regulate the upper-bound tolerable delays and do not exceed the remaining energy (e_i^t, e_n^t), which leads the proposed scheme to slice only 25% for these two sub-rewards. For *T-SDNC-UAV*, the result on energy sub-reward is peaked due to default SDNC configuration following the energy limitation constraints. For result on latency sub-reward, *SRL-SDNC-UAV* achieved higher immediate reward than the proposed scheme due to the single processing compared to double network architecture; however, the efficiency of configured action remained lower than the proposed scheme. Regarding reward efficiency perspective, this proposed scheme handles dynamic adjustment offloading paths with efficient computing capacities, and the selection/placement actions/decisions on UAV-MEC pair node are highly adequate for improving the resource-constrained or computation-intensive tasks and also for adaptability in UAV-MEC-assisted network environments.

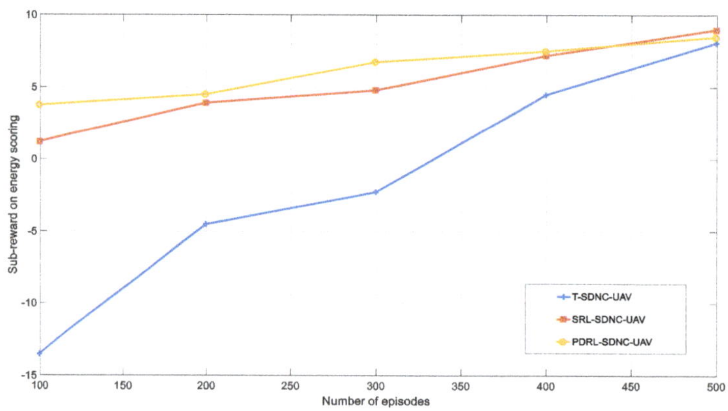

Figure 6. Performance metrics on sub-reward of energy scoring.

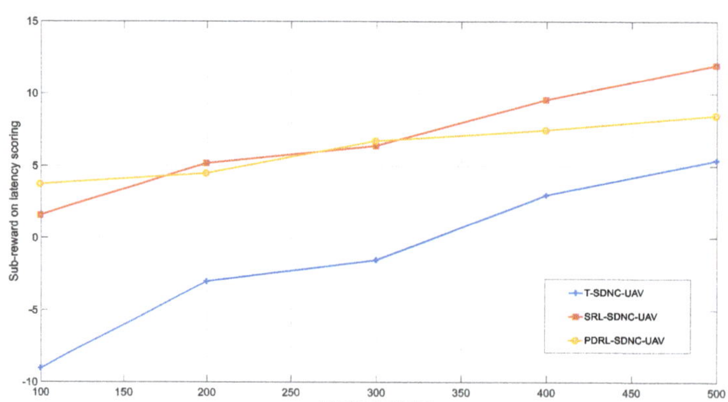

Figure 7. Performance metrics on sub-reward of latency scoring.

For task successful/failure ratios, the results are illustrated in Figures 8 and 9, where network states are consecutively configured by increasing congestion levels and application task complexity per 1000 simulation interval. We evaluate the agent flexibility by observing the result fluctuation differences compared to the diversity of heavy task and congestion states. The proposed scheme has the least fluctuation from 1000 to 5000 simulation times/conditions. *PDRL-SDNC-UAV* achieved the successful ratios (efficient offloading to adequate computational resources) from 99.91% to 99.98% (high to low intensity), which is (0.5%, and 0.03%) and (0.3% and 0.02%) better than *T-SDNC-UAV* and *SRL-SDNC-UAV*, respectively. In high computation-intensive and task generation rates, the failure rate of baseline approaches reached 0.59% and 0.39%, respectively. The proposed scheme indicates the policy reliability in achieving long-term adequacy, especially for multiple taxonomy-diversified IoT tasks. The resource-aware algorithm ensures the reliability and transferability of scalable UAV-MEC management systems by still efficiently considering the multi-aspect weighted sum reward on energy and delays.

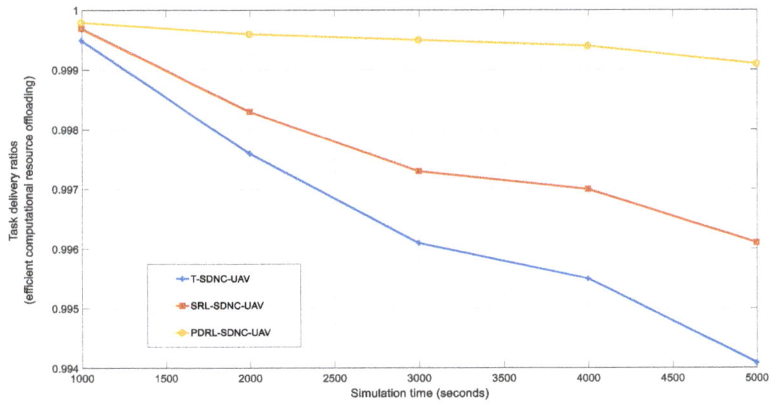

Figure 8. Performance metrics on task delivery ratios (successful rates from efficient computational resource control).

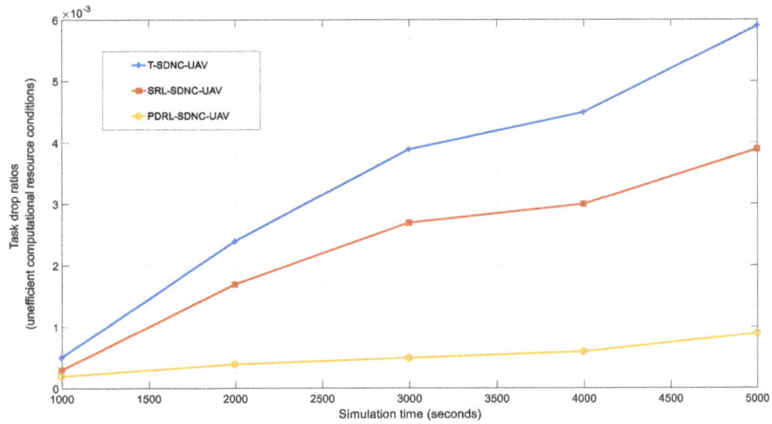

Figure 9. Performance metrics on task drop ratios (failure rates from inefficient selection/placement).

5. Conclusions and Future Work

This paper proposed a novel approach based on DRL for collaborating with SDN backbone control in UAV-assisted networks, termed PDRL-SDNC-UAV. The main objective was to develop an intelligent agent with primary sub-reward function on computational resource efficiency, and partial sub-rewards to meet the requirement on energy and delays. This paper presented the system architecture and interfaces, including the task processing in SDN-enabled 5G network systems. The components of DRL modeling are given, including the state observation, configurable actions in SDN policies and offloading decision, and reward objectives. Experimental simulations using OpenAI-TensorFlow and SDN emulator were conducted to evaluate the performance of the proposed approach compared to reference approaches in UAV-assisted networks. The evaluation considered reward measurements in terms of delay, energy, and resource. The task successful/failure is described to point out the stability of policy effectiveness over various network state conditions. *PDRL-SDNC-UAV* scheme outperformed the *SRL-SDNC-UAV* and *T-SDNC-UAV* schemes and achieved significantly higher rewards and more efficient resource utilization, indicating its effectiveness in optimizing UAV-assisted networking environments. Furthermore, the proposed scheme showed high reliability, even under varying levels of congestion and task complexity for scalable and long-term UAV-MEC management systems.

In future studies, an integrated softwarization on OpenAI-agent and SDN policy will be conducted and the extensive formulation on actor-critic will be used to advance this work for heterogeneous multi-UAV multi-MEC aspects. The state enhancement and interaction flow of each functionality will be extended for multi-objective awareness and demonstrated on improved DRL-based edge resource management.

Author Contributions: Conceptualization, I.S., P.T. and S.K. (Seokhoon Kim); methodology, I.S. and P.T.; software, P.T. and S.K. (Seungwoo Kang); validation, S.K. (Seungwoo Kang), I.S. and S.R.; formal analysis, S.K. (Seungwoo Kang), I.S. and S.R.; investigation, S.K. (Seokhoon Kim); resources, S.K. (Seokhoon Kim); data curation, P.T.; writing—original draft preparation, I.S. and P.T.; writing—review and editing, I.S. and P.T.; visualization, I.S.; supervision, S.K. (Seokhoon Kim); project administration, S.K. (Seokhoon Kim); funding acquisition, S.K. (Seokhoon Kim). All authors have read and agreed to the published version of the manuscript.

Funding: This work was supported by Institute of Information & communications Technology Planning & Evaluation (IITP) grant funded by the Korea government(MSIT) (No. RS-2022-00167197, Development of Intelligent 5G/6G Infrastructure Technology for The Smart City), in part by BK21 FOUR (Fostering Outstanding Universities for Research) under Grant 5199990914048, in part by the National Research Foundation of Korea (NRF), Ministry of Education, through Basic Science Research Program under Grant NRF-2020R1I1A3066543, and in part by the Soonchunhyang University Research Fund.

Data Availability Statement: Not applicable.

Conflicts of Interest: The authors declare no conflict of interest.

References

1. Mao, Y.; You, C.; Zhang, J.; Huang, K.; Letaief, K.B. A Survey on Mobile Edge Computing: The Communication Perspective. *IEEE Commun. Surv. Tutor.* **2017**, *19*, 2322–2358. [CrossRef]
2. Vhora, F.; Gandhi, J.C. A Comprehensive Survey on Mobile Edge Computing: Challenges, Tools, Applications. In Proceedings of the 2020 Fourth International Conference on Computing Methodologies and Communication (ICCMC), Erode, India, 11–13 March 2020; pp. 49–55.
3. Taleb, T.; Samdanis, K.; Mada, B.; Flinck, H.; Dutta, S.; Sabella, D. On Multi-Access Edge Computing: A Survey of the Emerging 5G Network Edge Cloud Architecture and Orchestration. *IEEE Commun. Surv. Tutor.* **2017**, *19*, 1657–1681. [CrossRef]
4. Duan, Q.; Wang, S.; Ansari, N. Convergence of Networking and Cloud/Edge Computing: Status, Challenges, and Opportunities. *IEEE Netw.* **2020**, *34*, 148–155. [CrossRef]
5. Sha, D.; Zhao, R. DRL-Based Task Offloading and Resource Allocation in Multi-UAV-MEC Network with SDN. In Proceedings of the 2021 IEEE/CIC International Conference on Communications in China (ICCC), Xiamen, China, 28–30 July 2021.
6. Do, Q.V.; Pham, Q.-V.; Hwang, W.-J. Deep Reinforcement Learning for Energy-Efficient Federated Learning in UAV-Enabled Wireless Powered Networks. *IEEE Commun. Lett.* **2022**, *26*, 99–103. [CrossRef]
7. Geraci, G.; Garcia-Rodriguez, A.; Azari, M.M.; Lozano, A.; Mezzavilla, M.; Chatzinotas, S.; Chen, Y.; Rangan, S.; Renzo, M.D. What Will the Future of UAV Cellular Communications Be? A Flight from 5G to 6G. *IEEE Commun. Surv. Tutor.* **2022**, *24*, 1304–1335. [CrossRef]
8. Li, B.; Fei, Z.; Zhang, Y. UAV Communications for 5G and Beyond: Recent Advances and Future Trends. *IEEE Internet Things J.* **2019**, *6*, 2241–2263. [CrossRef]
9. Kim, J.; Lee, J.; Yang, E.; Kang, S. Technology Forecasting from the Perspective of Integration of Technologies: Drone Technology. *KSII Trans. Internet Inf. Syst.* **2023**, *17*, 31–50.
10. Abdulghaffar, A.; Mahmoud, A.; Abu-Amara, M.; Sheltami, T. Modeling and Evaluation of Software Defined Networking Based 5G Core Network Architecture. *IEEE Access* **2021**, *9*, 10179–10198. [CrossRef]
11. Kiran, N.; Liu, X.; Wang, S.; Yin, C. VNF Placement and Resource Allocation in SDN/NFV-Enabled MEC Networks. In Proceedings of the 2020 IEEE Wireless Communications and Networking Conference Workshops (WCNCW), Seoul, Republic of Korea, 6–9 April 2020; pp. 1–6.

12. Hu, Y.; Zhu, L.; Zhang, J.; Cai, Z.; Han, J. Migration and Energy Aware Network Traffic Prediction Method Based on LSTM in NFV Environment. *KSII Trans. Internet Inf. Syst.* **2023**, *17*, 896–915.
13. Qiao, Q. Routing Optimization Algorithm for Logistics Virtual Monitoring Based on VNF Dynamic Deployment. *KSII Trans. Internet Inf. Syst.* **2022**, *16*, 1708–1734.
14. Tran, G.K.; Ozasa, M.; Nakazato, J. NFV/SDN as an Enabler for Dynamic Placement Method of mmWave Embedded UAV Access Base Stations. *Network* **2022**, *2*, 479–499. [CrossRef]
15. Zheng, J.; Tian, C.; Dai, H.; Ma, Q.; Zhang, W.; Chen, G.; Zhang, G. Optimizing NFV Chain Deployment in Software-Defined Cellular Core. *IEEE J. Sel. Areas Commun.* **2020**, *38*, 248–262. [CrossRef]
16. Tam, P.; Song, I.; Kang, S.; Ros, S.; Kim, S. Graph Neural Networks for Intelligent Modelling in Network Management and Orchestration: A Survey on Communications. *Electronics* **2022**, *11*, 3371. [CrossRef]
17. Ros, S.; Eang, C.; Tam, P.; Kim, S. ML/SDN-Based MEC Resource Management for QoS Assurances. In *Advances in Computer Science and Ubiquitous Computing*; Springer: Singapore, 2023; Volume 1028, pp. 591–597.
18. Ouamri, M.; Alkanhel, R.; Singh, D.; El-kenaway, E.S.M.; Ghoneim, S.S. Double Deep Q-Network Method for Energy Efficiency and Throughput in a UAV-Assisted Terrestrial Network. *Comput. Syst. Sci. Eng.* **2023**, *46*, 73–92. [CrossRef]
19. Ouamri, M.; Alkanhel, R.; Gueguen, C.; Alohali, M.; Ghoneim, S.S. Modeling and Analysis of UAV-Assisted Mobile Network with Imperfect Beam Alignment. *Comput. Mater. Contin.* **2022**, *74*, 453–467. [CrossRef]
20. Liu, Q.; Cheng, L.; Jia, A.L.; Liu, C. Deep Reinforcement Learning for Communication Flow Control in Wireless Mesh Networks. *IEEE Netw.* **2021**, *35*, 112–119. [CrossRef]
21. Tian, A.; Feng, B.; Zhou, H.; Huang, Y.; Sood, K.; Yu, S.; Zhang, H. Efficient Federated DRL-Based Cooperative Caching for Mobile Edge Networks. *IEEE Trans. Netw. Serv. Manag.* **2022**, *20*, 246–260. [CrossRef]
22. Tam, P.; Corrado, R.; Eang, C.; Kim, S. Applicability of Deep Reinforcement Learning for Efficient Federated Learning in Massive IoT Communications. *Appl. Sci.* **2023**, *13*, 3083. [CrossRef]
23. Yang, Z.; Pan, C.; Wang, K.; Shikh-Bahaei, M. Energy Efficient Resource Allocation in UAV-Enabled Mobile Edge Computing Networks. *IEEE Trans. Commun.* **2019**, *18*, 4576–4589. [CrossRef]
24. He, Y.; Zhai, D.; Huang, F.; Wang, D.; Tang, X.; Zhang, R. Joint Task Offloading, Resource Allocation, and Security Assurance for Mobile Edge Computing-Enabled UAV-Assisted VANETs. *Remote Sens.* **2021**, *13*, 1547. [CrossRef]
25. Tan, T.; Zhao, M.; Zeng, Z. Joint Offloading and Resource Allocation Based on UAV-Assisted Mobile Edge Computing. *ACM Trans. Sens. Netw. (TOSN)* **2022**, *18*, 1–21. [CrossRef]
26. Munaye, Y.Y.; Juang, R.-T.; Lin, H.-P.; Tarekegn, G.B.; Lin, D.-B. Deep Reinforcement Learning Based Resource Management in UAV-Assisted IoT Networks. *Appl. Sci.* **2021**, *11*, 2163. [CrossRef]
27. Yu, F.; Yang, D.; Wu, F.; Wang, Y.; He, H. Resource Optimization for UAV-Assisted Mobile Edge Computing System Based on Deep Reinforcement Learning. *Phys. Commun.* **2023**, *59*, 102107. [CrossRef]
28. Ren, Y.; Guo, A.; Song, C. Multi-Slice Joint Task Offloading and Resource Allocation Scheme for Massive MIMO Enabled Network. *KSII Trans. Internet Inf. Syst.* **2023**, *17*, 794–815.
29. Song, I.; Kang, S.; Tam, P.; Kim, S. Federated Logistic Regression for Reliable Prediction Models in Privacy-Preserving Healthcare Networks. In Proceedings of the 2022 6th International Conference on Interdisciplinary Research on Computer Science, Psychology, and Education (ICICPE' 2022), Pattaya, Thailand, 31 October 2022.
30. Zhang, P.; Su, Y.; Li, B.; Liu, L.; Wang, C.; Zhang, W.; Tan, L. Deep Reinforcement Learning Based Computation Offloading in UAV-Assisted Edge Computing. *Drones* **2023**, *7*, 213. [CrossRef]
31. Lin, C.; Han, G.; Shah, S.B.H.; Zou, Y.; Gou, L. Integrating Mobile Edge Computing into Unmanned Aerial Vehicle Networks: An SDN-Enabled Architecture. *IEEE Internet Things Mag.* **2021**, *4*, 18–23. [CrossRef]
32. Sellami, B.; Hakiri, A.; Ben Yahia, S.; Berthou, P. Deep Reinforcement Learning for Energy-Efficient Task Scheduling in SDN-Based IoT Network. In Proceedings of the 2020 IEEE 19th International Symposium on Network Computing and Applications (NCA), Cambridge, MA, USA, 24–27 November 2020; pp. 1–4.
33. Zhang, L.; Lu, Y.; Zhang, D.; Cheng, H.; Dong, P. DSOQR: Deep Reinforcement Learning for Online QoS Routing in SDN-Based Networks. *Secur. Commun. Netw.* **2022**, *2022*, 4457645. [CrossRef]
34. Tam, P.; Song, I.; Kang, S.; Kim, S. Privacy-Aware Intelligent Healthcare Services with Federated Learning Architecture and Reinforcement Learning Agent. In *Advances in Computer Science and Ubiquitous Computing*; Springer: Singapore, 2023; Volume 1028, pp. 583–590.
35. Ros, S.; Tam, P.; Kim, S. Modified Deep Reinforcement Learning Agent for Dynamic Resource Placement in IoT Network Slicing. *J. Internet Serv. Appl.* **2022**, *23*, 17–23.

Disclaimer/Publisher's Note: The statements, opinions and data contained in all publications are solely those of the individual author(s) and contributor(s) and not of MDPI and/or the editor(s). MDPI and/or the editor(s) disclaim responsibility for any injury to people or property resulting from any ideas, methods, instructions or products referred to in the content.

electronics

MDPI

Article

Intelligent Embedded Systems Platform for Vehicular Cyber-Physical Systems

Christopher Conrad *, Saba Al-Rubaye and Antonios Tsourdos

Centre for Autonomous and Cyberphysical Systems, Cranfield University, Cranfield MK43 0AL, UK; s.alrubaye@cranfield.ac.uk (S.A.-R.); a.tsourdos@cranfield.ac.uk (A.T.)
* Correspondence: christopher.conrad@cranfield.ac.uk

Abstract: Intelligent vehicular cyber-physical systems (ICPSs) increase the reliability, efficiency and adaptability of urban mobility systems. Notably, ICPSs enable autonomous transportation in smart cities, exemplified by the emerging fields of self-driving cars and advanced air mobility. Nonetheless, the deployment of ICPSs raises legitimate concerns surrounding safety assurance, cybersecurity threats, communication reliability, and data management. Addressing these issues often necessitates specialised platforms to cater to the heterogeneity and complexity of ICPSs. To address this challenge, this paper presents a comprehensive CPS to explore, develop and test ICPSs and intelligent vehicular algorithms. A customisable embedded system is realised using a field programmable gate array, which is connected to a supervisory computer to enable networked operations and support advanced multi-agent algorithms. The platform remains compatible with multiple vehicular sensors, communication protocols and human–machine interfaces, essential for a vehicle to perceive its surroundings, communicate with collaborative systems, and interact with its occupants. The proposed CPS thereby offers a practical resource to advance ICPS development, comprehension, and experimentation in both educational and research settings. By bridging the gap between theory and practice, this tool empowers users to overcome the complexities of ICPSs and contribute to the emerging fields of autonomous transportation and intelligent vehicular systems.

Keywords: intelligent system; cyber-physical system; vehicular; development platform; networking

1. Introduction

This section introduces intelligent cyber-physical systems (ICPSs) as a catalyst for advancing intelligent and autonomous transportation. For brevity, subsequent references to ICPSs will thereby imply intelligent vehicular cyber-physical applications. ICPS literature is first reviewed to determine the core methods used to develop ICPSs and summarise the barriers impeding their widespread adoption in vehicular applications. These challenges highlight the need for a simple yet comprehensive platform to facilitate ICPS development and promote research in intelligent and autonomous transportation systems. Specifically, field programmable gate arrays (FPGAs) are identified as a robust technology to realise an adaptable network of embedded systems to explore, develop and test ICPSs and intelligent vehicular algorithms. This work thereby prototypes an FPGA-based embedded system that serves as a versatile tool to advance ICPSs in both educational and research settings.

1.1. Intelligent Vehicular Cyber-Physical Systems

ICPSs are an advanced technology that bring together the physical and digital worlds through the use of sensors, actuators, and intelligent software. These systems are designed to interact with a physical environment, collect data from sensors, and analyse that data in real-time using advanced algorithms and machine learning (ML) techniques. Consequently, ICPSs can make intelligent decisions and execute actions based on incoming data [1]. ICPSs have the ability to transform various industries such as manufacturing, transportation, and energy. They can enhance efficiency, reduce costs, and improve safety in these industries by providing real-time data and intelligent insights to help optimise the underlying system processes. Nonetheless, Rani et al. [2] emphasize that the increased connectivity of ICPSs also brings about significant privacy and security concerns. As ICPSs become more prevalent, it is crucial to ensure that they are accurate, reliable, secure, and protected from cyberattacks. Despite these risks, ICPSs represent the next frontier of technology, with the potential to revolutionise the way we live, work and commute in the future [3].

Raj et al. [4] conclude that the integration of Internet of Things (IoT) with ICPSs is paramount for the advancement of transportation systems in smart cities. The authors assert that autonomous ground vehicles can significantly enhance the efficiency of traffic management systems, thereby improving road safety by establishing a smart and collaborative transportation infrastructure. Similarly, Mahrez et al. [5] confirm that real-time intelligent transportation systems (ITSs) safeguard vulnerable road users by employing collision warning systems, speeding alerts, safety indicators, and enhanced vision, radar and navigation systems. Moreover, they emphasize the potential of ITSs in predicting passenger, driver and traffic behaviour, which facilitates more efficient traffic management and routing algorithms. The authors also present ITSs as a key enabler of electric vehicles with varying degrees of autonomy, ranging from assisted driving to semi-automated or fully-autonomous vehicular systems. Liu [6] further presents a business case for autonomous vehicles equipped with infrastructure-to-vehicle communication capabilities, and concludes that a collaborative ITS is more efficient and cost-effective than relying solely on on-board autonomy. Similarly, Pan et al. [7] showcase a smart cloud-based commuting system that uses collaborative autonomous vehicles to improve passenger experience, vehicle utilisation and system efficiency within an ITS.

Nevertheless, the implementation of an ITS poses significant challenges. The transportation cyber-physical system (CPS) is a complex entity encompassing diverse physical components, including vehicles, road infrastructure, human drivers, machines, and sensors. In particular, the topology of the vehicular transportation network is in constant flux due to the high-speed movement of vehicles [8]. Moreover, human driver behaviour directly impacts the network topology, as drivers can join or exit the network at any given time. The density of vehicles within the transportation CPS also varies depending on the location. Furthermore, vehicular networks distinguish themselves from other wireless networks by boasting virtually unlimited power, storage, and computing capabilities [9]. Additionally, most existing wireless access technologies are not specifically designed to accommodate fast-moving vehicles. In the vehicular CPS, low latency is crucial for safety-critical applications that require the timely transmission of emergency messages, but bandwidth limitations arise, especially when considering multimedia and infotainment transmissions.

ICPSs play a critical role in enabling autonomous vehicles within the transportation CPS. Chen et al. [10] conduct a systematic review of computing and communication systems proposed for intelligent vehicles. Specifically, they divide the vehicular architecture into power and thermal management, storage management, security features, communication systems, computational hardware, and a variety of sensors and algorithms used to enable advanced perception, localisation, control and decision making. The authors identify light detection and ranging (LiDAR), radar, cameras, global navigation satellite systems (GNSS) and ultrasound sensors as the most commonly employed sensing systems within autonomous vehicles. They also emphasize the importance of robust communication protocols for vehicle-to-vehicle (V2V), vehicle-to-infrastructure (V2I), vehicle-to-pedestrian (V2P), and vehicle-to-cloud/network (V2C/N) communication. Similarly, Novak et al. [11] review the relationship between autonomous driving behaviours and the underlying sensing and computing technologies of an ICPS within an ITS framework. Additionally, Kliestik et al. [12] review state-of-the-art sensor fusion techniques and perception algorithms that enable autonomous vehicles in an ITS. Consequently, a development platform for ICPSs must remain compatible with typical sensing and communication systems, while facilitating research in each architectural layer of an intelligent vehicle.

The conclusions drawn for autonomous ground vehicles and self-driving cars can be extended to airborne vehicles, such as unmanned aircraft systems (UASs) and air taxis. In fact, the emerging fields of advanced air mobility (AAM) and UAS traffic management (UTM) envision a comprehensive aerial ITS for passenger and cargo deliveries. Pan et al. [13] systematically review advancements in flying car transportation systems and discuss the potential of AAM to revolutionise the transportation sector. Dense AAM operations, however, require advanced communication, navigation and surveillance (CNS) capabilities, robust cybersecurity systems and enhanced collaborative decision-making frameworks. Specifically, ICPSs for AAM impose stricter requirements compared to ground-based vehicles due to the greater safety risks associated with aerial operations. Nonetheless, the underlying system architecture remains analogous to that of traditional transportation systems. In fact, Zhou et al. [14] propose a computation and communication framework for UASs, and highlight the importance of collaborative V2V and V2I communication in improving the efficiency of the ITS. Furthermore, Shrestha et al. [15] emphasize the necessity of reliable communication protocols for widespread adoption of AAM, suggesting that 6G-enabled ecosystems will eventually supersede current 5G proposals.

Javaid et al. [16] review collaborative communication technologies for UASs as a key enabler of multi-UAS systems. Furthermore, Wilson et al. [17] review the sensors, communication technologies, and ML techniques commonly employed in UAS applications. Typical UASs incorporate camera, LiDAR, radar, ultrasonic, and inertial measurement unit (IMU) sensors, similar to ground-based vehicles. Additionally, wireless-fidelity (Wi-Fi), Bluetooth, long-term evolution (LTE) and long-range (LoRa) communication technologies are found to be prevalent in UAS applications. These findings further reinforce the hypothesis that a robust ICPS development platform must be compatible with a diverse range of common sensors and communication technologies.

1.2. Intelligent Vehicular Cyber-Physical System Methods

ICPS methods are employed to design, develop, and operate the systems that integrate physical processes with digital technologies. These methods are essential for creating safe, reliable, and efficient CPSs within the transportation CPS [18–21]. By integrating computing, communication, and control with physical tasks, CPSs can often optimise industrial operations to achieve improved reliability, efficiency and flexibility [22]. Moreover, the ability to monitor and control operations in real-time often results in quicker fault resolution and improved safety, especially in the context of autonomous vehicles [1]. In fact, CPSs serve as the foundation of Industry 4.0, a term used to describe increased automation, connectivity and data exchange in industrial processes. Some commonly used CPS methods are hereby presented:

- Model-Based Design: This approach involves developing mathematical models of the physical system and using them to design the control algorithms and software. The models can be simulated to assess system performance prior to implementation;
- Control Theory: Control theory is used to design control systems that regulate physical processes and maintain stability. It involves analysing the system dynamics and developing feedback control algorithms to adjust system behaviour;
- Real-Time Systems: Real-time systems ensure that a CPS responds to changes in the physical environment in a timely and predictable manner. This involves developing software that can execute in real-time and meet system timing requirements;
- Cybersecurity: Cybersecurity is essential for safeguarding CPSs from cyberattacks that can compromise their safety and reliability. CPS methods for cybersecurity include the development of secure communication protocols, intrusion detection systems, and access control mechanisms;
- Machine Learning: ML is used to develop ICPSs that can adapt to changing environments and optimise their performance. This involves training algorithms using data from sensors and other sources to make intelligent predictions and decisions.

Notably, on-going developments in artificial intelligence (AI), ML and cloud computing foster the design of intelligent systems capable of managing multiple intelligent sensors embedded in ICPSs, as depicted in Figure 1 [23]. These systems support fault prediction, increased autonomy and self-healing or self-adapting algorithms for safer and more reliable implementations. Additionally, embedded AI solutions can optimise energy usage and bandwidth utilisation of interconnected agents, leading to improved system performance and efficiency. Shuvo et al. [24] conduct a review of deep learning techniques applied to edge devices such as CPSs, IoT, autonomous systems, embedded hardware, and intelligent sensors. They highlight the challenges associated with optimising AI algorithms for embedded systems, often addressed through novel algorithms, improved optimisation, and more effective hardware acceleration or hardware–software co-design. Consequently, the evolution of ICPSs is inevitable to support the increased levels of autonomy and integration demanded by industry. A robust development platform for ICPSs must therefore facilitate the deployment of intelligent algorithms at both embedded and networked levels.

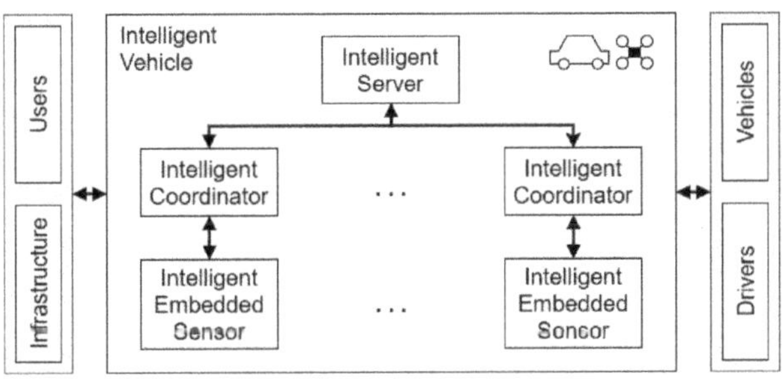

Figure 1. Generic architecture of an ICPS.

1.3. Cyber-Physical Challenges and Technical Barriers

Despite the benefits of ICPSs, several challenges and technical barriers hinder their widespread adoption, as outlined in Figure 2. Pundir et al. [25], in fact, suggest that the heterogeneity and complexity of the transportation CPS makes an ITS particularly susceptible to cyber vulnerabilities and technological challenges.

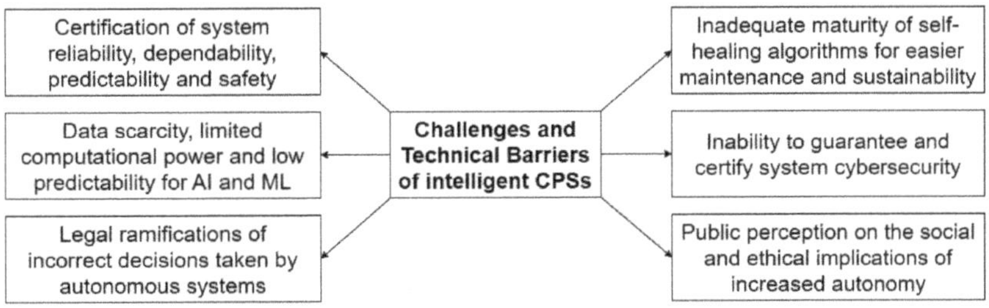

Figure 2. Challenges and technical barriers of ICPSs.

Reliability, dependability and predictability are fundamental requirements of any safety-critical system. In autonomous vehicles, for instance, unpredictable behaviour may lead to fatalities and significant damage to surrounding infrastructure. As technologies evolve, however, sensors, interfaces and CPS capabilities are becoming more diverse [26]. The lack of a unified set of standards makes it difficult to ascertain the interoperability and reliability of a system without undergoing numerous and potentially expensive in-field tests. Moreover, accurately modelling complex CPSs remains a considerable challenge for their widespread adoption. A unified set of cyber-physical standards must therefore be developed for different use-cases and applications.

AI algorithms introduce additional uncertainty, and may take erroneous decisions in scenarios for which they were not properly trained. This issue is particularly problematic when autonomous vehicles rely on external infrastructure for navigation, since road markings and landmarks often vary between different locations and tend to deteriorate over time. AI must therefore become more explainable and deterministic to support its use in safety-critical applications. Airbus, in fact, identifies the certification of AI as one of the biggest hurdles to introduce autonomous aerial vehicles into the airspace [27]. Additionally, regulations on data privacy may limit the data available to train intelligent algorithms. Consequently, system performance may deteriorate since AI models are trained over an insufficiently generic data pool [28]. Furthermore, the large computational power needed to train AI models poses a significant barrier to their adoption in low-cost systems [29].

The legal ramifications of incorrect decisions taken by ICPSs must also be properly defined. Legislation should clearly outline who has legal responsibility and insurance liability in the case of damage caused by autonomous robots or vehicles [30]. In fact, appropriate governmental policies are deemed necessary by Tran et al. [31] to develop and implement a successful ITS infrastructure with autonomous vehicles. Ethical dilemmas may also arise if, for example, an autonomous vehicle must choose between protecting the driver or safeguarding a passer-by in emergency situations.

Furthermore, technologies supporting the sustainability and maintainability of CPSs are insufficiently developed for many real-world applications. Several self-adapting and self-healing algorithms have been proposed in the literature, with most taking inspiration from biological and physical processes [32–34]. Such techniques, however, have yet to be adopted in many practical applications, and considerable testing is still needed to ensure their reliability in safety-critical scenarios. Additionally, Shakeri et al. [35] note that many existing data management and network technologies are inadequate to handle the increasing amounts of expected data traffic, highlighting the need for significant improvements in the scalability of CPSs.

Cybersecurity poses another significant challenge in CPSs, with Alguliyev et al. [36] suggesting that predicting, modelling and preventing cyberattacks are the most complex problems in any CPS. Al-Mhiqani et al. [37] comprehensively review historical cyberattacks on government infrastructure, transport systems and industrial companies and propose a risk analysis technique to evaluate the threat associated with each attack. While research on cybersecurity is ongoing, considerable work is still needed to cater for emerging hacking techniques and develop appropriate security standards. Cyberattacks on autonomous vehicles could otherwise cause catastrophic outcomes, serving as an entry point for cyber warfare or remote terrorist attacks.

The public perception of AI also presents a fundamental obstacle to the widespread adoption of ICPSs. Kurniawan et al. [38] argue that research on the social implications of autonomous vehicles is insufficiently developed. Understanding and shaping the social acceptance of ITSs can therefore be achieved by identifying the social driving forces behind autonomous vehicles. Redundancies due to automated labour, for instance, raise several social and ethical concerns. Community engagement is therefore critical to foster public acceptance and ensure that increased autonomy is mutually beneficial for both industry and society. Moreover, Pelau et al. [39] suggest that incorporating anthropomorphic features and emulating empathy in AI-based systems can enhance consumer acceptance. Denis et al. [40] also advocate for a socially equitable approach to ICPSs in urban mobility, and highlight the significant development required by developing countries to meet the infrastructural demands of ITSs.

A user-friendly development platform for ICPSs can help bridge the gap between theory and practice by providing a versatile tool to address the outstanding challenges of ICPSs. Technological, cybersecurity, and AI-related barriers can be directly addressed through exploratory research conducted within such a system. Furthermore, the platform can be utilised for educational purposes to facilitate discussions on regulations, social implications, and ethical concerns surrounding ICPSs, ITSs, and autonomy.

1.4. Field Programmable Gate Arrays

Overcoming the technical barriers of CPSs requires robust hardware platforms to support increasingly more complex intelligent algorithms. FPGAs are a promising candidate, and are defined by Xilinx as:

> semiconductor devices that are based around a matrix of configurable logic blocks (CLBs) connected via programmable interconnects, [that] can be reprogrammed to desired application or functionality requirements after manufacturing [41].

Boutros et al. [42] highlight the dynamic reconfiguration capabilities of this technology, thereby supporting self-healing and adaptive hardware algorithms. Moreover, the high degree of parallelism offered by FPGAs allows them to better replicate parallel physical processes and support quicker execution of AI algorithms.

Magyari et al. [43] analyse state-of-the-art FPGA applications in IoT networks, highlighting their use in ICPSs. They suggest that hardware designs can achieve different trade-offs between minimising latency, power consumption and form factor, and maximising throughput, security and data integrity. FPGAs, in fact, facilitate prototyping, and support a shorter time to market and a simpler design cycle when compared to traditional integrated circuit (IC) implementations. Consequently, FPGAs are particularly suitable for demonstrating, developing and testing new ICPS solutions.

1.5. Literature Gaps and Contributions

Existing literature on ICPSs reveals several gaps that must be addressed. Fundamentally, mechanical and cybernetic design philosophies must be better integrated in a single design approach, backed by reliable regulatory, legal and standardisation frameworks that support the certification of AI for increased autonomy. Additionally, ML algorithms must become more reliable to support their use in safety-critical applications such as self-driving cars and autonomous aerial vehicles. Similarly, advances in swarm intelligence [44], organic computing [45] and multiagent collaboration [46] will facilitate the integration of multiple cyber-physical agents into a cohesive ICPS with adaptive and self-sustaining characteristics.

Optimising both hardware and software components is crucial to improve the efficiency of embedded systems and support the transition towards a net-zero transportation industry [47]. Moreover, cloud computing must improve to cater for the high data management demands brought about by data-driven algorithms [48]. Cybersecurity protocols must also evolve to combat the cyber vulnerabilities of safety-critical ICPSs [49–51].

Research to overcome these technical barriers is ongoing, but addressing state-of-the-art concerns of ICPSs falls beyond the scope of this work. A gap, however, exists for a simple yet comprehensive ICPS platform that can be used for both educational purposes and exploratory cutting-edge research. Mohamed et al. [52] review engineering tools and languages for CPSs. Most existing platforms, however, tend to (1) require a large initial learning curve, making them unsuitable for educational purposes; (2) offer insufficient performance capabilities to support their use for prototyping state-of-the-art algorithms and ICPSs; or (3) solely focus on a single aspect of ICPSs, restricting their ability to demonstrate and explore all the elements that make up a complete ICPS [53].

This work introduces a simple yet robust platform, suitable for both beginners and experienced users, to explore the various elements of ICPSs. The platform supports the integration of multiple sensors, human–machine interfaces (HMIs) and data/communication links in an FPGA-based embedded system. Multiple embedded agents can also interface with a supervisory computer for high-level networking, data capture, processing and display. The entire system thereby allows for both standard and intelligent algorithms to be implemented at both embedded and networked levels, and remains compatible with a variety of sensor interfaces, hardware platforms, cybersecurity algorithms and communication protocols. This platform is suitable for research and educational endeavours, promoting discussions on ICPSs and ITSs while bridging the gap between ICPS theory and practice. The main contributions of this work are as follows:

- Design techniques and considerations for intelligent embedded systems in ICPSs are comprehensively discussed and summarised;
- A robust FPGA-based embedded system is proposed as an exploratory development platform for ICPSs, supporting a range of hardware and software implementations for intelligent vehicular applications;
- The basic functionality of the proposed platform is validated by designing, implementing and testing a simple FPGA-based embedded system with a soft microprocessor core. This interfaces with an IMU and rotary encoder, as typical elements of a vehicular CPS, using embedded software to process the collected data and transmit the system location/orientation and encoder position to a supervisory computer whenever the encoder shaft is rotated. These data are captured and plotted in real-time by a Matlab application running on the host computer;
- The developed prototype is reviewed to discuss the suitability of a simple FPGA-based platform for educational and research endeavours concerning ICPSs and ITSs.

2. System Design and Implementation Methodologies

This section discusses design considerations that are crucial when developing an intelligent embedded system for ICPSs, and summarises the design objectives and requirements of this work. It subsequently presents the designed system architecture, including all hardware and software elements. This is complemented by an overview of all data processing algorithms and ancillary HMI features included within the developed prototype.

2.1. Design Considerations for Intelligent Vehicular Embedded Systems

Lee [54] suggests that existing design abstractions for hardware and software systems are insufficiently compatible to support the development of CPSs, and must be redefined to better complement the realities of physical processes. Nonetheless, Khaitan et al. [55] provide a comprehensive analysis of design techniques and applications of CPSs, highlighting common approaches found in the literature. They propose that effective hardware and software co-design is a well-established approach that can streamline and parallelise system development. Similarly, Segura et al. [56] use a distributed co-simulation protocol to propose a generic interface for CPS simulation and development.

A co-design methodology is particularly important for an embedded system, where hardware and firmware must be seamlessly integrated in a single product. An example of such a co-design flow is illustrated in Figure 3 [57] and promotes a simple yet effective methodology for the design of intelligent embedded systems. The system description is first used to identify the primary objectives of the embedded platform. This is partitioned into unique software and hardware requirements, such that both elements can be developed in parallel. Sufficient inter-communication is further required to guarantee hardware–software compatibility. At each development milestone, remaining tasks are re-partitioned, and the process is repeated until the final embedded system is complete.

When adopting a co-simulation design approach, several design trade-offs need to be considered based on the specific system under development. Some key considerations for intelligent embedded systems are depicted in Figure 4 and hereby discussed.

Primarily, an embedded system should maximise its performance in terms of data throughput, accuracy and speed, particularly in real-time systems [58]. This often imposes a set of memory and processing requirements on the hardware platform, especially when complex AI algorithms must be implemented. Notably, autonomous vehicles require high CNS performance capabilities to remain connected to the ITS, while safely navigating their environment and avoiding collisions. Bijjahalli et al. [59], in fact, suggest that accurate navigation is critical for self-driving cars, and propose a low-cost solution to meet this requirement. Additionally, Namuduri et al. [60] systematically review the current research on CNS systems for AAM, and highlight the challenges that must be addressed for wide-scale AAM operations. These include secure V2V communications, enhanced navigation, and advanced detect and avoid (DAA) capabilities in hostile environments.

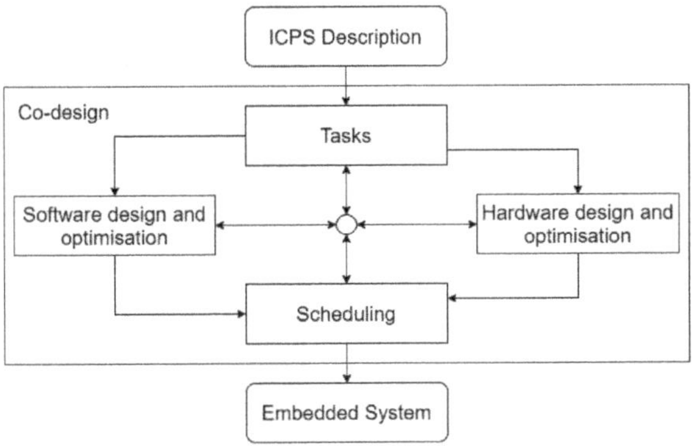

Figure 3. Overview of hardware–software co-design for embedded systems.

In addition to performance, embedded systems must also strive for low power consumption, particularly in battery-operated vehicles and devices [61]. Electric vehicles, for example, require efficient power usage to minimise the frequency of charging. Moreover, Swaminathan et al. [62] affirm that the battery life directly impacts the range, utility and business case of AAM, underscoring the significance of power efficiency in vehicular systems. A trade-off between performance and power efficiency must therefore be achieved. Furthermore, embedded solutions must respect the limitations of available hardware resources, which may further constrain circuit design and optimisation requirements [63]. Timilsina et al. [64], for instance, demonstrate that batteries in electric vehicles have a shorter lifespan than other on-board components, imposing the most stringent design requirements on the vehicular architecture.

Safety and reliability are also critical design factors, especially when devices are embedded in an ICPS. Compliance with industrial standards for electromagnetic interference (EMI), electromagnetic compatibility (EMC), and electrostatic discharge (ESD) protection is particularly important to ensure reliable operation in noisy environments, including industrial machinery and autonomous vehicles. Zhang et al. [65], in fact, highlight the potential damage to an autonomous vehicle's transceiver caused by a high-power microwave pulse, while Arandhakar et al. [66] emphasize the impact of EMI and temperature effects on electric vehicle performance. Printed circuit boards (PCBs) must also be well-designed to withstand harsh environments in which they may be deployed. PCBs within moving vehicles on rough terrain, for instance, require greater resilience to vibrations, while those embedded close to high-power components require a higher thermal tolerance. Fault tolerant designs and hardware redundancies may also enhance system reliability, particularly in high-risk environments with limited accessibility, such as satellites in outer space [67].

Hardware reliability must be complemented by good cybersecurity features, such that the embedded system is well-protected from invasive attacks that may compromise system integrity. Sharma et al. [68] review state-of-the-art cybersecurity developments for connected autonomous vehicles, stressing the importance of AI-based solutions to combat cyberattacks. Moreover, Kukkala et al. [69] discuss the cybersecurity challenges surrounding autonomous vehicles, and propose a roadmap to address outstanding concerns. Open challenges include ensuring data privacy, developing tamper-proof AI solutions and evolving cybersecurity protocols to adapt to emerging cyberattack methodologies. Establishing public acceptance and trust in autonomous vehicles and ITSs is also crucial. Consequently, Khan et al. [70] suggest the use of cybersecurity digital labels to inform consumers and passengers on the cybersecurity health status of intelligent vehicles.

Furthermore, a well-designed embedded system should support future upgrades and modifications, thereby promoting flexibility and adaptability. Proper documentation is therefore necessary to ensure that future developers can readily understand the existing system design and implementation. Moreover, embedded systems must offer a simple and effective interface to interact with other agents in the ICPS. In particular, networked channels and communication buses must be efficiently managed to prevent a single embedded system from saturating a common bus and stalling the entire system. To address this concern, Choi et al. [71] propose technical solutions to meet the data transfer and latency requirements of ICPSs using a communication bus topology.

Apart from technical considerations, an embedded system must conform to a specific form factor according to its allocated enclosure. The practicality of the system should also be taken into account by minimising costs, predicting a reasonable time-to-market, and considering potential supply chain shortages [72]. Additionally, the design should incorporate a suitable HMI that augments human capabilities within the ICPS. Tan et al. [73] systematically review the challenges and opportunities surrounding HMIs for intelligent vehicular applications. They highlight the complexity and heterogeneity of human intentions as a significant challenge for predictive HMI systems, and suggest that recognition performance must be improved to enable more advanced HMIs. They further observe a lack of datasets to adequately train HMI models for autonomous vehicle applications.

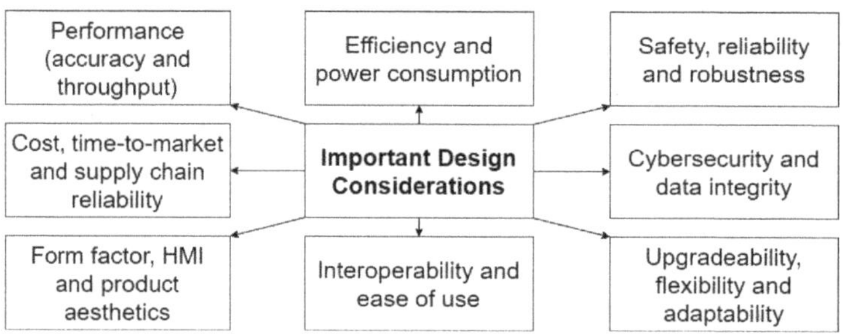

Figure 4. Design considerations for intelligent embedded systems.

2.2. Design Requirements and Objectives

This work focuses on the development of a simple FPGA-based embedded system that serves as a versatile tool for educational and research purposes in the field of ICPSs. The objective is to create a platform that remains compatible with various CNS technologies while accurately representing the unique requirements of intelligent vehicular systems. To demonstrate its capabilities, a simple sensory use-case is considered, showcasing how the FPGA-based embedded system can integrate different aspects of an ICPS and bridge the gap between theory and practice.

The selected approach involves using an FPGA development board, which provides a simple solution for interfacing the embedded system with various commercial off-the-shelf (COTS) components. The FPGA is connected to an IMU and rotary encoder, representing common sensors and interfaces found in intelligent vehicles. Whenever the encoder shaft is rotated, the device transmits its location/orientation and the shaft position to a supervisory computer over a serial communication interface. A high-level application subsequently captures and plots incoming data in real-time. The prototype thereby encapsulates typical components of a vehicular architecture. It serves as a preliminary demonstration on how the proposed platform can be extended to realise a complete ICPS for educational or research purposes, by integrating a wide variety of different sensors, actuators, processor implementations, communication links and HMIs. Consequently, this prototype does not aim to meet specific performance, safety, or security requirements. Instead, its purpose is to present a simple, robust, flexible, and upgradeable approach for deploying and testing ICPS solutions and intelligent vehicular algorithms.

Flexibility and upgradeability are identified as critical design features for an experimentation platform which aims to support future ICPS designs and algorithms. To achieve these characteristics, a COTS FPGA development board is used, allowing for hardware reconfiguration and interfacing with different COTS sensors and modules. Additionally, a soft microcontroller core with limited memory is employed to run embedded firmware. This offers a familiar coding platform for users that are unfamiliar with hardware description languages (HDLs), while simulating the memory and processing constraints of real embedded systems. Nonetheless, HDL development is still supported, ensuring that the platform remains relevant for low-level implementations of embedded ICPS solutions.

To ensure a user-friendly CPS, an on-board HMI is implemented with common features like a reset switch and light emitting diode (LED) indicator. The system also establishes a simple data link between the embedded agent and a supervisory computer, which can be replaced with more complex wired or wireless communication links for testing communication protocols and cybersecurity algorithms. The supervisory computer can further act as an intelligent coordinator or server, facilitating communication with multiple embedded agents to realise a complex multi-agent ICPS. Mizutani et al. [74] even propose a network solution for inter-FPGA communication in collaborative networking applications. This confirms that an FPGA-based embedded solution can be readily extended to collaborative multi-agent vehicular applications.

In summary, the prototyped architecture involves a single FPGA-based embedded agent interfacing with on-board HMIs, displays, and sensors. In the considered use-case, the FPGA controls an LED display and communicates with an IMU and rotary encoder. Additionally, a soft microcontroller core is implemented within the FPGA to enable the deployment of custom embedded software within the ICPS agent. Each agent is connected to a supervisory computer via a wired data link, which can be upgraded to support wireless communication systems. The supervisory computer captures, processes, and displays incoming data, and can also transmit commands to the embedded agent for controlling data transmission. Furthermore, the computer can be connected to multiple agents to create a networked set of embedded ICPS agents, resembling connected vehicles or vehicle components within an ITS. This is illustrated in Figure 5.

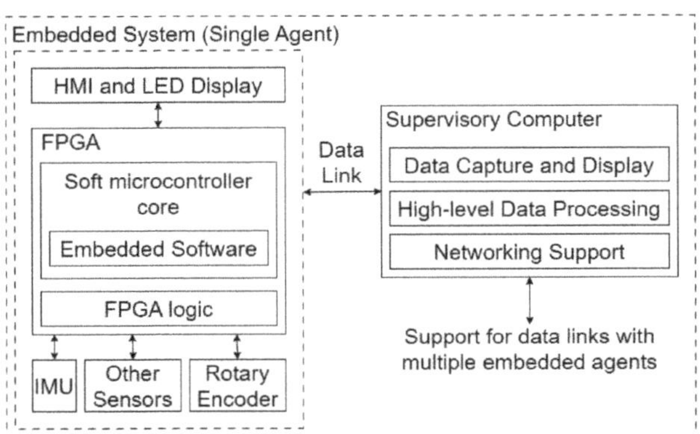

Figure 5. High-level overview of the required system architecture.

2.3. Hardware Design

The embedded system utilises the COTS ARTY A7 development board, housing an XC7A35TICSG324-1L FPGA with 88,280 logic cells, four 6.6 GB/s transceivers and 250 input/output (I/O) slices. This offers a robust yet low-cost development platform that can readily interface with several sensors, particularly through the peripheral module interface (PMOD) standard. Digilent offers well-documented PMOD packages that enable the quick connection of a wide range of sensors, actuators, memory systems, transceivers, and displays to the ARTY A7 board. These include sensors such as time-of-flight, infrared, colour, IMU, humidity, pressure, temperature, and ultrasonic sensors commonly found in intelligent vehicles. The PMOD packages also offer secure digital (SD) or micro SD card slots, flash memory extensions, wireless communication modules or transceivers, as well as a variety of wired communication interfaces, facilitating the deployment and testing of embedded algorithms for ICPSs with different memory and communication constraints. Additionally, there are various button, keypad, joystick, LED, and display boards available to enable simple HMIs within the embedded system. The wide range of COTS modules ensures that the platform can be easily customised for different vehicular requirements without the need to develop application-specific hardware from scratch. The proposed solution thereby remains an effective tool for facilitating ICPS education and expediting exploratory ICPS research. Moreover, the ARTY A7 development board offers extensive community support and resources, including online forums, tutorials, and example projects. This ensures that users have access to a wealth of knowledge and can easily seek assistance when developing their ICPS solutions. Nonetheless, custom extension boards can also be developed and interface with the ARTY A7 board if specific components must be included for cutting-edge research on a specific vehicular system.

To implement a softcore Microblaze processor on the FPGA, Xilinx intellectual property (IP) cores are used. The Microblaze processor has 128 kbits of block random-access memory (BRAM) and incorporates the necessary interrupt, clock, and peripheral controllers in FPGA logic. This approach allows the microcontroller to be dynamically synthesized and mapped onto the FPGA's configurable logic fabric, rather than being introduced as a physical hardware component. The softcore microcontroller thereby ensures a flexible and re-configurable design that is suitable for different vehicular applications. All the implemented cores are inter-connected through the advanced extensible interface (AXI) protocol that forms part of the ARM advanced microcontroller bus architecture (AMBA), commonly used in system on chip (SoC) designs. An AXI bus is also employed to connect the microcontroller to Xilinx IP cores designed to interface with the rotary encoder and NAV PMOD devices, connected to the JA and JB headers of the development board, respectively. Moreover, the lite Xilinx IP core for UART communication is included to establish a serial communication link with the host computer. A set of on-board red–green–blue (RGB) LEDs are also connected to the microcontroller peripheral controller, to act as a simple on-board HMI. The use of Xilinx IP cores provides a well-established and reliable foundation for implementing various functionalities within the FPGA. These IP cores are rigorously tested and validated, ensuring compatibility and reducing the risk of errors or design flaws. Consequently, only customised settings, logic blocks and embedded firmware that differ from default block implementations are detailed throughout this work.

A summary of the hardware design is illustrated in Figure 6. The default clocking wizard IP core is used to clock the soft-core Microblaze processor and ancillary peripheral and interrupt controllers. Additionally, an on-board reset switch is used to enable a hard reset of the clocking and processor logic. A memory module is also connected to the soft-core microcontroller, with the specified 128 kbits of BRAM, and the default debug module is implemented to facilitate embedded firmware development. The default AXI peripheral controller is subsequently used as an interface between the processor core and all peripheral controller IP cores. Specifically, the default IP cores developed for the PMOD encoder, PMOD NAV IMU, on-board GPIOs and UART communication are employed and connected to external peripherals through the ARTY A7 connection headers. Additionally, the on-board UART-Universal Serial Bus (USB) bridge is used to facilitate communication with the host computer through a simple micro-USB to USB connection.

The flexibility of the FPGA-based embedded system allows for iterative development and rapid prototyping. Users can easily modify and reconfigure the hardware and firmware to accommodate the evolving requirements and experiment with different design approaches, accelerating the development cycle. This platform can also be modified to include different processor implementations. Khairullah [75], for example, showcases the possibility of realising a reduced instruction set computing (RISC) processor within an FPGA. The prototype developed in this work, however, solely aims to showcase the potential of a COTS FPGA board to deploy complete ICPS systems. More complex developments must also consider additional FPGA design concepts, such as pipelining, occupancy and datapath restrictions, as reviewed by Monmasson et al. [76].

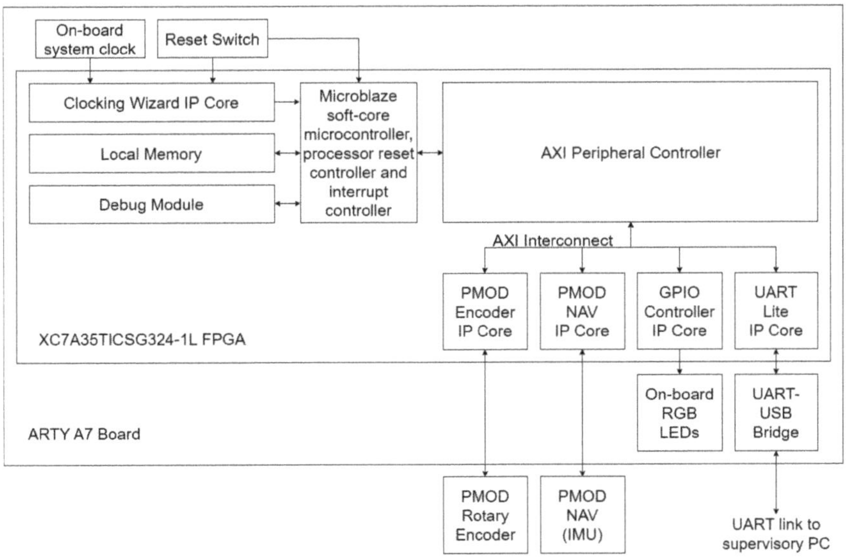

Figure 6. Overview of the implemented hardware design.

2.4. Sensor Interfaces and Communication Protocols

Selecting appropriate communication protocols is crucial in any hardware design. Most ICs support serial peripheral interface (SPI) and/or inter-integrated circuit (I2C) protocols for intra-board communication, while UART or Ethernet protocols are commonly used for communication with external devices. Considerable literature [77] exists describing the operation of such protocols. In general, however, SPI tends to support faster data transfer than I2C, but requires more connections between peripherals. Additionally, Ethernet offers higher data rates and better noise immunity than UART, at the expense of non-deterministic behaviour and higher protocol complexity. A summary of this comparison is presented in Table 1. SPI is thereby used to communicate with the PMOD NAV, ensuring swift IMU data transfer. Conversely, the rotary encoder is monitored through a set of general purpose I/O (GPIO) pins, and UART is used to communicate with the host computer. Consequently, the system can be readily monitored using open source UART virtual terminals. Nonetheless, future platform upgrades may incorporate controllers for additional communication protocols to better support diverse system requirements.

Table 1. Comparison of typical communication protocols.

Feature	UART	SPI	I2C	Ethernet
Data rate	<460 kbps	<20 Mbps	<1 Mbps	>1 Gbps
Type of communication	Asynchronous	Synchronous	Synchronous	Synchronous
Hardware complexity	Low	Moderate	Moderate	High
Benefits	Simple and widely supported	Full duplex communication and low power consumption	Supports multiple masters and only needs a two-line bus	Good quality and high data rates over long distances
Disadvantages	Restricted to two devices at a pre-defined data rate	Requires a large number of connections for multiple slaves	Half duplex communication with low overhead	Performs poorly in high data traffic and is non-deterministic

2.5. Embedded Software Design and Data Processing

All sensor interfacing, data collection, low-level processing and data transmission algorithms are implemented on the softcore microcontroller using the C programming language. These are optimised to respect the memory limitations of the Microblaze processor, ensuring a more realistic prototype scenario. After traversing a custom start-up sequence, the embedded system has four modes of operation, configured by the switch on the PMOD encoder board. A high-level overview of the embedded firmware is shown in Figure 7.

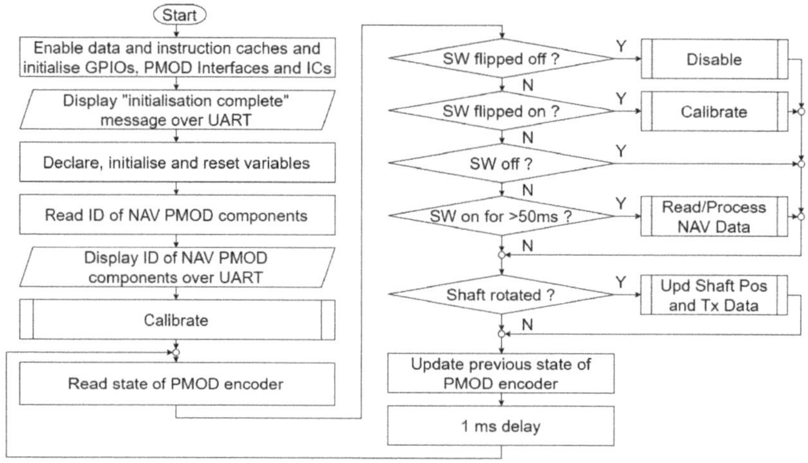

Figure 7. Flow-chart detailing a high-level overview of the implemented embedded firmware.

The system is designed to appropriately initialise all interfaces, peripherals and caches on start-up, according to the datasheet of each respective device. Specifically, the default libraries for SPI communication with the LSM9DS1 IC on the PMOD NAV were modified to initialise the device with a sampling frequency of 50 Hz, supporting a quicker data update rate that is more conducive to the requirements of autonomous vehicles. Moreover, the internal accelerometer filters were enabled by appropriately configuring the device registers, reducing the need for subsequent digital filtering in software. Upon completion of this startup sequence, an appropriate message is transmitted to the supervisory computer over UART. The code subsequently initialises and resets all run-time variables, configures the PMOD NAV and reads the IDs of all on-board sensors. These are transmitted over UART for validation by the supervisory computer.

Calibration offsets for each sensor are determined, by averaging the readings obtained over 100 successive sampling instances, while instructing the user to not touch the device. Throughout this process, appropriate messages are transmitted over UART, and the on-board LEDs are set to blue to indicate that the user should avoid moving the board. At 1 ms intervals, the processor polls the state of the encoder PMOD modules, and disables the platform when the switch is flipped off. This involves resetting all run-time variables (effectively resetting the reference frame of the system and the zero position of the encoder shaft) and setting the on-board LEDs to red. Whenever the switch is flipped back on, the system re-collects calibration offsets to account for possible temperature variations or device movements, corresponding to a different effect of gravity along each accelerometer axis. This calibration procedure ensures that the prototype retains reliable and accurate results throughout all stages of its operation. Trinh et al. [78] discuss more sophisticated IMU calibration procedures using Kalman filter techniques. Such algorithms, however, were deemed beyond the scope of this prototype.

Upon shaft rotation, the number and direction of discrete ticks or rotary motions are recorded and converted to the shaft position in degrees using:

$$P = \frac{\#\ ticks}{max\ \#\ ticks\ in\ 1\ revolution} \times 360^\circ = \frac{\#\ ticks}{20} \times 360^\circ, \quad (1)$$

where P is the angular position of the shaft. The result is transmitted over UART to the host computer, together with the current 3D location and orientation of the board. The encoder state is polled every 1 ms to ensure that all tick rotations are detected by the device. A summary of these procedures is illustrated in Figure 8.

If the switch is on, the processor sets the on-board LEDs to green and reads IMU data at 50 ms intervals. This avoids stalling the processor while being frequent enough to reliably capture user movements. Incoming accelerometer and gyroscope data are subsequently processed, as shown in Figure 9. On-board processing eliminates the need for continuous data transmission to a host computer, thereby improving the communication bus efficiency. While computationally intensive sensor fusion algorithms could not be implemented on the limited memory available, several data processing techniques were employed to reliably and accurately obtain the device position and orientation from the raw sensor data.

Data measurements are calibrated by removing zero offsets and accounting for any scale factor deviations. Since integration is used to derive position and orientation data from the accelerometer and gyroscope readings, small numerical offset errors will cause output saturation and system instability. A mechanical filtering window is thereby implemented, whereby readings below a particular threshold are set to zero to prevent integrator saturation. A numerical integrator is subsequently used to determine the device orientation along each axis, with the result manipulated to lie between -360° and 360° and prevent register saturation. Similarly, accelerometer readings are converted to metres per second squared, and a double integration process is used to determine the 3D device position.

For ease of testing, position and orientation are computed in the body frame of the inertial sensor. The algorithm, however, can be readily used to compute the absolute device position and orientation in an external reference frame, as described in Section 3.5 below. Moreover, numerical integration is implemented according to the trapezoidal rule for computational efficiency. Nonetheless, more accurate yet memory-intensive numerical integration techniques can be used when higher accuracy is needed.

Finally, NXP [79] suggests that small residual velocities after integrating acceleration data will cause the output of the second position integrator to quickly saturate and yield large output errors. Consequently, an end-of-movement check is performed, such that the velocity output of the first integrator is reset to zero if several successive accelerometer readings indicate that the device is stationary. This is a reasonable assumption, unless the vehicle is expected to move at a perfectly constant velocity for long periods of time.

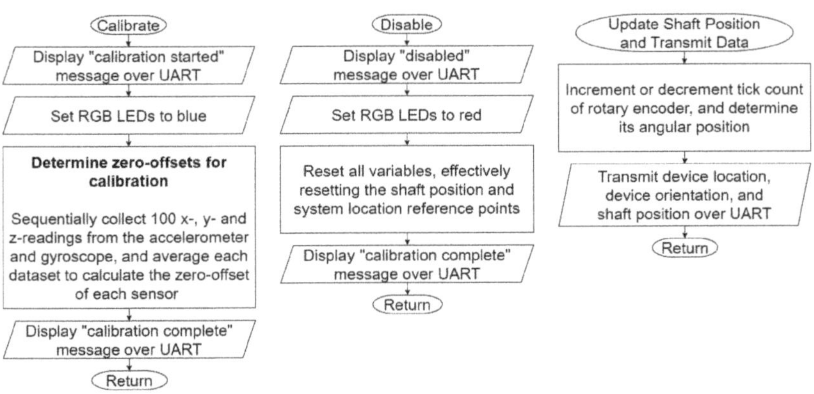

Figure 8. Flow-charts summarising device operation during calibration (**left**), disable operations (**centre**), and upon rotation of the encoder shaft (**right**).

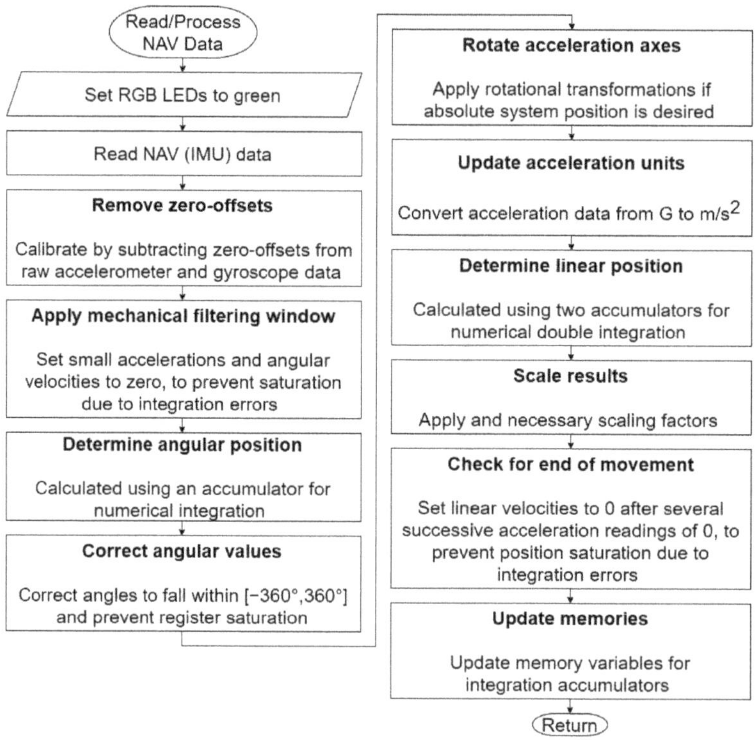

Figure 9. Flow-chart summarising the algorithm used to process incoming sensor data.

2.6. Data Capture and Human Machine Interface

To complement the on-board LED indicators, a user-friendly HMI is implemented on the host computer. The serial data transmitted by the embedded system may be viewed through a virtual serial port (configured for a baud rate of 9600 bps), or captured and displayed on a custom Matlab application. The latter plots the board location and orientation against the shaft position, using the algorithm summarised in Figure 10. This feature serves as a simple demonstration on how the developed prototype supports high-level data capture and display, and can be readily upgraded to represent frameworks that are more conducive to vehicular applications.

A call-back function is used to read incoming data whenever a termination character is detected on the serial port. Each incoming string is subsequently processed and used to control a real-time display. If the incoming message contains any data, it is extracted and used to update the plot in real-time. If not, incoming messages are simply displayed in the Matlab user terminal. Moreover, if a disable message is detected, suggesting that the on-board reset switch was flipped, the plot is reset in anticipation of a new set of datapoints. Additionally, the system halts execution if the user presses the Enter key, ensuring that the application does not become stuck in an infinite loop.

This application can be extended to support bi-directional control of the embedded system, by transmitting commands over UART to the embedded agent. With appropriate firmware development, the soft-core microcontroller can poll and decipher incoming UART commands. Alternatively, standard FPGA logic blocks may be employed to receive and interpret UART data. Moreover, networked operations across a set of embedded agents can be realised by simultaneously communicating with multiple embedded agents. Various online articles and tutorials exist to implement such upgrades, ensuring that the platform remains accessible to inexperienced ICPS users.

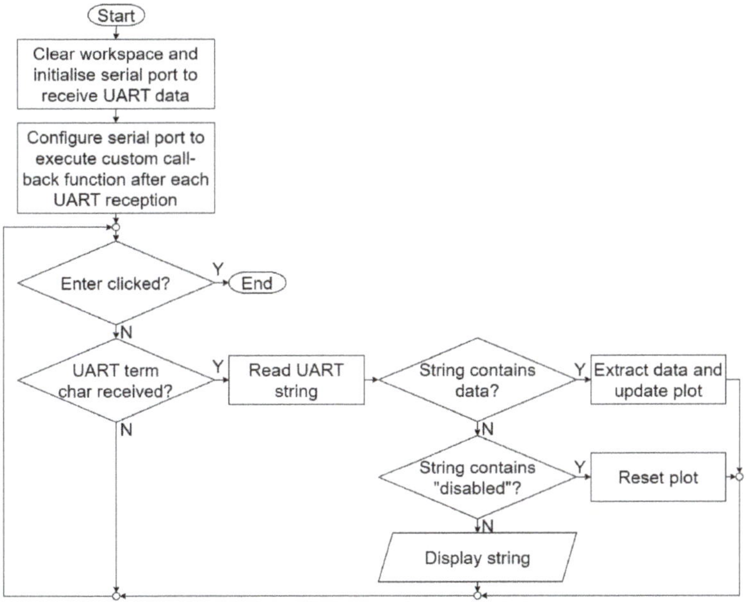

Figure 10. Flow-chart summarising the implemented Matlab application.

2.7. Implemented Prototype

The implemented prototype is shown in Figure 11, supported by several HDL, constraint, C and Matlab files available as an addendum to this work.

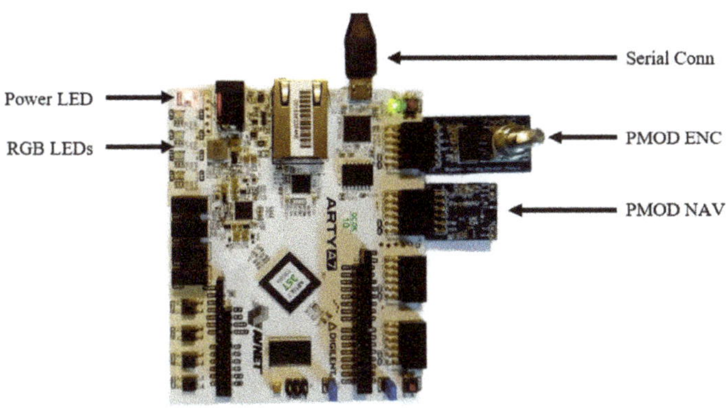

Figure 11. Layout of the implemented hardware prototype with an Arty A7 development board.

3. Testing, Results and Discussion

This section briefly reviews testing procedures for CPSs and FPGA-based embedded systems, before describing the testing approach used throughout this work. It subsequently presents the results obtained when validating and characterising the implemented prototype. Finally, potential design improvements and future modifications are highlighted, including techniques to improve system security, efficiency, accuracy and reliability. Moreover, the flexibility of the developed platform is reviewed to assess its suitability as a development platform for ICPSs and intelligent vehicular algorithms.

3.1. Testing Procedure

Any intelligent embedded system or CPS must be rigorously verified, tested and validated to ensure it meets project requirements. Zhou et al. [80] systematically review testing methodologies and testbeds for CPSs, and conclude that the heterogeneity and complexity of CPSs pose considerable challenges to existing testing procedures. Nonetheless, the V-model [81] has been widely adopted as an industry standard for software development, and can be extended to outline a sequential testing procedure for simple ICPSs. This involves individual testing of system elements against their respective requirements, followed by integration testing of multiple elements to assess the system as a whole. Moreover, the final system must be validated in the expected operation environment, under reasonable extreme conditions (such as radiation, temperature and humidity levels), and considering all edge cases and boundary conditions, when possible. While this may be impossible when considering random physical processes and stochastic intelligent algorithms, every effort must be made to validate system performance in all anticipated operating scenarios, particularly in safety-critical applications. Several intelligent algorithms have also been proposed for automatic optimisation and testing of CPSs [82].

In applications like spacecraft design, physical testing is not always possible, such that simulations are often used for the initial system validation. In particular, FPGA-based embedded designs often use behavioural, post-synthesis and post-implementation simulations to sequentially verify the logical, synthesized and implemented designs through a well-defined test bench, respectively. Additionally, timing, memory and layout constraints must be met to ensure a reliable implementation.

Zheng et al. [83] review the perception of researchers on CPS verification and validation techniques, and conclude that existing formal methodologies are insufficient to support the development of robust ICPSs. While simulations or controlled environments may help mimic the physical process associated with a CPS, these often fall short in accurately representing all the factors affecting the system in its true operating environment. When possible, in fact, in-field tests are preferable to fully validate system operation. The ASTREA project undertaken by the University of Malta, for instance, aims to provide a low-cost pico-satellite platform to directly test the reliability of electronic components and embedded systems in space, rather than relying on a series of controlled laboratory tests. Similarly, intelligent vehicles will benefit from physical tests that replicate the physical ITS environment. Moreover, safety-critical systems such as autonomous vehicles should adhere to relevant safety standards to guarantee system reliability.

Rigorous testing and characterisation was deemed unnecessary for the implemented system, since its main purpose is to showcase the flexibility of a platform for CPS research. Moreover, test-bench simulations for FPGA validation were considered unnecessary since standard Xilinx IP cores were utilised throughout the hardware design. Nonetheless, all elements of the CPS were validated against the project requirements, as depicted in Figure 12. The implemented hardware was first validated by ensuring that the microcontroller could successfully communicate with all sensor and communication modules. This was confirmed by successfully reading the IDs of each sensor and transmitting them over UART to the host computer. All embedded firmware and data processing algorithms were subsequently validated by confirming that the system operated correctly in all four modes of operation. Additionally, the high-level data processing and display algorithms were validated by ensuring correct processing and display of incoming UART messages. Finally, all components were tested within the integrated system, and a preliminary characterisation of the implemented use-case was performed. The primary objective of this research, however, remains to demonstrate that a simple commercially available FPGA platform can be utilised for educational and research purposes in ICPSs.

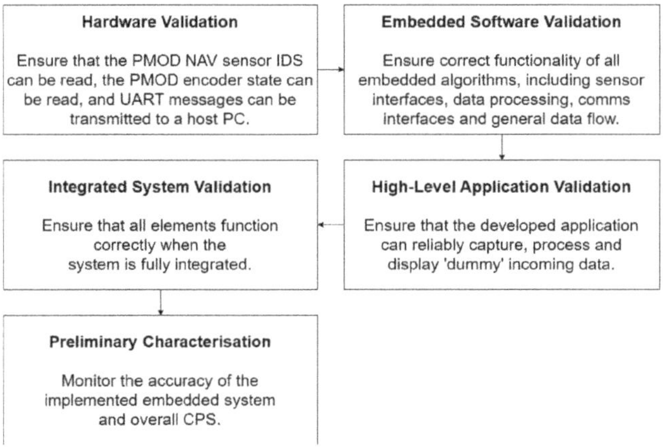

Figure 12. Overview of the adopted testing procedure.

3.2. System Validation

During system testing, both the individual components and unified CPS exhibited correct operation. The implemented prototype successfully integrated an IMU, switch and rotary encoder in an FPGA-based embedded system. Moreover, a softcore microcontroller was successfully implemented and used to collect, process and manipulate IMU data to determine the device position and orientation. This information, together with the position of the encoder shaft, was successfully transmitted over a serial UART interface on shaft rotation, captured by a Matlab application, and displayed through a series of real-time plots. Examples showing the observed plots when rotating the encoder shaft and moving the device along all six degrees of freedom (DOFs) are displayed in Figures 13 and 14.

The implemented HMI was also observed to function correctly, with the on-board LEDs appropriately indicating the mode of system operation, as shown in Figure 15. The embedded system further operated correctly when interfaced with an open-source virtual serial terminal, as demonstrated in Figure 16. Moreover, the custom application successfully displayed all system status messages, reset the displayed plots on a user reset, and halted execution upon clicking the Enter key, as shown in Figure 17.

These results affirm the successful design and implementation of a robust and flexible ICPS development platform. The prototype also demonstrates strong scalability and extensibility, allowing for seamless integration of additional components and modules as the complexity of ICPSs increases. With its modular design and flexible architecture, researchers and developers can easily incorporate new sensors, actuators, or communication interfaces to address specific application requirements. Specifically, all components of a simple CPS were successfully developed using an FPGA development board interfaced with COTS components. These included a sensor, visual LED display, rotary encoder, embedded processor with custom firmware, user-activated switches, and a data and communication link with a supervisory controller. A high-level application was also successfully deployed to collect, process and display data received from the embedded agent.

These claims substantiate the platform's capability to efficiently implement customised ICPSs without the need for specialised hardware or software expertise. Consequently, this platform holds significant value for educational purposes by simplifying the complexity of real-world heterogeneous ICPSs and bridging the gap between ICPS theory and practical implementation. Notably, sensors and communication interfaces can be readily substituted with alternative COTS or customised options to fulfil specific application requirements. Similarly, custom embedded firmware, FPGA logic, and high-level applications can be developed for different research scenarios, ensuring that the platform remains relevant for cutting-edge ICPS research. Furthermore, the CPS can incorporate customised communication and cybersecurity protocols, including intelligent algorithms.

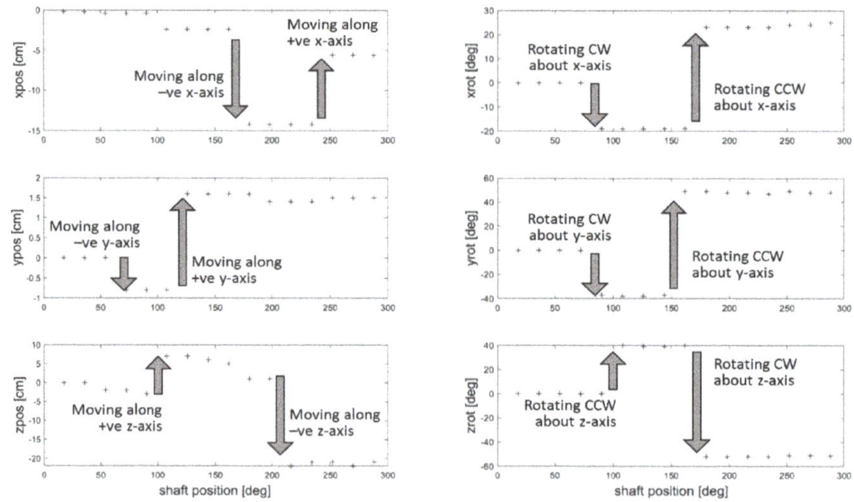

Figure 13. Summary of plots from six independent validation tests, showing correct system operation when moving or rotating the device along all 6 DOFs while rotating the encoder shaft.

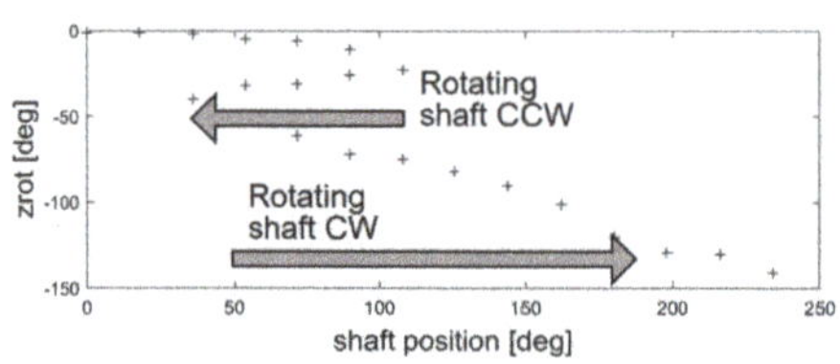

Figure 14. Plot showing correct system operation when rotating the encoder shaft in both directions, while rotating the entire board clockwise about the z-axis.

Figure 15. Correct LED display in disabled (**left**), calibration (**middle**) and enabled (**right**) modes.

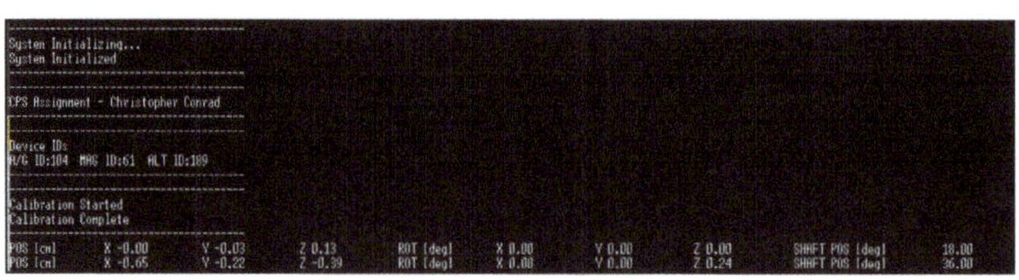

Figure 16. Example of the HMI when using a virtual serial port.

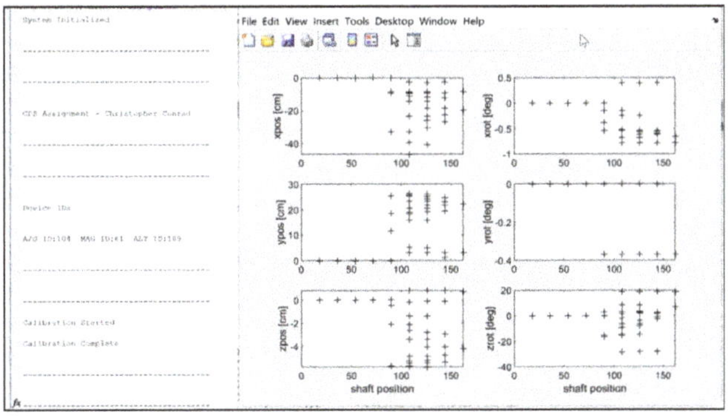

Figure 17. Example of the HMI when using the custom Matlab application.

3.3. Prototype Accuracy

Characterising the IMU-based system falls beyond the scope of this work, which solely aims to demonstrate that an FPGA-based system can be used to develop a complete ICPS. Notably, the prototype cannot be reliably characterised without an appropriate test setup, owing to a large degree of human error. Specifically, it is impossible to guarantee that device motion is perfectly aligned along a particular axis. Nonetheless, crude measurements demonstrated reasonable accuracy along all six DOFs, as summarised in Figure 18. To limit inaccuracies, these readings were taken while slowly orienting the platform. This approach, however, is not conducive of high-speed vehicular applications, and solely aims to demonstrate correct operation of the implemented use-case. Angular readings were observed to be more accurate than position readings, as expected due to the additional integrator needed to extract positional data from the accelerometer readings. These observations further verified the correct system operation.

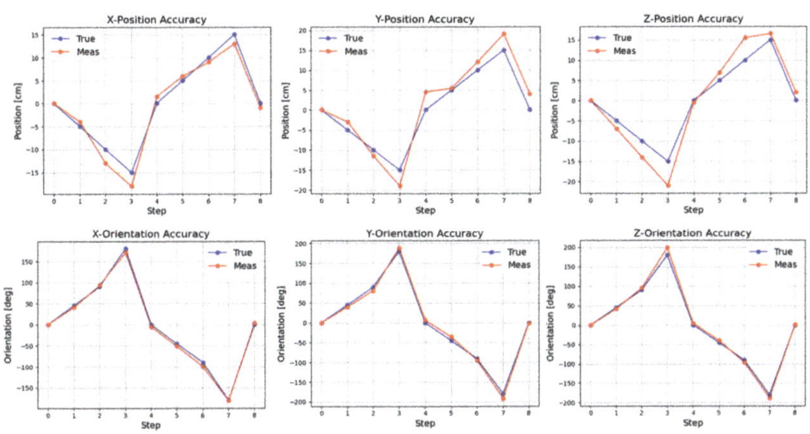

Figure 18. Preliminary showcase of prototype accuracy from a crude proof-of-concept test.

When conducting research on the proposed platform in an academic setting, it is imperative to employ rigorous testing and characterisation methodologies to validate the hypothesis or research question under investigation. This becomes particularly critical in the context of autonomous vehicle applications, where stringent testing protocols are essential to ensure the safety of drivers, vehicle occupants, pedestrians, and other users of the transportation infrastructure. In fact, Naufal et al. [84] assert that without robust safety assessments and comprehensive test data, the realisation of autonomous vehicles and ITSs remains unattainable.

3.4. Accuracy Improvements

Inertial systems implemented in real ICPSs will likely require considerable accuracy improvements. While more expensive and accurate IMUs can be used, several processing techniques may also be employed to improve position and orientation accuracy:

- Thermal compensation algorithms may be included, using the temperature sensor on board the PMOD NAV to monitor the ambient system temperature;
- Software-based timing algorithms can be replaced with interrupt-based hardware timers, thereby ensuring more accurate and consistent integration time steps;
- More complex algorithms may be used to fuse magnetometer, gyroscope and accelerometer data and obtain better position and orientation estimates. Several sensor fusion techniques have been proposed to handle this problem, often based on extended Kalman filtering [85]. Mahony and Madgwick filters [86] have also been shown to achieve good accuracy while incurring less computational overhead;
- IMU data can be fused with data obtained from other sensors [87];
- Intelligent embedded algorithms can be used to improve the estimation accuracy [88];
- If real-time data capture is not required, bi-directional digital filters can be implemented after data collection for better estimation accuracy.

3.5. Absolute 3D Positioning

While relative positioning in the device body frame was suitable for demonstrative purposes, absolute positioning in an external inertial frame is often required. An inertial system mounted on an UAS or self-driving car, for example, should support position and orientation calculations in a pre-defined external reference frame. This can be accomplished by rotating the acceleration axes using standard rotational matrices. Alternatively, quaternions offer a more robust representation of the 3D orientation of an object and are commonly used to translate IMU readings to a defined reference frame. Algorithms using this hyper-complex number system tend to be more computationally efficient, making them particularly suitable for embedded firmware within CPSs. Patel et al. [89], for example, use quaternions to deploy and evaluate IMU fusion algorithms for UAS navigation.

3.6. Efficiency

Embedded algorithms were implemented on a softcore microcontroller to demonstrate the flexibility of an FPGA-based platform. FPGA resources, however, could be used more efficiently to achieve better performance with lower power consumption. Custom VHDL scripts or IP cores, for instance, could be used to manage sensor interfaces and implement the data signal conditioning blocks in hardware. In fact, Seng et al. [90] systematically review embedded intelligence on FPGA systems and conclude that high computational power and low power consumption must be enabled through efficient FPGA designs.

3.7. Cybersecurity Considerations

Since the implemented prototype solely aimed to showcase the robust capabilities of the proposed platform, no cybersecurity features were designed or implemented. Nonetheless, encryption and decryption algorithms can be readily implemented and tested at both an embedded and supervisory level. Simple linear-feedback shift registers (LFSRs), for instance, may be used for pseudo-random number generation within a stream cypher [91]. More advanced block cypher and hash function-based encryption may also be implemented and tested for more advanced cybersecurity protocols. In fact, developing new and resilient cryptographic techniques remains an open research problem [92–94] and this platform offers a simple yet effective way to test such algorithms over a one or more data and communication links. This is especially important for autonomous vehicles and ICPSs within an ITS, where cyberattacks can have catastrophic and fatal results.

3.8. Flexibility for ICPS Experimentation

The primary objective of this research is to investigate the potential of using an FPGA development board with COTS or custom modules to facilitate the exploration, development, and testing of ICPSs and intelligent vehicular algorithms. The deployed use-case demonstrates that this approach offers a viable solution to advance ICPS research, foster ICPS education, and stimulate discourse surrounding the fundamental challenges associated with the deployment of ICPSs in autonomous vehicles and ITSs. As opposed to the platform developed by Zhang et al. [95], this solution explicitly considers the requirements of vehicular applications, including collaborative and networked ICPSs, stricter safety and performance criteria, and the need for better cybersecurity protocols. Additionally, the Raspberry Pi platform proposed by Garcia et al. [96] offers less flexibility than the developed FPGA-based embedded system, which enables dynamic hardware reconfiguration.

Isakovic et al. [97] suggest that custom hardware can reduce the costs of CPS platforms for research and education. Their approach, however, does not offer an open-source solution that is independent of specific hardware and software systems. In contrast, this work promotes the adoption of COTS development boards and modules to enable a broader audience to engage in ICPS experimentation and advance the emerging fields of autonomous vehicles and ITSs. Landolfi et al. [98] also propose a platform to support automotive CPS deployment, but focus on the marketplace architecture rather than emphasizing the educational and research potential of a technical ICPS experimentation platform.

Several simulation tools have also been developed to aid ICPS research and education. Schmittle et al. [99] and Iannino et al. [100], for instance, respectively propose simulation environments for UAS and factory systems. Nonetheless, the hardware-based experimentation platform prototyped in this work offers significant advantages over a purely software-based implementation. Simulation platforms often lack the ability to simulate low-level electronics, making it challenging to conduct accurate testing of intelligent embedded systems. In vehicular applications, ensuring proper power and thermal management at an embedded level is particularly crucial, highlighting the suitability of a hardware-based experimentation platform for ICPSs. Similarly, a simulated system may struggle to precisely model all the elements of the physical system in which an ICPS will be deployed.

Additionally, the proposed platform addresses the need for interdisciplinary collaboration in ICPS research. By providing a platform that integrates hardware and software components, researchers from diverse backgrounds, such as computer science, engineering, and transportation, can collaborate effectively to tackle the complex challenges of ICPSs in autonomous vehicles and ITSs. This interdisciplinary approach fosters a holistic understanding of ICPSs, promotes cross-pollination of ideas, and encourages the development of innovative solutions that consider both technical and societal aspects. Ultimately, it enables a more comprehensive and integrated approach towards building safe and efficient ITSs.

3.9. Limitations

While the FPGA-based platform presented in this research offers numerous advantages, it is important to acknowledge its limitations in addressing ICPS research questions. Primarily, cost-effective FPGA development boards are often limited in terms of logic elements, memory, and processing capabilities. This restricts the complexity and scale of the solutions that can be deployed within a single embedded agent. Although more powerful development boards are available, these will significantly increase the cost of such an experimentation tool. Additionally, the platform is limited by the hardware connections available on the ARTY A7 development board. Adding more connections or deploying custom sensors may require custom hardware design and development.

The platform may also face limitations in accurately simulating all aspects of a real-world environment. Physical factors, such as power management, thermal effects, and environmental conditions, may not be fully replicated in controlled hardware-based experimentation, potentially impacting the fidelity of the ICPS evaluation. Nonetheless, this solution remains a valid and versatile tool for initial exploratory research.

4. Conclusions

This work successfully demonstrates the potential of an FPGA-based embedded platform for exploring, developing, and testing ICPSs for autonomous vehicles and ITSs. The developed platform uses COTS components and a customisable FPGA development board to offer a cost-effective and accessible solution for ICPS research and education. This is showcased through a simple use-case that demonstrates the seamless integration of various sensors, communication modules, and HMIs within an embedded FPGA system. A high-level application further highlights the platform's ability to support real-time data collection, processing and display by a supervisory computer. The proposed solution explicitly addresses the specific requirements of intelligent vehicular applications, to encourage ICPS research and promote discussions on the fundamental issues surrounding their deployment in autonomous vehicles and ITSs. Moreover, the platform supports the deployment of low-level embedded firmware, intelligent algorithms, and custom sensor modules, ensuring that it remains relevant for state-of-the-art exploratory research.

This research offers a valuable resource for researchers, educators, and practitioners in the fields of ICPSs and ITSs. By fostering interdisciplinary collaboration, addressing domain-specific requirements, and supporting real-world experimentation, the platform opens up new avenues for advancing the understanding, development, and deployment of intelligent and efficient transportation systems of the future.

Future work will focus on enhancing and testing the capabilities of the platform, such as incorporating advanced ML algorithms for intelligent decision-making, deploying cybersecurity protocols over wired and wireless communication channels, and investigating techniques for ensuring the resilience and reliability of ICPSs in dynamic and unpredictable environments. Additionally, the applicability of this tool for CPS experimentation in different domains will be explored. Furthermore, empirical studies and user evaluations should be conducted to assess the effectiveness and usability of this tool in educational and research settings. This would provide valuable insights to optimise its research or educational potential. Finally, different use-cases will be deployed on the developed prototype, to better understand the capabilities and limitations of an FPGA-based embedded system for ICPS experimentation and exploration.

Author Contributions: Conceptualization, C.C. and S.A.-R.; methodology, C.C.; software, C.C.; validation, C.C.; formal analysis, C.C.; investigation, C.C.; resources, S.A.-R.; data curation, C.C.; writing—original draft preparation, C.C.; writing—review and editing, S.A.-R.; visualization, C.C.; supervision, S.A.-R. and A.T.; project administration, A.T. All authors have read and agreed to the published version of the manuscript.

Funding: This research received no external funding.

Institutional Review Board Statement: Not applicable.

Informed Consent Statement: Not applicable.

Data Availability Statement: Data sharing not applicable.

Conflicts of Interest: The authors declare no conflict of interest.

Abbreviations

The following abbreviations are used in this manuscript:

AAM	Advanced air mobility
AI	Artificial intelligence
AMBA	Advanced microcontroller bus architecture
AXI	Advanced extensible interface
BRAM	Block random access memory
CLB	Configurable logic blocks
CNS	Communication, navigation and surveillance
COTS	Commercial off-the-shelf

CPS	Cyber-physical system
DAA	Detect and avoid
DOF	Degree of freedom
EMC	Electromagnetic compatibility
EMI	Electromagnetic interference
ESD	Electrostatic discharge
FPGA	Field programmable gate array
GNSS	Global navigation satellite systems
GPIO	General purpose input / output
HDL	Hardware description language
HMI	Human machine interface
IC	Integrated circuit
ICPS	Intelligent cyber-physical system
IMU	Inertial measurement unit
I/O	Input/output
IoT	Internet of things
IP	Intellectual property
ITS	Intelligent transportation system
I2C	Inter-integrated circuit
LED	Light emitting diode
LFSR	Linear-feedback shift registers
LiDAR	Light detection and ranging
LoRa	Long range
LTE	Long-term evolution
ML	Machine Learning
PCB	Printed circuit board
PMOD	Peripheral module interface
RGB	Red-green-blue
RISC	Reduced instruction set computing
SD	Secure digital
SoC	System on chip
SPI	Serial peripheral interface
UART	Universal asynchronous receiver transmitter
UAS	Unmanned aerial system
USB	Universal serial bus
UTM	UAS traffic management
V2C/N	Vehicle-to-cloud/network
V2I	Vehicle-to-infrastructure
V2P	Vehicle-to-pedestrian
V2V	Vehicle-to-vehicle
Wi-Fi	Wireless fidelity

References

1. Gao, Y.; Tian, F.; Li, J.; Fang, Z.; Al-Rubaye, S.; Song, W.; Yan, Y. Joint Optimization of Depth and Ego-Motion for Intelligent Autonomous Vehicles. *IEEE Trans. Intell. Transp. Syst.* **2022**, 1–14. [CrossRef]
2. Rani, S.; Kataria, A.; Chauhan, M.; Rattan, P.; Kumar, R.; Kumar Sivaraman, A. Security and Privacy Challenges in the Deployment of Cyber-Physical Systems in Smart City Applications: State-of-Art Work. *Mater. Today Proc.* **2022**, *62*, 4671–4676. [CrossRef]
3. Gunes, V.; Peter, S.; Givargis, T.; Vahid, F. A survey on concepts, applications, and challenges in Cyber-Physical Systems. *KSII Trans. Internet Inf. Syst.* **2014**, *8*, 134–159.
4. Fantin Irudaya Raj, E.; Appadurai, M. Internet of Things-Based Smart Transportation System for Smart Cities. In *Intelligent Systems for Social Good: Theory and Practice*; Springer Nature: Singapore, 2022; pp. 39–50. [CrossRef]
5. Mahrez, Z.; Sabir, E.; Badidi, E.; Saad, W.; Sadik, M. Smart Urban Mobility: When Mobility Systems Meet Smart Data. *IEEE Trans. Intell. Transp. Syst.* **2022**, *23*, 6222–6239. [CrossRef]
6. Liu, S. The Business Case for Infrastructure-Vehicle Cooperative Autonomous Driving. *IEEE Eng. Manag. Rev.* **2022**, *50*, 189–194. [CrossRef]
7. Pan, M.; Li, Y.; Zhang, Z.L.; Luo, J. SCCS: Smart Cloud Commuting System With Shared Autonomous Vehicles. *IEEE Trans. Big Data* **2022**, *8*, 1301–1311. [CrossRef]

8. Shi, X.; Wei, H. An Embedded Vehicular Integrative Platform Based on CDMA. In Proceedings of the 2006 2nd IEEE/ASME International Conference on Mechatronics and Embedded Systems and Applications, Beijing, China, 13–16 August 2006; pp. 1–5. [CrossRef]
9. Rawat, D.B.; Bajracharya, C.; Yan, G. Towards intelligent transportation Cyber-Physical Systems: Real-time computing and communications perspectives. In Proceedings of the SoutheastCon 2015, Fort Lauderdale, FL, USA, 9–12 April 2015; pp. 1–6. [CrossRef]
10. Chen, L.; Li, Y.; Huang, C.; Xing, Y.; Tian, D.; Li, L.; Hu, Z.; Teng, S.; Lv, C.; Wang, J.; et al. Milestones in Autonomous Driving and Intelligent Vehicles—Part 1: Control, Computing System Design, Communication, HD Map, Testing, and Human Behaviors. *IEEE Trans. Syst. Man, Cybern. Syst.* **2023**, 1–17. [CrossRef]
11. Novak, A.; Novak Sedlackova, A.; Vochozka, M.; Popescu, G.H. Big Data-driven Governance of Smart Sustainable Intelligent Transportation Systems: Autonomous Driving Behaviors, Predictive Modeling Techniques, and Sensing and Computing Technologies. *Contemp. Readings Law Soc. Justice* **2022**, *14*, 100–117. [CrossRef]
12. Kliestik, T.; Musa, H.; Machova, V.; Rice, L. Remote Sensing Data Fusion Techniques, Autonomous Vehicle Driving Perception Algorithms, and Mobility Simulation Tools in Smart Transportation Systems. *Contemp. Readings Law Soc. Justice* **2022**, *14*, 137–152. [CrossRef]
13. Pan, G.; Alouini, M.S. Flying Car Transportation System: Advances, Techniques, and Challenges. *IEEE Access* **2021**, *9*, 24586–24603. [CrossRef]
14. Zhou, J.; Tian, D.; Sheng, Z.; Duan, X.; Shen, X. Joint Mobility, Communication and Computation Optimization for UAVs in Air-Ground Cooperative Networks. *IEEE Trans. Veh. Technol.* **2021**, *70*, 2493–2507. [CrossRef]
15. Shrestha, R.; Bajracharya, R.; Kim, S. 6G Enabled Unmanned Aerial Vehicle Traffic Management: A Perspective. *IEEE Access* **2021**, *9*, 91119–91136. [CrossRef]
16. Javaid, S.; Saeed, N.; Qadir, Z.; Fahim, H.; He, B.; Song, H.; Bilal, M. Communication and Control in Collaborative UAVs: Recent Advances and Future Trends. *IEEE Trans. Intell. Transp. Syst.* **2023**, *24*, 5719–5739. [CrossRef]
17. Wilson, A.N.; Kumar, A.; Jha, A.; Cenkeramaddi, L.R. Embedded Sensors, Communication Technologies, Computing Platforms and Machine Learning for UAVs: A Review. *IEEE Sens. J.* **2022**, *22*, 1807–1826. [CrossRef]
18. Prag, K.; Woolway, M.; Celik, T. Toward Data-Driven Optimal Control: A Systematic Review of the Landscape. *IEEE Access* **2022**, *10*, 32190–32212. [CrossRef]
19. Tantawy, A.; Abdelwahed, S.; Erradi, A. Cyber LOPA: An Integrated Approach for the Design of Dependable and Secure Cyber-Physical Systems. *IEEE Trans. Reliab.* **2022**, *71*, 1075–1091. [CrossRef]
20. De Klerk, M.L.; Saha, A.K. A Comprehensive Review of Advanced Traction Motor Control Techniques Suitable for Electric Vehicle Applications. *IEEE Access* **2021**, *9*, 125080–125108. [CrossRef]
21. Ding, D.; Han, Q.L.; Wang, Z.; Ge, X. A Survey on Model-Based Distributed Control and Filtering for Industrial Cyber-Physical Systems. *IEEE Trans. Ind. Inform.* **2019**, *15*, 2483–2499. [CrossRef]
22. Al-Dulaimi, A.; Al-Rubaye, S.; Cosmas, J. Adaptive congestion control for mobility in cognitive radio networks. In Proceedings of the 2011 Wireless Advanced, London, UK, 20–22 June 2011; pp. 273–277. [CrossRef]
23. Cogliati, D.; Falchetto, M.; Pau, D.; Roveri, M.; Viscardi, G. Intelligent cyber-physical systems for industry 4.0. In Proceedings of the 2018 First International Conference on Artificial Intelligence for Industries (AI4I), Laguna Hills, CA, USA, 26–28 September 2018; pp. 19–22.
24. Shuvo, M.M.H.; Islam, S.K.; Cheng, J.; Morshed, B.I. Efficient Acceleration of Deep Learning Inference on Resource-Constrained Edge Devices: A Review. *Proc. IEEE* **2023**, *111*, 42–91. [CrossRef]
25. Pundir, A.; Singh, S.; Kumar, M.; Bafila, A.; Saxena, G.J. Cyber-Physical Systems Enabled Transport Networks in Smart Cities: Challenges and Enabling Technologies of the New Mobility Era. *IEEE Access* **2022**, *10*, 16350–16364. [CrossRef]
26. Xing, G.; Simplicio, M.; Pillai, A. *Autonomous Vehicles: Cyber-Physical Risk on a Massive Scale;* Technical Report; IEEE Transmitter: Manhattan, NY, USA, 2022.
27. Wilson, P. *Certifying AI for Safety Critical Aircraft Systems;* Technical Report; Acubed: Sunnyvale, CA, USA, 2020.
28. Qazi, S.; Khawaja, B.A.; Farooq, Q.U. IoT-Equipped and AI-Enabled Next Generation Smart Agriculture: A Critical Review, Current Challenges and Future Trends. *IEEE Access* **2022**, *10*, 21219–21235. [CrossRef]
29. Demigha, S. The impact of Big Data on AI. In Proceedings of the 2020 International Conference on Computational Science and Computational Intelligence (CSCI), Las Vegas, NV, USA, 16–18 December 2020; pp. 1395–1400. [CrossRef]
30. Rodrigues, R. Legal and human rights issues of AI: Gaps, challenges and vulnerabilities. *J. Responsible Technol.* **2020**, *4*, 100005. [CrossRef]
31. Tran, C.N.N.; Tat, T.T.H.; Tam, V.W.Y.; Tran, D.H. Factors affecting intelligent transport systems towards a smart city: A critical review. *Int. J. Constr. Manag.* **2023**, *23*, 1982–1998. [CrossRef]
32. Khairullah, S.S.; Elks, C.R. Self-repairing hardware architecture for safety-critical cyber-physical-systems. *IET Cyber-Phys. Syst. Theory Appl.* **2019**, *5*, 92–99. [CrossRef]
33. Zhang, Y.; Qian, C.; Lv, J.; Liu, Y. Agent and cyber-physical system based self-organizing and self-adaptive intelligent shopfloor. *IEEE Trans. Ind. Inform.* **2017**, *13*, 737–747. [CrossRef]
34. Dutt, N.; Jantsch, A.; Sarma, S. Self-aware cyber-physical systems-on-chip. In Proceedings of the 2015 IEEE/ACM International Conference on Computer-Aided Design (ICCAD), Austin, TX, USA, 2–6 November 2015; pp. 46–50. [CrossRef]

35. Shakeri, R.; Al-Garadi, M.A.; Badawy, A.; Mohamed, A.; Khattab, T.; Al-Ali, A.K.; Harras, K.A.; Guizani, M. Design Challenges of Multi-UAV Systems in Cyber-Physical Applications: A Comprehensive Survey and Future Directions. *IEEE Commun. Surv. Tutorials* **2019**, *21*, 3340–3385. [CrossRef]
36. Alguliyev, R.; Imamverdiyev, Y.; Sukhostat, L. Cyber-physical systems and their security issues. *Comput. Ind.* **2018**, *100*, 212–223. [CrossRef]
37. Al-Mhiqani, M.N.; Ahmad, R.; Abidin, Z.Z.; Yassin, W.M.; Hassan, A.; Mohammad, A.N.; Clarke, N.L. A new taxonomy of insider threats; an initial step in understanding authorized attack. *Int. J. Inf. Syst. Manag.* **2018**, *1*, 343–359.
38. Kurniawan, J.H.; Chng, S.; Cheah, L. The Social Acceptance of Autonomous Vehicles. *IEEE Potentials* **2021**, *40*, 39–44. [CrossRef]
39. Pelau, C.; Dabija, D.C.; Ene, I. What makes an AI device human-like? The role of interaction quality, empathy and perceived psychological anthropomorphic characteristics in the acceptance of artificial intelligence in the service industry. *Comput. Hum. Behav.* **2021**, *122*, 106855. [CrossRef]
40. Denis, A.S.; Moskolai Ngossaha, J.; Ari, A. Cyber-Physical Urban Mobility Systems: Opportunities and Challenges in Developing Countries. *Int. J. Softw. Innov.* **2023**, *11*, 1–21. [CrossRef]
41. Xilinx. What Is an FPGA? Available online: https://www.xilinx.com/products/silicon-devices/fpga/what-is-an-fpga.html (accessed on 10 April 2023).
42. Boutros, A.; Betz, V. FPGA Architecture: Principles and Progression. *IEEE Circuits Syst. Mag.* **2021**, *21*, 4–29. [CrossRef]
43. Magyari, A.; Chen, Y. Review of state-of-the-art FPGA applications in IOT Networks. *Sensors* **2022**, *22*, 19. [CrossRef]
44. Chakraborty, A.; Kar, A.K. Swarm Intelligence: A Review of Algorithms. In *Nature-Inspired Computing and Optimization*; Springer: Berlin/Heidelberg, Germany, 2017. [CrossRef]
45. Steghöfer, J.P.; Kiefhaber, R.; Bee, K.; Bernard, Y.; Klejnowski, L.; Reif, W.; Ungerer, T.; André, E.; Hähner, J.; Müller-Schloer, C. Trustworthy Organic Computing Systems: Challenges and Perspectives. In *Autonomic and Trusted Computing, Proceedings of the 7th International Conference, ATC 2010, Xi'an, China, 26–29 October 2010*; Springer: Berlin/Heidelberg, Germany, 2010. [CrossRef]
46. Nguyen, T.T.; Nguyen, N.D.; Nahavandi, S. Deep Reinforcement Learning for Multiagent Systems: A review of challenges, solutions, and applications. *IEEE Trans. Cybern.* **2020**, *50*, 3826–3839. [CrossRef]
47. Handayani, K.; Anugrah, P. Assessing the implications of net-zero emissions pathways: An analysis of the Indonesian power sector. In Proceedings of the 2021 International Conference on Technology and Policy in Energy and Electric Power (ICT-PEP), Jakarta, Indonesia, 29–30 September 2021; pp. 270–275. [CrossRef]
48. Tomar, D.; Tomar, P. Integration of Cloud Computing and Big Data Technology for Smart Generation. In Proceedings of the 2018 8th International Conference on Cloud Computing, Data Science & Engineering (Confluence), Noida, India, 11–12 January 2018; pp. 1–6. [CrossRef]
49. Al-Rubaye, S.; Rodriguez, J.; Fragonara, L.Z.; Theron, P.; Tsourdos, A. Unleash Narrowband Technologies for Industrial Internet of Things Services. *IEEE Netw.* **2019**, *33*, 16–22. [CrossRef]
50. Gopalakrishnan, S.K.; Al-Rubaye, S.; Inalhan, G. Adaptive UAV Swarm Mission Planning by Temporal Difference Learning. In Proceedings of the 2021 IEEE/AIAA 40th Digital Avionics Systems Conference (DASC), San Antonio, TX, USA, 3–7 October 2021; pp. 1–10. [CrossRef]
51. Aloseel, A.; Al-Rubaye, S.; Zolotas, A.; Shaw, C. Attack-Detection Architectural Framework Based on Anomalous Patterns of System Performance and Resource Utilization—Part II. *IEEE Access* **2021**, *9*, 87611–87629. [CrossRef]
52. Mohamed, M.A.; Kardas, G.; Challenger, M. Model-Driven Engineering Tools and Languages for Cyber-Physical Systems–A Systematic Literature Review. *IEEE Access* **2021**, *9*, 48605–48630. [CrossRef]
53. Escobar, L.; Moyano, C.; Aguirre, G.; Guerra, G.; Allauca, L.; Loza, D. Multi-robot platform with features of Cyber-physical systems for education applications. In Proceedings of the 2020 IEEE ANDESCON, Quito, Ecuador, 13–16 October 2020; pp. 1–6. [CrossRef]
54. Lee, E.A. Cyber Physical Systems: Design Challenges. In Proceedings of the 2008 11th IEEE International Symposium on Object and Component-Oriented Real-Time Distributed Computing (ISORC), Orlando, FL, USA, 5–7 May 2008; pp. 363–369. [CrossRef]
55. Khaitan, S.K.; McCalley, J.D. Design Techniques and Applications of Cyberphysical Systems: A Survey. *IEEE Syst. J.* **2015**, *9*, 350–365. [CrossRef]
56. Segura, M.; Poggi, T.; Barcena, R. A Generic Interface for x-in-the-Loop Simulations Based on Distributed Co-Simulation Protocol. *IEEE Access* **2023**, *11*, 5578–5595. [CrossRef]
57. Hubert, H.; Stabernack, B. Profiling-Based Hardware/Software Co-Exploration for the Design of Video Coding Architectures. *IEEE Trans. Circuits Syst. Video Technol.* **2009**, *19*, 1680–1691. [CrossRef]
58. Luyan, W.; Zhangguo, S.; Long, C. The performance analysis for embedded systems using statistics methods. *Telkomnika Indones. J. Electr. Eng.* **2013**, *11*, 4099–4103. [CrossRef]
59. Bijjahalli, S.; Sabatini, R. A High-Integrity and Low-Cost Navigation System for Autonomous Vehicles. *IEEE Trans. Intell. Transp. Syst.* **2021**, *22*, 356–369. [CrossRef]
60. Namuduri, K.; Fiebig, U.C.; Matolak, D.W.; Guvenc, I.; Hari, K.; Määttänen, H.L. Advanced Air Mobility: Research Directions for Communications, Navigation, and Surveillance. *IEEE Veh. Technol. Mag.* **2022**, *17*, 65–73. [CrossRef]
61. Schmitz, M.T.; Al-Hashimi, B.M.; Eles, P. *System-Level Design Techniques for Energy-Efficient Embedded Systems*, 1st ed.; Springer: New York, NY, USA, 2006. [CrossRef]

62. Swaminathan, N.; Reddy, S.R.P.; RajaShekara, K.; Haran, K.S. Flying Cars and eVTOLs—Technology Advancements, Powertrain Architectures, and Design. *IEEE Trans. Transp. Electrif.* **2022**, *8*, 4105–4117. [CrossRef]
63. Panda, P.R.; Catthoor, F.; Dutt, N.D.; Danckaert, K.; Brockmeyer, E.; Kulkarni, C.; Vandercappelle, A.; Kjeldsberg, P.G. Data and memory optimization techniques for embedded systems. *ACM Trans. Des. Autom. Electron. Syst.* **2001**, *6*, 149–206. [CrossRef]
64. Timilsina, L.; Badr, P.R.; Hoang, P.H.; Ozkan, G.; Papari, B.; Edrington, C.S. Battery Degradation in Electric and Hybrid Electric Vehicles: A Survey Study. *IEEE Access* **2023**, *11*, 42431–42462. [CrossRef]
65. Zhang, D.; Zhou, X.; Cheng, E.; Wan, H.; Chen, Y. Investigation on Effects of HPM Pulse on UAV's Datalink. *IEEE Trans. Electromagn. Compat.* **2020**, *62*, 829–839. [CrossRef]
66. Arandhakar, S.; Jayaram, N.; Shankar, Y.R.; Gaurav.; Kishore, P.S.V.; Halder, S. Emerging Intelligent Bidirectional Charging Strategy Based on Recurrent Neural Network Accosting EMI and Temperature Effects for Electric Vehicle. *IEEE Access* **2022**, *10*, 121741–121761. [CrossRef]
67. Shen, Q.; Yue, C.; Goh, C.H.; Wang, D. Active fault-tolerant control system design for spacecraft attitude maneuvers with actuator saturation and faults. *IEEE Trans. Ind. Electron.* **2019**, *66*, 3763–3772. [CrossRef]
68. Sharma, P.; Gillanders, J. Cybersecurity and Forensics in Connected Autonomous Vehicles: A Review of the State-of-the-Art. *IEEE Access* **2022**, *10*, 108979–108996. [CrossRef]
69. Kukkala, V.K.; Thiruloga, S.V.; Pasricha, S. Roadmap for Cybersecurity in Autonomous Vehicles. *IEEE Consum. Electron. Mag.* **2022**, *11*, 13–23. [CrossRef]
70. Khan, W.Z.; Khan, M.K.; Arshad, Q.u.A. Cybersecurity Digital Labels for Connected and Autonomous Vehicles. *IEEE Consum. Electron. Mag.* **2023**, *12*, 87–93. [CrossRef]
71. Choi, E.; Song, H.; Kang, S.; Choi, J.W. High-Speed, Low-Latency In-Vehicle Network Based on the Bus Topology for Autonomous Vehicles: Automotive Networking and Applications. *IEEE Veh. Technol. Mag.* **2022**, *17*, 74–84. [CrossRef]
72. Yassine, A.; Souweid, S. Time-to-Market and Product Performance Tradeoff Revisited. *IEEE Trans. Eng. Manag.* **2021**, 1–16. [CrossRef]
73. Tan, Z.; Dai, N.; Su, Y.; Zhang, R.; Li, Y.; Wu, D.; Li, S. Human–Machine Interaction in Intelligent and Connected Vehicles: A Review of Status Quo, Issues, and Opportunities. *IEEE Trans. Intell. Transp. Syst.* **2022**, *23*, 13954–13975. [CrossRef]
74. Mizutani, K.; Yamaguchi, H.; Urino, Y.; Koibuchi, M. OPTWEB: A Lightweight Fully Connected Inter-FPGA Network for Efficient Collectives. *IEEE Trans. Comput.* **2021**, *70*, 849–862. [CrossRef]
75. Khairullah, S.S. Realization of a 16-bit MIPS RISC pipeline processor. In Proceedings of the 2022 International Congress on Human-Computer Interaction, Optimization and Robotic Applications (HORA), Ankara, Turkey, 9–11 June 2022; pp. 1–6. [CrossRef]
76. Monmasson, E.; Cirstea, M.N. FPGA Design Methodology for Industrial Control Systems—A Review. *IEEE Trans. Ind. Electron.* **2007**, *54*, 1824–1842. [CrossRef]
77. Chen, J.; Huang, S. Analysis and Comparison of UART, SPI and I2C. In Proceedings of the 2023 IEEE 2nd International Conference on Electrical Engineering, Big Data and Algorithms (EEBDA), Changchun, China, 24–26 February 2023; pp. 272–276. [CrossRef]
78. Trinh, X.D.; Le, M.C.; Huy, T. IMU Calibration Methods and Orientation Estimation Using Extended Kalman Filters. In *AETA 2019—Recent Advances in Electrical Engineering and Related Sciences: Theory and Application*; Springer International Publishing: Berlin/Heidelberg, Germany, 2021; pp. 541–551. [CrossRef]
79. NXP. AN3397—Implementing Positioning Algorithms Using Accelerometers—NXP, 2007. Available online: https://www.nxp.com/docs/en/application-note/AN3397.pdf (accessed on 1 April 2023).
80. Zhou, X.; Gou, X.; Huang, T.; Yang, S. Review on Testing of Cyber Physical Systems: Methods and Testbeds. *IEEE Access* **2018**, *6*, 52179–52194. [CrossRef]
81. Allouis, E.; Blake, R.; Gunes-Lasnet, S.; Jorden, T.; Maddison, B.; Schroeven-Deceuninck, H.; Stuttard, M.; Truss, P.; Ward, K.; Ward, R.; et al. A Facility for the Verification and Validation of Robotics and Autonomy for Planetary Exploration. In Proceedings of the 44th Annual Lunar and Planetary Science Conference, The Woodlands, TX, USA, 18–22 March 2013.
82. Deshmukh, J.V.; Horvat, M.; Jin, X.; Majumdar, R.; Prabhu, V.S. Testing cyber-physical systems through Bayesian optimization. *ACM Trans. Embed. Comput. Syst.* **2017**, *16*, 1–18. [CrossRef]
83. Zheng, X.; Julien, C.; Kim, M.; Khurshid, S. Perceptions on the State of the Art in Verification and Validation in Cyber-Physical Systems. *IEEE Syst. J.* **2017**, *11*, 2614–2627. [CrossRef]
84. Naufal, J.K.; Camargo, J.B.; Vismari, L.F.; de Almeida, J.R.; Molina, C.; González, R.I.R.; Inam, R.; Fersman, E. A2CPS: A Vehicle-Centric Safety Conceptual Framework for Autonomous Transport Systems. *IEEE Trans. Intell. Transp. Syst.* **2018**, *19*, 1925–1939. [CrossRef]
85. Kownacki, C. Optimization approach to adapt Kalman filters for the real-time application of accelerometer and gyroscope signals' filtering. *Digit. Signal Process.* **2011**, *21*, 131–140. [CrossRef]
86. Ludwig, S.A.; Burnham, K.D. Comparison of Euler Estimate using Extended Kalman Filter, Madgwick and Mahony on Quadcopter Flight Data. In Proceedings of the 2018 International Conference on Unmanned Aircraft Systems (ICUAS), Dallas, TX, USA, 12–15 June 2018; pp. 1236–1241. [CrossRef]
87. Zhang, P.; Gu, J.; Milios, E.E.; Huynh, P. Navigation with IMU/GPS/digital compass with unscented Kalman filter. In Proceedings of the IEEE International Conference Mechatronics and Automation, Niagara Falls, ON, Canada, 29 July–1 August 2005; Volume 3, pp. 1497–1502.

88. Steinbrener, J.; Brommer, C.; Jantos, T.; Fornasier, A.; Weiss, S. Improved state propagation through AI-based pre-processing and down-sampling of high-speed inertial data. In Proceedings of the 2022 International Conference on Robotics and Automation (ICRA), Philadelphia, PA, USA, 23–27 May 2022; pp. 6084–6090.
89. Patel, U.N.; Faruque, I.A. Multi-IMU Based Alternate Navigation Frameworks: Performance & Comparison for UAS. *IEEE Access* **2022**, *10*, 17565–17577. [CrossRef]
90. Seng, K.P.; Lee, P.J.; Ang, L.M. Embedded Intelligence on FPGA: Survey, Applications and Challenges. *Electronics* **2021**, *10*, 895. [CrossRef]
91. Datta, D.; Datta, B.; Dutta, H.S. Design and implementation of multibit LFSR on FPGA to generate pseudorandom sequence number. In Proceedings of the 2017 Devices for Integrated Circuit (DevIC), Kalyani, India, 23–24 March 2017; pp. 346–349. [CrossRef]
92. Sharma, D.K.; Singh, N.C.; Noola, D.A.; Doss, A.N.; Sivakumar, J. A review on various cryptographic techniques & algorithms. *Mater. Today Proc.* **2022**, *51*, 104–109. [CrossRef]
93. Sohal, M.; Sharma, S. BDNA-a DNA inspired symmetric key cryptographic technique to secure cloud computing. *J. King Saud Univ.—Comput. Inf. Sci.* **2022**, *34*, 1417–1425. [CrossRef]
94. Qian, Y.; Feng, Y.; Chen, H.H. Cryptographic techniques. In *Security in Wireless Communication Networks*; John Wiley and Sons, Ltd.: Chichester, UK, 2022; pp. 51–76.
95. Zhang, J.; Prabakar, K.; Hasandka, A.; Alam, S.M.S.; Jiang, Y.; Hodge, B.M.; Gao, D.W. Power and Communications Hardware-In-the-Loop CPS Architecture and Platform for DER Monitoring and Control Applications. In Proceedings of the 2021 IEEE International Conference on Communications, Control, and Computing Technologies for Smart Grids (SmartGridComm), Aachen, Germany, 25–28 October 2021; pp. 170–174. [CrossRef]
96. García, M.V.; Pérez, F.; Calvo, I.; Morán, G. Building industrial CPS with the IEC 61499 standard on low-cost hardware platforms. In Proceedings of the 2014 IEEE Emerging Technology and Factory Automation (ETFA), Barcelona, Spain, 16–19 September 2014; pp. 1–4. [CrossRef]
97. Isakovic, H.; Ratasich, D.; Hirsch, C.; Platzer, M.; Wally, B.; Rausch, T.; Nickovic, D.; Krenn, W.; Kappel, G.; Dustdar, S.; et al. CPS/IoT Ecosystem: A platform for research and education. In *Cyber Physical Systems. Model-Based Design, Proceedings of the 8th International Workshop, CyPhy 2018, and 14th International Workshop, WESE 2018, Turin, Italy, 4–5 October 2018*; Springer International Publishing: Berlin/Heidelberg, Germany, 2019; pp. 206–213.
98. Landolfi, G.; Barni, A.; Menato, S.; Cavadini, F.A.; Rovere, D.; Dal Maso, G. Design of a multi-sided platform supporting CPS deployment in the automation market. In Proceedings of the 2018 IEEE Industrial Cyber-Physical Systems (ICPS), St. Petersburg, Russia, 15–18 May 2018; pp. 684–689. [CrossRef]
99. Schmittle, M.; Lukina, A.; Vacek, L.; Das, J.; Buskirk, C.P.; Rees, S.; Sztipanovits, J.; Grosu, R.; Kumar, V. OpenUAV: A UAV Testbed for the CPS and Robotics Community. In Proceedings of the 2018 ACM/IEEE 9th International Conference on Cyber-Physical Systems (ICCPS), Porto, Portugal, 11–13 April 2018; pp. 130–139. [CrossRef]
100. Iannino, V.; Colla, V.; Denker, J.; Göttsche, M. A CPS-Based Simulation Platform for Long Production Factories. *Metals* **2019**, *9*, 1025. [CrossRef]

Disclaimer/Publisher's Note: The statements, opinions and data contained in all publications are solely those of the individual author(s) and contributor(s) and not of MDPI and/or the editor(s). MDPI and/or the editor(s) disclaim responsibility for any injury to people or property resulting from any ideas, methods, instructions or products referred to in the content.

electronics

MDPI

Article

Video Blockchain: A Decentralized Approach for Secure and Sustainable Networks with Distributed Video Footage from Vehicle-Mounted Cameras in Smart Cities

Kasun Moolikagedara, Minh Nguyen, Wei Qi Yan * and Xue Jun Li *

School of Engineering, Computer and Mathematical Sciences, Auckland University of Technology, Auckland 1010, New Zealand; kasun.moolikagedara@autuni.ac.nz (K.M.); minh.nguyen@aut.ac.nz (M.N.)
* Correspondence: weiqi.yan@aut.ac.nz (W.Q.Y.); xuejun.li@aut.ac.nz (X.J.L.)

Abstract: In this paper, we explore video blockchain for establishing connectivity among vehicles in a smart city through utilizing blockchain technology. By leveraging intelligent vehicular systems that provide location-based visualization through multiple deployed cameras in vehicles, we expand the scope of collecting video surveillance data for observation, thereby enhancing overall situational awareness. We utilize the decentralized nature of blockchain to implement a vehicle-based surveillance system across a smart city. To ensure reliability, the integration of two cryptographic functions, hashing and signing, with the blockchain is employed. This integration ensures secure and tamper-proof solutions for the existing intelligent surveillance system. In this paper, our primary focus is on combining blockchain technology to achieve sustainable and robust smart solutions for intelligent vehicular distributed video networks while eliminating the need for third-party intermediaries. Through extensive experiments and analysis, we demonstrate the effectiveness and feasibility of our proposed video blockchain approach. The results indicate that this innovative framework provides enhanced security, privacy, and scalability for intelligent vehicular distributed networks in smart cities, paving the way for a connected and efficient urban environment.

Keywords: video blockchain; distributed network; intelligent vehicular

1. Introduction

The rapid advancement of blockchain has significantly impacted multiple facets of society, necessitating the careful consideration of the complexities that play a vital role in community development and that enhance our daily lives. As we navigate this area, addressing both its positive and negative aspects becomes imperative. Therefore, in the context of contemporary technologies, ensuring security is of paramount importance. Particularly in smart cities, where sensitive information is abundant, the establishment of a highly secure repository is necessary.

The concept of a "smart city" represents an ambitious goal pursued by cities worldwide, driven by the intricate nature of urban environments and the potential offered by intelligent technologies. Within this context, it is crucial to thoroughly examine digital surveillance systems, as they hold significant importance today. Furthermore, prioritizing the safety and security of individuals by reducing crime rates and accidents stands as one of the primary objectives of smart cities [1,2].

In video surveillance [3], the primary objective is to capture video footage using vehicle cameras and securely store it onsite, leveraging blockchain to ensure data communication integrity and resistance against tampering and cyberattacks. Video surveillance serves multiple purposes, including crime solving, anomalous event detection, and privacy policy enforcement, making recorded video footage highly valuable for diverse applications. However, there exists a potential risk of malicious attackers, savvy hackers, or unauthorized third parties manipulating cameras and video repositories, rendering them ineffective in crime scenarios. These attacks may include false frame injection, data tampering, and privacy violations within a surveillance system. To address these challenges and enhance the security of video data captured using intelligent surveillance systems, we propose the following hypotheses:

- The utilization of a video blockchain framework will significantly enhance the security and integrity of video data;
- The integration of enhanced cryptographic functions, such as the Schnorr signature and hash (SHA256), within a video blockchain, along with the implementation of Merkle trees, will establish a comprehensive security mechanism that ensures the robust storage, retrieval, and data integrity of video records within a network;
- The combination of vehicle cameras, blockchain, and third-party certificate authority (CA) verification will ensure the secure and sequential storage of video data and protect it from tampering and unauthorized access;
- The proposed video blockchain will effectively mitigate the risks associated with malicious attacks, data tampering, and privacy violations in "intelligent vehicular distributed video networks", thereby improving its effectiveness in crime scenarios.

In this paper, our focus was driven by our intention to combine video blockchain with an enhancement of intelligent vehicular systems. We aimed for an evaluation of a novel method for capturing and storing video data using vehicle-mounted cameras and blockchain. Building upon existing work and implementing enhanced cryptographic functions, our objective was to enhance the security of vehicle videos in smart cities.

The rest of this paper is organized as follows. Section 2 provides a survey of related work, followed by a detailed presentation of our proposed method in Section 3, as well as a demonstration of the experimental results in Section 4 to assess the validity and effectiveness of our hypotheses. In Section 5, we discuss the results and conclude this paper, with the limitations and future work in Section 6.

2. Related Work

2.1. Theoretical Frameworks for Blockchain

In this paper, we explore how computational methods can be applied to video surveillance, focusing on the video blockchain approach that predominantly protects user identification by using hashed public keys [4]. Previous work [3] stored video footage from surveillance cameras and ensured the integrity of video data. However, the work did not cover the storage of real videos in the blockchain, which plays a significant role in simplifying human lifestyles across various sectors, including energy, healthcare, and supply chains [5,6].

The key feature of blockchain is to enable a decentralized community and reach consensus on a transparent transaction history without relying on pre-established trust, thus mitigating double-spending attacks [7,8]. The use of hash functions, particularly in the context of proof-of-work, has proven effective in resisting distributed denial-of-service (DDoS) attacks, as demonstrated by bitcoin.

In the process of calculating blockchain data, cryptographic encryption is employed to establish connections between preceding and succeeding blocks [9]. Each block has a block header, which contains the information necessary to establish the link between adjacent blocks and a block body that typically stores transaction details and records [10].

In a blockchain, each block contains specific information, including a local number indicating the block position in the chain, the identity of the source or creator, the hash value of the previous block, a timestamp, and a Merkle root. The hash value of the preceding block is critical for maintaining the integrity and continuity of the blockchain. The timestamp serves multiple purposes, ensuring the reliability, authenticity, and traceability of the data while preventing tampering [11].

2.2. Surveillance System for Smart Cities Approaches

The concept of smart cities [1] emerged during the 2008 economic crisis when the world sought to leverage information and communication technologies (ICT) to become more intelligent and resource-efficient. Smart cities aim to achieve cost and energy savings, improved service delivery and quality of life, and reduced environmental footprints [12].

A smart city includes a surveillance system that captures videos and aims to enhance the integrity of the recorded data. In [13], using Blocksee, through the method of using the event of a car accident, detected through built-in accelerometers, relevant videos were cryptographically hashed and recorded through distributed storage on the blockchain. To ensure the integrity of the recorded videos, we leveraged the unique features of the distributed and tamper-proof characteristics offered by blockchain technology. In the blockchain, timestamping features are applied to verify and transfer unaltered data to a distributed repository. Similarly, Ref. [7] explored the use of blockchain-based systems to guarantee the storage of recorded data from closed-circuit television (CCTV) cameras in smart cities, preventing alteration or tampering. This mechanism assists law enforcement and clients in securing data recordings from digital surveillance systems by utilizing the metadata recorded in the blockchain. To ensure data security in a smart city, it is crucial to adhere to the principles of confidentiality, integrity, and availability (CIA) [14,15] in both video surveillance and video blockchain. Although video surveillance lacks built-in mechanisms to ensure secure data transfer, we can sort the videos in the correct order [16] and ensure their proper storage in a video blockchain without tampering. The video blockchain mechanism organizes the videos accurately in ascending or descending order from the video website and stores large data in the blockchain [17–19]. Additionally, blockchain addresses data integrity in systems such as medical record keeping and intelligent gas monitoring in smart cities [20–22].

2.3. Extracting Cryptographic Functions: Advances and Limitations

While considering smart cities worldwide, infrastructure security and data privacy emerge as two critical aspects. A distributed ledger, such as blockchain, offers enhanced security for connected electric vehicles [23]. Blockchain eliminates single points of failure and employs various cryptographic algorithms to ensure data integrity. Bitcoin, implemented with blockchain since 2009 [13,24], has extended its support to sophisticated sectors like smart cities. Blockchain has also eliminated the need for third-party involvement in transactions, providing a secure and transparent mechanism [25].

Hashing, a one-way function, guarantees transaction information without tampering. By converting plain text to irreversible hash data, the blockchain employs hashing algorithms such as SHA for data encryption. The blockchain encrypts transaction data in blocks using a hash algorithm and saves a unique string of 32-bit numbers mixed with an arbitrary three-column string. The integrity of the stored data is ensured through comparisons between hash values. To create a high-confidence smart city, we propose an approach that combines a (k, n) secret-sharing mechanism and software-defined networking (SDN) framework to secure data transportation among smart cities [26]. Data security is guaranteed by using the (k, n) secret-sharing scheme, while SDN-based transmission strategies leverage SDN's advantages in network management and are scheduled to overcome challenges posed by unstable network states. Extensive experiments demonstrated that this approach significantly reduces attack success rates with reasonable overhead.

A cryptographic hash function plays a crucial role in a blockchain by mapping arbitrary data to a fixed-size string. The security requirements for hash functions include one-way characteristics and collision resistance [16]. To ensure a minimum of 80-bit security, the output length of hash functions should be at least 160 bits. In blockchains, the general hash function is SHA256, which belongs to the SHA (secure hash algorithms) family of cryptographic hash functions. Hash functions in blockchains are utilized for various purposes, such as proof-of-work (PoW), address generation, block generation (as part of the Merkle-tree paradigm), signature validation, pseudorandom number generation, and other essential components, like the Fiat–Shamir mechanism.

A Merkle tree is employed to organize and represent the primary transaction data in a bottom-up manner. Each leaf node corresponds to a transaction, and the hash value of two transactions is computed to obtain the hash value of the intermediate node. This process continues until a final hash value, known as the Merkle root, is derived. Each set of transaction data has a unique Merkle root associated with it.

The Schnorr signature scheme reduces the computation required for signature generation, especially during idle time. The main computational work for generating signatures is independent of the message, and it involves multiplying a 2n-bit integer with an n-bit integer. The scheme is based on a prime modulus, p, where $p - 1$ has a prime factor, q, of an appropriate size, satisfying the condition $p - 1 = 1 \pmod q$ [16]. Typically, p is chosen to be approximately 21,024, and q is approximately 2160, resulting in a 1024-bit number for p and a 160-bit number, which matches the length of the SHA-1 hash value [24].

In the field of cryptography, the Schnorr signature scheme is well known for its simplicity and security based on the intractability of discrete logarithm problems [27]. This scheme minimizes the computational effort required to generate a signature, and it is widely utilized in public-key cryptography for digital signatures. Digital signatures provide message authentication among communicating parties, protecting against fraudulent message creation or denial. Digital signatures involve the use of public-key algorithms among the communicating parties. The sender's private key is harnessed to encrypt either the entire message or a hash code of the message, forming the digital signature. Confidentiality can be further achieved by encrypting the entire message along with the signature using public or private key schemes. To resolve disputes, the signature function and the message must be accessible to a third party. However, the security of these approaches relies on the protection of the sender's private key against forgery, loss, or theft. Digital certificates and certificate authorities, along with timestamps and key revocation mechanisms, are commonly employed to address these security threats.

3. Proposed Video Blockchain

In this paper, a descriptive research design is employed to investigate a secure mechanism for linking vehicle-mounted cameras in a smart city. Both primary and secondary video data are utilized to achieve the research objectives. For primary data collection, videos from vehicle-mounted cameras were collected, and the images extracted from these videos were used for experimentation. This implementation was part of our studies in the visual blockchain [28]. During the image encryption process, the raw image data were converted into a sequence of bytes, and the original data of the raw image could not be accessed once the encryption process was completed [29,30].

The main objective of this paper was to create a chain of blocks using vehicle videos and connect them in a distributed manner to ensure the correct video sequence without tampering, while minimizing the involvement of third parties in a smart city. To attain secure surveillance data communication using blockchain technology, a combination of a SHA-256 hash function and Schnorr signature was employed. The hash function ensures collision resistance, while the Schnorr signature scheme enhances the security of the communication.

Let (G', S', V') represent a signature scheme, and $H(\cdot)$ denotes a hash function with collision resistance properties. We utilized these components to achieve the desired security requirements for the communication of surveillance data over our blockchain.

Key Generation (G'):

- Generate a private key, s_k, randomly from the set G.
- Compute the corresponding public key v_k as $v_k = g^{sk}$.

Signing (S'):

- Take a message, m, as input.

- Generate a random value, b, from the set of integers modulo, $| G$.
- Compute u as $u = g^b$;
- Calculate a by applying the random oracle $H(\cdot)$ to u and the message m: $a = H(u, m)$.
- Compute s as $s = a.\ s_k + b \bmod q$, where q represents the modulus of the group G.
- Output the signature (a, s).

Verification (V'):

- Take a message, m, and a signature (a, s) as input.
- Retrieve the public key v_k.
- Compute u' as $u' = g^s \cdot v_k^{(-a)}$.
- Calculate a' by applying the random oracle $H(\cdot)$ to u' and the message m: $a' = H(u', m)$;
- If a' is equal to a, the signature is considered valid. Otherwise, it is invalid.

In the context of blockchain data processing, the calculation involves generating both hashing and signature functions. This dual process ensures the integrity and security of the data. In this paper, we calculated the hash value using a specific hash function, such as SHA-256. This hash value acts as a unique identifier for the data, helping to ensure its integrity and prevent tampering.

To enhance the data protection of the vehicle-based video surveillance system in the smart city, we employed a third-party certification process. This involved collaborating with a certificate authority (CA) to verify the authority and authenticity of the data. The involvement of a trusted third party, such as a CA, adds an extra layer of security and credibility to the data.

Moreover, a private blockchain was employed to transmit the video data. By utilizing a private blockchain, we ensured that the data sent through the CA were received without any tampering. The private blockchain serves as a secure and decentralized platform for transmitting and storing video frame data, ensuring its integrity and protection. This mechanism, which combines the calculation of hash values, third-party certification, and the use of a private blockchain, effectively enhances the security and integrity of data in the vehicle-based video surveillance system in the smart city. It offers a reliable and trustworthy approach to guarantee the authenticity and integrity of data originating from vehicles.

In Algorithm 1, we assume the availability of a blockchain web service (Blockchain Web Service) and aim to process incoming requests (R_i) from authenticated nodes (N_i). The steps involved in the process include:

Algorithm 1: Blockchain-Based Image Encryption Process

Require: Blockchain Web Service
Ensure: Genesis Block
while T has not expired do
    if node N_i is authenticated == **True** then
        if request R_i is matched == **True** then
            if Ri is identified as a processed request == **False** then
                process the response to C_i
                Hash (video_frame_metadata)
                Sign (video_frame_metadata)
                Update chain
            else
                Response to N_i that R_i is not a valid request
            end if
        else
            Deny the request
        end if
        Validate and add block into chain
    end if
end while

In this updated version of Algorithm 2, a certificate authority (CA) has been introduced to provide certificates for the hashed features. The algorithm outlines a process for image encryption using blockchain and assumes the availability of a blockchain web service. Authenticated nodes can submit requests for image encryption, which are then vali-dated and processed according to specific conditions. Valid requests undergo the encryption process, which includes generating a response for the requesting node, hashing the video frame metadata, and signing it for integrity. The algorithm updates the blockchain by adding a new block containing the encrypted image or relevant information. Invalid requests are responded to accordingly, either by notifying the node of the issue or denying the request outright. The algorithm ensures the integrity and security of the blockchain by validating and adding processed blocks. It continues processing requests until a specified time has not expired. A certificate authority has also been introduced to generate certificates for the hashed features, enhancing security and verification capabilities.

Algorithm 2: Hashed Features Authentication Procedures

```
Function: hash_feature (features_set)
    string_features [empty_string]
    for feature_vector in features_set.items do
        feature_string = Convert_to_string (feature_vector)
        string_features = (string_features | | feature_string)
    end for
    bytes_features = Convert_to_bytes (string_features)
    feature_hash = Convert_to_hash (bytes_features)
    feature_id = features_set.name
    return {feature_id: feature_hash}
Procedure: record_hashed_feature (features_set)
    feature_tx = hash_feature (features_set)
    certificate = CertificateAuthority.generate_certificate (feature_tx)
    tx_result = transaction_commit(feature_tx, certificate)
    return tx_result
Procedure: verify_hashed_feature (features_set)
    feature_tx = hash_feature (features_set)
    query_feature_tx = transaction_query (feature_tx)
    if query_feature_tx == feature_tx then
        verify_hash = True
    else
        verify_hash = False
    end if
    return verify_hash
```

4. Results

The process involves converting captured videos into individual video frames, each ranging from 50 KB to 1024 KB. These frames were utilized to establish a private blockchain system using Microsoft Windows 11 Pro 64 bit. An experimental setup was executed to assess the efficacy of the proposed approach against various attacks. The experimentation led to the creation of innovative computational techniques for video blockchains, combining specific cryptographic algorithms with video blockchain technology. This article serves as a valuable contribution to the realm of video blockchain by introducing a fresh method for safeguarding video data in surveillance setups. The study's outcomes can serve as a cornerstone for forthcoming endeavors in both video blockchain and cryptographic algorithms.

Each root structure within the Merkle tree ensures the integrity of the connection between video frames and their respective hashing order. This safeguards against any alteration to the image sequence without modifying the entire root structure of the tree. Overall, these findings offer valuable insights into the advancement of web interfaces for video blockchain systems, and their implications can be harnessed to augment the dependability and security of such systems in the future.

The results in Figure 1 reflect Merkle tree-based approaches for data verification. The evaluation was conducted by measuring the computational time and amount of data for each experiment, with each experiment being repeated 100 times to minimize errors caused by outliers.

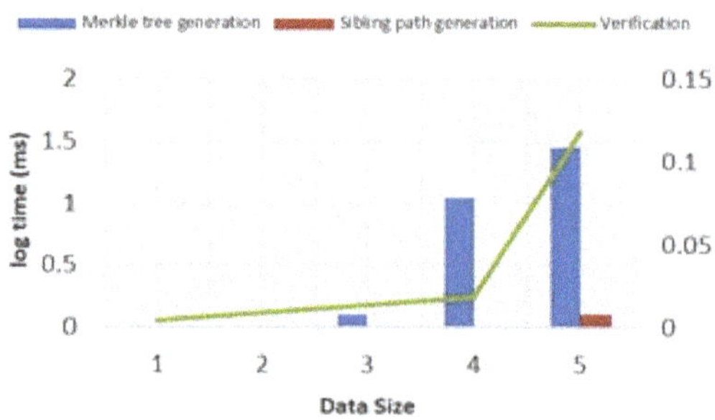

Figure 1. Average computational time (millisecond) for authentication based on Merkle tree by data size.

However, beyond 100 KB of data, these approaches exhibit only marginal differences. For instance, while generating a Merkle tree for a 1 MB data file, the prover's computational time is dominated by the process, accounting for approximately 99.9% of the total time, and the results are presented in a Figure 1. The execution time varies based on different input sizes and block sizes, and we analyzed this variability to ascertain the function's time complexity. Figure 2 illustrates the relationship between execution time (on the y-axis) and input size or block size (on the x-axis). This comparison encompasses various input sizes and block sizes, providing insights into the function's performance.

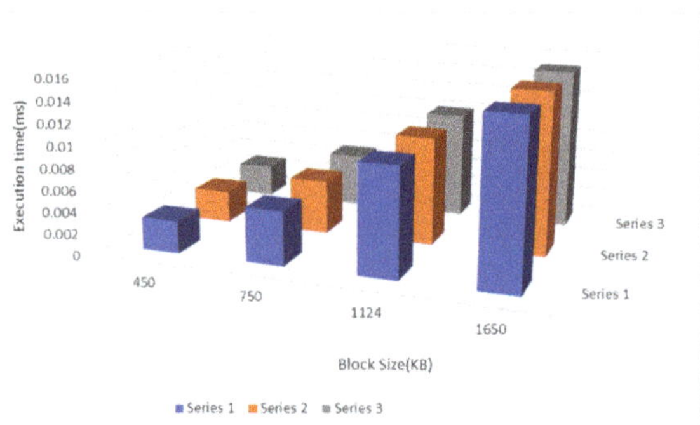

Figure 2. Execution times for different input sizes and block sizes.

In Figure 3, we also compare the computation time and data size metrics with those of other similar work. Our results show a significant change after a data size of 100 KB, indicating that the study's outcomes are comparatively reliable. These findings can be applied to improve the efficiency and accuracy of Merkle tree-based approaches for data verification.

The use of blockchain technology in intelligent surveillance faces many challenges, such as scalability, interoperability, and regulatory compliance. Scaling blockchain to handle large volumes of data and transactions, integrating it with the existing systems, and navigating regulatory frameworks are crucial areas requiring attention in this domain. In our future work, we plan to address these issues.

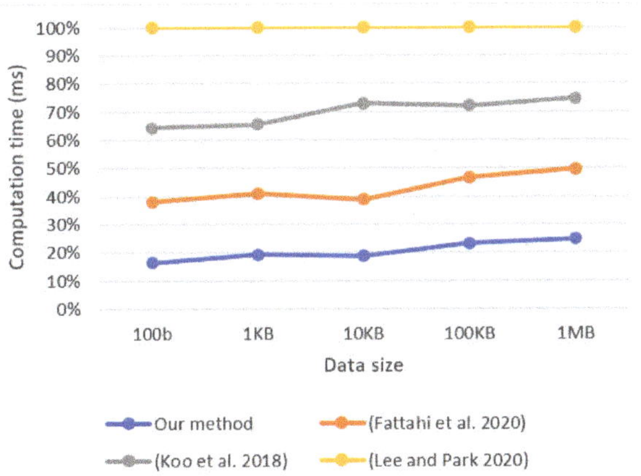

Figure 3. Comparison of the computational time between ours and other similar work [17,27,31].

5. Discussion and Result Analysis

In this paper, our primary findings are rooted in videos captured from vehicles. To evaluate our video blockchain system, we established a blockchain environment using visual coding and the Hyperledger Fabric private blockchain platform. Hyperledger Fabric, designed for distributed ledger research and experimentation, facilitated the creation of our private blockchain, enabling us to conduct comprehensive analyses. Additionally, we harnessed the integrated development environment (IDE) in MATLAB to gather computational results and compare them against video frame encryption. The MATLAB IDE was also employed to measure the success rates of our proposed method.

This paper contributes valuable insights into the evolution of web interfaces for video blockchain systems, outlining avenues for bolstering their reliability and security in forthcoming developments. Our recorded videos yielded results for the number of normalized pixels change rate (NPCR) of 99.6021% and the unified averaged changing intensity (UACI) of 33.4065%. Table 1 shows a comparison with other results, indicating our method outperforms others. In Equation (1), M denotes the width (i.e., number of columns) of the images, and N signifies the height (i.e., number of rows) of the images.

$$P = M * N. \tag{1}$$

Table 1. Comparisons of deferential attack.

Type of Test	This Work	Belazi et al. [32]	Muhammad et al. [33]
NPCR	99.6021	99.6098	99.61
UACI	33.40	33.4384	33.44

Let $|1(i, j)$ and $|2(i, j)$ represent the pixel values at position (i, j) in the first and second images, respectively. Equation (1) calculates the normalized pixel difference rate ($NPCR$) as a percentage. It quantifies the proportion of pixels that differ between the two images and is computed by summing the absolute differences between corresponding pixel values not equal ($|I1(i, j)$ - $|2(i, j)|$)and dividing it by the total number of pixels (M $*$ N), multiplied by 100.

$$NPCR = \frac{\Sigma|I1(i, j) \neq I2(i, j)|}{(M * N)} \times 100 \tag{2}$$

where $UACI$ provides an indication of the average changingintensity difference between the two images. Higher $UACI$ values indicate greater average changing intensity differences, while lower $UACI$ values suggest that the images have more similar average intensities.

In this context, Σ represents the summation over all pixel positions (i, j). Additionally, $|I_1(i, j) - I_2(i, j)|$ denotes the absolute difference in pixel values between corresponding positions (i, j) in the two images.

$$UACI = \frac{\Sigma|I_1(i,j) - I_2(i,j)|}{(M * N * L)} \times 100 \quad (3)$$

We also established a peer-to-peer (P2P) network to interconnect the vehicle cameras and created a blockchain network using the required vehicles category in the smart city. For instance, the blockchain model can be applied to monitor city activities. Compared to the traditional client–server model, this P2P model offers numerous benefits. On the other hand, current surveillance systems heavily depend on service availability. However, a client–server architecture cannot guarantee high availability [31] because of its inherent nature. This structured P2P network continues to function even in the event of a node failure because of its distributed nature. Thus, we ensure the system's availability for users and maintain the security of the system's data.

It is noteworthy that while the results largely align with our expectations, there exist slight discrepancies when compared to the results of other studies. These variations can be attributed to several factors. Firstly, the choice of blockchain platform and encryption algorithms can lead to nuances in the results because of their inherent characteristics. Additionally, differences in the types of videos, varying frame rates, and even environmental conditions during recording could contribute to these deviations. These discrepancies are not unexpected within the realm of research, as the uniqueness of each approach and experimental setup can inevitably lead to divergent outcomes.

In conclusion, our discussion of the results highlights the alignment between our anticipated outcomes and the actual findings of our study. Moreover, the comparative analysis against other related research lends credence to the effectiveness of our video blockchain system. The minor discrepancies identified can be attributed to a combination of methodological choices and contextual variations. This discussion, thus, reinforces the value of our research in advancing the field of video blockchain systems while also acknowledging the nuanced nature of the experimental results in a diverse research landscape.

6. Conclusions

This paper presents a video blockchain framework that validates the hypotheses formulated in the study. The results underscore the significant enhancement in security and data integrity achievable through the utilization of a video blockchain within smart city surveillance systems. This outcome posits that the integration of a video blockchain establishes a robust security mechanism for the storage and retrieval of surveillance camera video records. Moreover, the inclusion of Merkle trees in the video blockchain architecture further bolsters the security and data integrity of the surveillance system.

The combination of vehicle cameras, blockchain technology, and third-party certification authority (CA) verification ensures the secure and sequential storage of video data, protecting it against tampering and unauthorized access. Lastly, the proposed video blockchain approach effectively mitigates the risks associated with malicious attacks, data tampering, and privacy violations in surveillance systems, enhancing their effectiveness in crime scenarios.

This paper contributes to the field by establishing a link between video frames captured by intelligent surveillance systems and blockchain. By leveraging cryptographic functions and decentralized storage platforms, the security of vehicle camera transferring video data is significantly improved. The proposed blockchain-based approach not only enhances the security and integrity of vehicle video data but also promotes trust, reliability, and controlled disclosure in smart cities. Throughout this paper, one of the identified limitations pertains to the necessity of maintaining a minimum number of connected nodes to enhance the system's availability. Additionally, our method is not resistant to quantum computer attacks. Taking into account these limitations, we plan to extend our research in future endeavors to enhance the system's resistance against attacks from quantum computers. This approach is of significant value to law enforcement monitoring, autonomous vehicles, insurance providers, and traffic management systems, providing increased security and adaptability for an advanced vehicular distributed video network in smart urban environments.

Author Contributions: Conceptualization, K.M. and W.Q.Y.; methodology, K.M.; software, K.M.; validation, K.M. formal analysis, K.M.; investigation, K.M.; resources, K.M. and W.Q.Y.; data curation, K.M.; writing—original draft preparation, K.M. writing—review and editing, K.M., W.Q.Y., M.N. and X.J.L.; visualization, K.M. supervision, W.Q.Y. and M.N.; project administration, X.J.L., W.Q.Y. and M.N. All authors have read and agreed to the published version of the manuscript.

Funding: This research received no external funding.

Data Availability Statement: Data sharing not applicable.

Conflicts of Interest: The authors declare no conflict of interest.

Abbreviations

The following abbreviations are used in this manuscript.

NPCR	Normalized pixels change rate
UACI	Unified average changing intensity

References

1. Aldairi, A.; Tawalbeh, L. Cyber security attacks on smart cities and associated mobile technologies. *Procedia Comput. Sci.* **2017**, *109*, 1086–1091. [CrossRef]
2. Alketbi, A.; Nasir, Q.; Abu Talib, M. Novel blockchain reference model for government services: Dubai government case study. *Int. J. Syst. Assur. Eng. Manag.* **2020**, *11*, 1170–1191. [CrossRef]
3. Deepak, K.; Badiger, A.N.; Akshay, J.; Awomi, K.A.; Deepak, G.; Harish Kumar, N. Blockchain-based management of video surveillance systems: A survey. In Proceedings of the International Conference on Advanced Computing and Communication Systems 2020, Coimbatore, India, 6–7 March 2020; pp. 1256–1258. [CrossRef]
4. Khan, M.A.; Salah, K. IoT security: Review, blockchain solutions, and open challenges. *Future Gener. Comput. Syst.* **2018**, *82*, 395–411. [CrossRef]
5. Engelhardt, M.A. Hitching healthcare to the chain: An introduction to blockchain technology in the healthcare sector. *Technol. Innov. Manag. Rev.* **2017**, *7*, 22–34. [CrossRef]
6. Fill, H.; Haerer, F. Knowledge blockchains: Applying blockchain technologies to enterprise modeling. In Proceedings of the Hawaii International Conference on System Sciences 2018, Waikoloa Village, HI, USA, 3–6 January 2018; pp. 4045–4054.
7. Fitwi, A.; Chen, Y. Secure and privacy-preserving stored surveillance video sharing atop permissioned blockchain. In Proceedings of the International Conference on Computer Communications and Networks 2021, Athens, Greece, 19–22 July 2021; pp. 1–8.
8. Fu, J.; Qiao, S.; Huang, Y.; Si, X.; Li, B.; Yuan, C. A study on the optimization of blockchain hashing algorithm based on PRCA. *Secur. Commun. Netw.* **2020**, *2020*, 8876317. [CrossRef]
9. Kullig, N.; Lämmel, P.; Tcholtchev, N. Prototype implementation and evaluation of a blockchain component on IoT devices. *Procedia Comput. Sci.* **2020**, *175*, 379–386. [CrossRef]
10. Hasan, O.; Brunie, L.; Bertino, E. Privacy-Preserving Reputation Systems Based on Blockchain and Other Cryptographic Building Blocks: A Survey. Ph.D. Thesis, University of Lyon, Lyon, France, 2023.
11. Hou, L.; Zheng, K.; Liu, Z.; Xu, X.; Wu, T. Design and prototype implementation of a blockchain-enabled lora system with edge computing. *IEEE Internet Things J.* **2021**, *8*, 2419–2430. [CrossRef]
12. Khrais, L.T. The combination of iot-sensors in appliances and block-chain technology in smart cities energy solutions. In Proceedings of the International Conference on Advanced Computing and Communication Systems 2020, Coimbatore, India, 6–7 March 2020; pp. 1373–1378. [CrossRef]
13. Gallo, P.; Pongnumkul, S.; Nguyen, U.Q. BlockSee: Blockchain for IOT video surveillance in smart cities. In Proceedings of the IEEE International Conference on Environment and Electrical Engineering and IEEE Industrial and Commercial Power Systems Europe, Palermo, Italy, 12–15 June 2018. [CrossRef]
14. Chen, Y.C.; Chou, Y.P.; Chou, Y.C. An image authentication scheme using Merkle tree mechanisms. *Future Internet* **2019**, *11*, 149. [CrossRef]
15. Gřivna, T.; Drápal, J. Attacks on the confidentiality, integrity and availability of data and computer systems in the criminal case law of the Czech Republic. *Digit. Investig.* **2019**, *28*, 1–13. [CrossRef]
16. Khan, P.W.; Byun, Y.C.; Park, N. A data verification system for CCTV surveillance cameras using blockchain technology in smart cities. *Electronics* **2020**, *9*, 484. [CrossRef]

17. Lee, D.; Park, N. Blockchain based privacy preserving multimedia intelligent video surveillance using secure Merkle tree. *Multimed. Tools Appl.* **2020**, *80*, 34517–34534. [CrossRef]
18. Li, T.; Chen, Y.; Wang, Y.; Wang, Y.; Zhao, M.; Zhu, H.; Tian, Y.; Yu, X.; Yang, Y. Rational protocols and attacks in blockchain system. *Secur. Commun. Netw.* **2020**, *2020*, 8839047. [CrossRef]
19. Majdoubi, D.E.L.; El Bakkali, H.; Sadki, S. Towards smart blockchain-based system for privacy and security in a smart city environment. In Proceedings of the International Conference on Cloud Computing and Artificial Intelligence: Technologies and Applications (2020), Marrakesh, Morocco, 24–26 November 2020. [CrossRef]
20. Chen, J.; Ruan, Y.; Guo, L.; Lu, H. BCVehis: A blockchain-based service prototype of vehicle history tracking for used-car trades in China. *IEEE Access* **2020**, *8*, 214842–214851. [CrossRef]
21. Chen, X.; Xing, Z.; Karki, B.; Li, Y.; Chen, Z. Blockchain simulation: A web application for it education. In Proceedings of the IEEE Annual Computing and Communication Workshop and Conference 2021, Virtual, 27–30 January 2021; pp. 486–491. [CrossRef]
22. Chukwu, E.; Garg, L. A systematic review of blockchain in healthcare: Frameworks, prototypes, and implementations. *IEEE Access* **2020**, *8*, 21196–21214. [CrossRef]
23. Gabay, D.; Akkaya, K.; Cebe, M. Privacy-preserving authentication scheme for connected electric vehicles using blockchain and zero knowledge proofs. *IEEE Trans. Veh. Technol.* **2020**, *69*, 5760–5772. [CrossRef]
24. George, R.V.; Harsh, H.O.; Ray, P.; Babu, A.K. Food quality traceability prototype for restaurants using blockchain and food quality data index. *J. Clean. Prod.* **2019**, *240*, 118021. [CrossRef]
25. Hartwig, M. ECDSA Security in bitcoin and Ethereum: A research survey. *Blog. Coinfabrik* **2016**, 1–10.
26. Gergely, A.M.; Crainicu, B. Randadminsuite: A new privacy-enhancing solution for private blockchains. *Procedia Manuf.* **2020**, *46*, 562–569. [CrossRef]
27. Koo, D.; Shin, Y.; Yun, J.; Hur, J. Improving security and reliability in Merkle tree-based online data authentication with leakage resilience. *Appl. Sci.* **2018**, *8*, 2532. [CrossRef]
28. Gedara, K.M.; Nguyen, M.; Yan, W.Q. Enhancing privacy protection in intelligent surveillance: Video blockchain solutions. In *BLOCKCHAIN'23 Proceedings*; Springer LNNS: Berlin/Heidelberg, Germany; Available online: https://cerv.aut.ac.nz/wp-content/uploads/2023/05/BLOCKCHAIN23.pdf (accessed on 23 August 2023).
29. Gedara, K.; Nguyen, M.; Yan, W. Visual blockchain for intelligent surveillance in a smart city. In *Blockchain Technologies for Sustainable Development in Smart Cities*; IGI Global: Hershey, PA, USA, 2021.
30. Kumar, M.; Kaur, G. High performance scalable recursive block matrix inverse for multicore architectures. In Proceedings of the International Conference on Parallel, Distributed and Grid Computing 2022, Solan, India, 25–27 November 2022; pp. 45–49.
31. Fattahi, S.; Makanju, A.; Fard, A. SIMBA: An efficient simulator for blockchain applications. In Proceedings of the 2020 50th Annual IEEE-IFIP International Conference on Dependable Systems and Networks-Supplemental Volume (DSN-S), Valencia, Spain, 29 June–2 July 2020; pp. 51–52.
32. Belazi, A.; Abd El-Latif, A.A.; Belghith, S. A novel image encryption scheme based on substitution-permutation network and chaos. *Signal Process.* **2016**, *128*, 155–170. [CrossRef]
33. Muhammad, K.; Hamza, R.; Ahmad, J.; Lloret, J.; Wang, H.; Baik, S.W. Secure surveillance framework for IoT systems using probabilistic image encryption. *IEEE Trans. Ind. Inform.* **2018**, *14*, 3679–3689. [CrossRef]

Disclaimer/Publisher's Note: The statements, opinions and data contained in all publications are solely those of the individual author(s) and contributor(s) and not of MDPI and/or the editor(s). MDPI and/or the editor(s) disclaim responsibility for any injury to people or property resulting from any ideas, methods, instructions or products referred to in the content.

electronics

MDPI

Article

QoS-Driven Slicing Management for Vehicular Communications

Prohim Tam [1], Seyha Ros [1], Inseok Song [1] and Seokhoon Kim [1,2,*]

1 Department of Software Convergence, Soonchunhyang University, Asan 31538, Republic of Korea; prohimtam@gmail.com (P.T.); seyha@sch.ac.kr (S.R.); sis5041@sch.ac.kr (I.S.)
2 Department of Computer Software Engineering, Soonchunhyang University, Asan 31538, Republic of Korea
* Correspondence: seokhoon@sch.ac.kr

Abstract: Network slicing is introduced for elastically instantiating logical network infrastructure isolation to support different application types with diversified quality of service (QoS) class indicators. In particular, vehicular communications are a trending area that consists of massive mission-critical applications in the range of safety-critical, intelligent transport systems, and on-board infotainment. Slicing management can be achieved if the network infrastructure has computing sufficiency, a dynamic control policy, elastic resource virtualization, and cross-tier orchestration. To support the functionality of slicing management, incorporating core network infrastructure with deep learning and reinforcement learning has become a hot topic for researchers and practitioners in analyzing vehicular traffic/resource patterns before orchestrating the steering policies. In this paper, we propose QoS-driven management by considering (edge) resource block utilization, scheduling, and slice instantiation in a three-tier resource placement, namely, small base stations/access points, macro base stations, and core networks. The proposed scheme integrates recurrent neural networks to trigger hidden states of resource availability and predict the output of QoS. The intelligent agent and slice controller, namely, RDQ3N, gathers the resource states from three-tier observations and optimizes the action on allocation and scheduling algorithms. Experiments are conducted on both physical and virtual representational vehicle-to-everything (V2X) environments; furthermore, service requests are set to massive thresholds for rendering V2X congestion flow entries.

Keywords: deep q-networks; quality of service; recurrent neural networks; slicing management; vehicle-to-everything

1. Introduction

The advent of vehicular communications has led to a new era of vehicle-to-everything (V2X) with enhanced safety, intelligent transportation systems (ITS), improved traffic management, and the realization of autonomous driving [1,2]. The user-centric service demands of the Internet of Vehicles (IoV) in 5G and beyond networks necessitate the deployment of a paradigm shift in logical network architecture and resource-efficient mechanisms such as intelligent slicing orchestration to cater to the diverse quality of service (QoS) requirements in V2X communications. The multi-class requirements are within safety-critical applications, non-safety services in ITS, and on-board infotainment [3,4]. However, the heterogeneous features of vehicular mobility and protocols pose significant challenges to ensuring the applicability of slicing federation and orchestration in real-world use cases.

Before commencing the augmentation of network slicing (NS), it is crucial to have a comprehensive multi-aspect awareness of potential challenges in conventional approaches. There are primary challenges for slicing orchestration in vehicular communications, such as scalability [5], end-to-end (E2E) management and orchestration (MANO) [5], slice optimality [6], core architecture evolutions and APIs [7], mobility management [8], and virtualization sharing [8]. E2E MANO confronts complexities arising from the proliferation of IoV applications from V2X [9], such as vehicle-to-network (V2N), vehicle-to-infrastructure (V2I), vehicle-to-vehicle

(V2V), and vehicle-to-pedestrian (V2P), leading to an increased number of dedicated network slices. To handle the scalability, slice grouping and prioritization have to be either well-fixed or autoscaling. Core architecture evolutions and APIs need rich feature observability and elastic reconfigurability to the interfaces for enabling on-demand, autonomous, predictability, and dynamic configurations. Therefore, the convergence of well-known enabling NS paradigms with prediction-based deep learning and autonomy-based deep reinforcement learning (DRL) introduces innovative schemes (e.g., resource sharing, elastic virtualization, and QoS-driven isolation between slices), which can be evaluated by the reward functions of QoS-aware agents [10–12].

QoS metrics play a significant role in IoV services by dictating the performance and reliability required to support a diverse range of applications [13,14]. These metrics can be captured either reactively or proactively. Reactive capturing, while convenient for analyzing past performance, is less suitable for intelligent NS. In contrast, proactive capturing involves assessing QoS performance prior to data transmission to ensure compliance with predefined class indicators. In the context of vehicular communications, the proactive approach entails evaluating factors such as bandwidth allocation, resource availability, and computational resources before deploying a service. To support proactivity, recurrent neural networks (RNN) and long short-term memory (LSTM) have been implemented by several studies [15–18]; furthermore, integrating with DRL [19] brings promising approaches for long-term self-organizing and adaptive slicing control policies.

Therefore, in this paper, we first gather well-known enabling paradigms in E2E NS for supporting the applicability of our proposed system architecture, and then we emphasize our contributions with QoS-driven system models and reliable environment representation to support the integration of our proposed RNN and deep Q-networks (DQN), namely, the RDQ3N-assisted intelligent slice controller. V2X communications consist of numerous natural parameters; therefore, we normalize the input states, following our sole priority in logical resource slice sharing. We input the processing features to obtain the hidden states of resource availability and allocability in small base stations (SBSs) as edge nodes and macro base stations (MBS) as resource-assisted nodes, later considered DQN states. The output determined the proactive QoS performances of the processing V2X features. RDQ3N has a group-based fixed slice instance of (1) safety-critical, (2) ITS, and (3) on-board infotainment. The reward function jointly considers resource utilization and QoS performances in evaluating the efficiency of action configurations in each timeslot, namely, slice instantiation, resource placement, and scheduling. Experiments are conducted to substantiate the RDQ3N-assisted intelligent slice controller in our built V2X environment, comparing key metrics including average delays per slice, resource efficiency, and reliability with reference schemes. Table 1 presents the acronyms used in this paper.

Table 1. Primary acronyms and descriptions.

Acronym	Description	Acronym	Description
DQN	Deep Q-networks	QoS	Quality of service
DNN	Deep neural networks	RNN	Recurrent neural networks
DRL	Deep reinforcement learning	SDN	Software-defined networking
E2E	End-to-end	SBS	Small base stations
EPC	Evolved packet core	SLAs	Service-level agreements
IoV	Internet of Vehicles	TD	Temporal difference
ITS	Intelligent transport systems	V2I	Vehicle-to-infrastructure
LSTM	Long short-term memory	V2N	Vehicle-to-network
MBS	Macro base stations	V2P	Vehicle-to-pedestrian
MANO	Management and orchestration	V2V	Vehicle-to-vehicle
MEC	Multi-access edge computing	V2X	Vehicle-to-everything
NFV	Network functions virtualization	VNFs	Virtual network functions
NS	Network slicing	VNFM	VNFs manager

2. Preliminary of Intelligent NS in Vehicular Communications

2.1. Primary Enabling Paradigms

Software-defined networking (SDN) [20], network function virtualization (NFV) [21], multi-access edge/cloud computing [22], and virtualization techniques [23] are undeniably the primary enabling paradigms for high-applicability infrastructure in E2E NS. In the operational phase, SDN is dynamically responsible for creating and managing network slices in response to application demands. SDN architecture supports resource allocation, traffic steering, and service-level agreements (SLAs) enforcement. Additionally, SDN ensures the security and isolation of network slices, which requires virtualization from NFV to support efficient and reliable operation of logical isolation networks in a 5G and beyond environment [24,25]. NFV complements SDN by providing software-based network functions that instantiate and orchestrate service function chaining across the E2E network [26]. Together, SDN and NFV provide a holistic approach in terms of new instance creation, management, and optimization of logical network slices. The operational phase can be leveraged closer to edge networks using multi-access edge computing (MEC), which enables better decisions in compute-intensive task offloading [27], therefore reducing backbone latency and improving response times for IoV applications.

Despite the popularity of these paradigms, a converged platform that fully supports joint functionality and operability can be challenging to deploy as a reliable system architecture. The positions of SDN resources, controllers, and applications in NFV are varied in different scenarios [28], which leads to more complexity at the core orchestration level. The cross-tier management of edge-core collaboration is difficult to design, particularly for supporting rich feature state observability in resource-constrained environments.

2.2. Add-on Paradigms for Intelligence

On top of primary enabling paradigms, self-organizing flexibility and autonomy have been added to NS MANO capability using artificial intelligence [29], blockchain [30], federated learning [31], and edge intelligence [32]. In terms of artificial intelligence choices, many studies [33–38] have introduced intelligent NS design for V2X services with RNN/LSTM as early feature prediction and DRL as a policy maker. In [34], the authors reviewed how different research studies apply RNN and its variant, LSTM, to predict traffic and manage resources. RNN/LSTM outperforms other techniques in traffic prediction for vehicular-specific slices and supports predicting user mobility for inter-slice resource allocation. However, the computational cost of RNN/LSTM has led researchers to explore lighter variants like simplified or soft-gated recurrent units, which demonstrated comparable accuracy with reduced computation time. Moreover, RNN-based QoS prediction in IoV was presented in [35]. The proposed RNN utilized a multi-layer gated recurrent unit structure to capture QoS variations and model the evolving IoV environment. By leveraging experience features and objective factors, the proposed RNN and collaborative filtering can handle data sparsity issues and enhance the precision of QoS predictions.

Beyond RNN/LSTM, DRL can emerge as a pivotal add-on paradigm in the realization of intelligent and efficient resource management for 5G/6G NS, where the significant challenges of tasks such as admission control, resource allocation, scheduling, and orchestration demand autonomous decision making and policy making [36,37]. In [38], the authors introduced a novel multi-agent-based dynamic slicing framework for vehicular networks. The proposed DRL-assisted approach leveraged a multi-agent deep deterministic policy gradient to enable base stations to cooperatively allocate spectrum and computing resources. Experimental results demonstrated the effectiveness of the framework in achieving significant cost reductions and high QoS satisfaction rates. Furthermore, the DRL-based approach for resource optimization in UAVs and IoV networks necessitates utilizing 5G/6G infrastructure [39–41]. An online MEC-based scheme was also developed for slice selection, coverage selection, resource block, and power allocation [42].

Building upon the promising advancements in intelligent NS using RNN/LSTM and DRL, the role of the core orchestrator should be expanded and integrated into cross-tier management of the edge-core framework to achieve E2E NS for reliable vehicular communications.

2.3. Slice Instances in Vehicular Communications

NS is employed to facilitate diverse QoS requirements, particularly in massive use cases such as V2X services. However, the creation of slice instances necessitates adherence to a standardized set of specific criteria. With well-defined slice instances, scalability can be effectively covered and extended to a wide range of applications. There are several well-known studies that categorize slice instances and use cases differently. Table 2 presents the expected performances and critical requirements of use cases and possible slices of V2X services [5,43].

Table 2. Expected performances of slice instances from complementary studies [5,43].

Slice Instances	Latency	Data Rate	Scalability (Per Vehicle)	Ref.
Autonomous vehicles	1 ms	10 Mb/s	50–200	[5]
Tele-operated vehicles	20 ms	1Mb/s or 25 Mb/s	50–100	
Remote diagnostic	Not a concern	Not a concern	10–20	
Vehicle infotainment system	100 ms	0.5–15 Mb/s	5–15	
Slice Instances	**Latency**	**Data Rate**	**Reliability**	**Ref.**
Localization and navigation	10–100 ms	1 Mb/s	Low to high	[43]
Driving/transportation safety	50–100 ms	1 Mb/s	99.9%	
Autonomous driving	1 ms	10 Mb/s	Close to 100%	
Infotainment services	Up to 100 ms	15 Mb/s	Fair	

Each study classifies the slice criteria differently based on its own system models and target optimization objectives. In [44], the authors proposed two slices as follows: (1) slice 1 (security services) aims for low latency and high reliability (high-quality V2V links for safety information), and (2) slice 2 (infotainment service) focuses on high transmission rates (the roadside unit provides infotainment services via high-quality V2I links).

The primary target of optimization in this study [44] was to maximize infotainment slice throughput while ensuring latency constraints for the security slice. The authors considered traffic admission control, spectrum allocation, and transmit power allocation. In [45], the authors proposed a network clustering and slicing algorithm to group vehicles based on traffic demands. Slice grouping was based on use case requirements that prioritize safety messages for autonomous driving and video streaming for infotainment. Furthermore, in [46], the authors presented a road-state-based adaptive NS framework for vehicular networks, which prioritizes emergency traffic during incidents while maintaining acceptable QoS for non-emergency services. The study employed a Mininet-WIFI emulator to create a simulated vehicular network, the ONOS SDN controller to manage network resources, and the OpenVirteX NS tool (Open Network Foundation (ONF), Korea University, Seoul, Republic of Korea) to implement the proposed scheme.

3. Proposed System Framework

3.1. Working Flow

Our proposed framework runs on SDN and NFV platforms from edge to core. The network slice request originates from the NFV orchestrator, which acts as the central control plane for managing and orchestrating network slices. The NFV orchestrator receives the request from the V2X service provider or application, which specifies the desired network slice characteristics, such as latency, bandwidth, and QoS requirements. Figure 1 illustrates our proposed system framework and its working flow. Overall, there are three tiers in this edge-core framework, as follows:

- ***Tier-1***. We represent SBSs with three possible general entities, denoted as $S = \{1, 2, \ldots s\}$, to symbolize eNodeB, gNodeB, or access points as vehicular NFV-enabled edge computing and network connectivity. eNodeB or gNodeB, the base station component of LTE and 5G networks, can be deployed with NFV infrastructure. Traditionally, eNodeB and gNodeB have been implemented as dedicated hardware appliances, but NFV enables the virtualization of eNodeB functions onto general-purpose hardware, such as servers or (edge) cloud platforms. In our framework, tier-1 is prioritized for low-latency slicing services since it is a computing resource-constrained environment. Vehicle nodes are denoted as $V = \{1, 2, \ldots v\}$. Upon receiving the service request at timeslot-t, denoted as $E_t = \{1_t, 2_t, \ldots e_t\}$, the SDN controller collaborates with the NFV orchestrator to determine the optimal network path for V2X traffic within the requested service e_t. The controller observes and considers the capabilities of the SBS resources, current topology states, and traffic patterns, denoted as $X_t = \{1_t, 2_t, \ldots x_t\}$. X_t is input to the main RNN execution, part of RDQ3N. The SDN controller translated the determined path into flow rules, which are instructions to direct traffic in tier-1 enabling slicing-aware traffic forwarding.
- ***Tier-2***. We denote MBS as $M = \{1, 2, \ldots m\}$ with higher resource capabilities. In this tier, the SDN controller continuously monitors the network load and traffic patterns within the network slice. The controller adjusts traffic flow across MBS-m in tier-2, which ensures that the load is distributed evenly and that V2X traffic receives the QoS-driven priority based on the sliced modelling. As node-v moves across the network, the controller coordinates with the mobility management entity to manage handover between MBSs within the network slice, while ensuring that V2X communication remains uninterrupted and that QoS expectation is maintained during handover.
- ***Tier-3***. We refer to the evolved packet core (EPC), 5G core networks, or core data center. The SDN controller establishes and manages connectivity to external entities for the network slice, therefore enabling V2X communication with entities outside the slice. The core entity collects hidden states h_t and QoS metrics QoS_t proactively from our proposed RDQ3N (main RNN) to build a virtual representational environment. The NFV orchestrator, in conjunction with the virtual network functions manager (VNFM), orchestrates the allocation of NFV resources, such as virtual network functions (VNFs) and computing resources, to the defined slice instances. The VNFM deploys and configures the required VNFs onto the allocated SBS resources in tier-1, while ensuring that the required network functionalities are provided within the slice.

To enhance the realm of NS, the core orchestration serves as the backbone of the network and orchestrates the allocation of resources and services across the various slices. Hence, interface N2 establishes a robust communication channel between the core network controller and access point, and is responsible for orchestrating slicing, often utilizing a protocol such as the next-generation application protocol (NGAP), enabling communication control and user-plane critical information exchange, such as slicing configuration, resource allocation, and user mobility management. Thus, it allows the core network to manage and direct traffic control, ensuring the optimal allocation of resources for diverse applications of slicing. Simultaneously, the interface connects the core network to SBS, which extends network convergence and capacity. For the xn interface, we aim to provide the corresponding slice-specific configurations, resource allocations, and traffic prioritization across the network, ensuring that each slice receives the necessary resource and operates. Lastly, access points to the SBS interface ensure handovers and consist of service delivery within network, which handles data forwarding and load balancing across multiple SBSs within a slice.

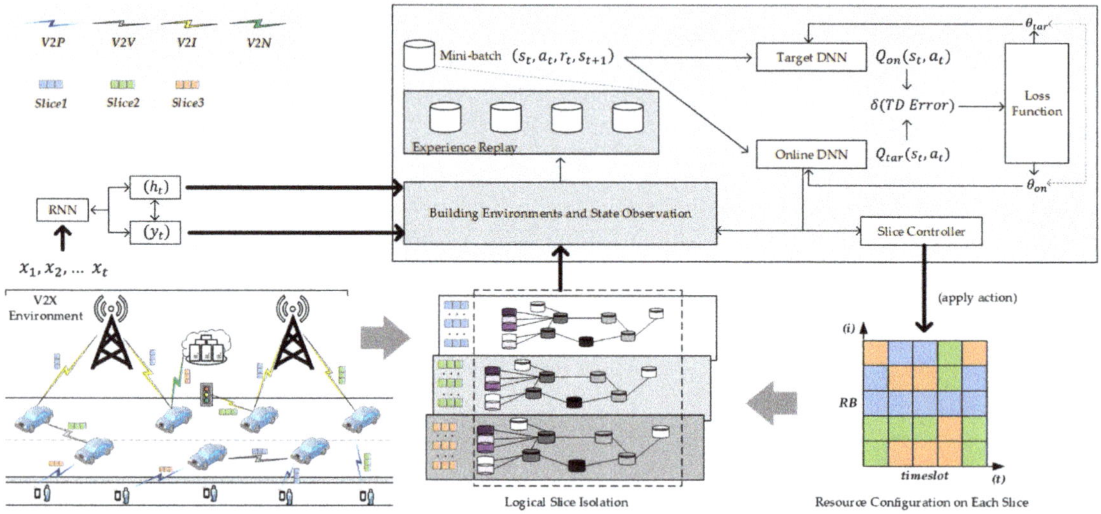

Figure 1. System framework between the V2X environment and RDQ3N components.

This subsection presents the collaborative working flow from the infrastructure layer to SDN and NFV (tier-1, tier-2, and tier-3) in enabling the dynamic provisioning and management of E2E network slices. The details of the main RNN and DQN to achieve an RDQ3N-assisted intelligent slice controller will be presented in Section 4. The extension of Figure 1 description will be given in Section 4 on the target and online deep neural networks (DNN), as well as how it associates with the slice controller to adjust resource configuration for each logical slice isolation.

3.2. Slice Instances

Based on our preliminary studies in Section 2, we propose three slice instances: (1) safety-critical service, (2) ITS-supported service, and (3) on-board infotainment services. These instances correspond to slices 1, 2, and 3, respectively. The expected performance of each slice is given in Table 3. The slicing indices are denoted as L_i, where i ranges from 1 to 3. However, due to multi-service involvement, we cannot establish fixed and nonduplicated criteria for each slice. For instance, autonomous driving can be considered both safety-critical and ITS-supported, depending on the use cases. Autonomous systems require planning a safe path that avoids collisions and adheres to traffic rules, which necessitates their consideration as a safety-critical slice instance. Simultaneously, autonomous systems also coordinate vehicle movements to reduce congestion and improve traffic flow, which makes it an ITS-supported slice instance as well. Therefore, our slicing model is a balanced trade-off that determines as follows:

- ***Slice 1***. The safety-critical slice instance is prioritized with ultra-reliable transmission of mission-critical services, while ensuring rapid latency with a maximum delay of 1–20 ms. The data rate ranges from 1 Mb/s to 10 Mb/s, which emphasizes vehicle-cooperative awareness messages that require low bandwidth but more dense requests. Scalability can be between 50 and 200. Reliability expectation is targeted as high. If the agent outputs actions on slice identification as L_1, the allocation metrics in MBS-m and scheduling in SBS-s will be higher priority compared to L_2 and L_3.
- ***Slice 2***. The ITS-supported slice instance acknowledges the importance of assisting applications that enhance traffic management and efficiency. With a maximum delay of 1–50 ms, slice 2 balances low-latency requirements with broader application coverage. The data rate is in the same range as with safety-critical applications. Scalability is specified between 50 and 100. The reliability level for ITS-supported applications is

classified as medium compared to slice 1. The guaranteed bit rates can be arranged less than L_1 in congestion states.

- ***Slice 3.*** The on-board infotainment slice instance caters to non-safety-related services that prioritize high data rates for non-mission-critical streaming entertainment. With a maximum delay of up to 100 ms, slice 3 allows for a more relaxed transmission timeframe. The data rate spans from 0.5 Mb/s to 15 Mb/s, accommodating the varying bandwidth needs of heavy-traffic streaming data for infotainment content. Scalability is less, ranging from 5 to 15. The reliability level for this slice is classified as low since it can afford the unavoidable loss in a congested and constrained timeslot.

Table 3. Slice instances for our approach.

Slice Instances	Max. Delay	Data Rate	Scalability (Per Vehicle)	Reliability
Safety-critical	1–20 ms	1–10 Mb/s	50–200	High
ITS-supported	1–50 ms	1–10 Mb/s	50–100	Low–High
On-board infotainment	Up to 100 ms	0.5–15 Mb/s	5–15	Low

4. RDQ3N-Assisted Intelligent Slice Controller

4.1. Environment Components

In our study, we synchronized the physical and logical infrastructure to support the integration of RNN and DQN as RDQ3N for the intelligent slice controller. We gathered the parameters that matter to our virtual environment for building RDQ3N agent policies. Let x_t denote the state of vehicular edge networks that later input to the main RNN component. h_t is a partial feature in state observations that is input to build the virtual environment. Let s_t denote the state of the vehicular core networks at time t, including resource block conditions and V2X service request demands. The action a_t is a three-tuple adjustment on slice identification $L_i(e_t)$ of the requested services, resource allocation $\rho_t(m|e_t)$, and scheduling capacity $\varphi_t(s|e_t)$ in that timeslot-t. Equation (1) presents the action spaces of the RDQ3N framework. The proposed agent at time t obtains the output of QoS predictions as QoS_t from the main RNN component. After deploying a_t to s_t, the agent evaluates a reward r_t that considering the performances of QoS metrics, denoted as QoS_t, including expected latency L_t, expected throughput T_t, expected reliability (loss ratio) P_t, and resource efficiency R_t.

$$a_t \leftarrow \{L_i(e_t), \rho_t(m|e_t),\ \varphi_t(s|e_t)\} \tag{1}$$

4.2. QoS-Driven Models

The primary target is to evaluate the actions from RDQ3N that guide SDN control and NFV orchestration to achieve better QoS performance. The QoS-driven model is considered as the goal of our approach, represented as G_t in Equation (2), which aims to optimize policy π_t in maximizing the cumulative discounted reward. k denotes the possible future timesteps in the learning process. As k-step grow larger, the discount factor γ^k and the future rewards (r_{t+k+1}) have less impact on the RDQ3N policy making; therefore, our RDQ3N agent gains the capability of balancing the short- and long-term decision weights. The reward r_t, as given in Equation (3), reflects the performances of a_t on slice identifications and resource placement, including (1) sub-reward on throughput r_t^{TP}, (2) sub-reward on latency r_t^{LC}, and (4) sub-reward on resource efficiency r_t^{RP}.

$$G_t = \sum_{k=0}^{N} \gamma^k r_{t+k+1},\ \forall k \in N \tag{2}$$

$$r_t = r_t^{TP} + r_t^{LC} + r_t^{RP} \tag{3}$$

4.3. Algorithm Flows

There are three primary phases in our study, including (1) RDQ3N predictions on hidden state and slicing QoS, (2) action selections in RDQ3N, and (3) slice controller operations.

In the first phase, in the context of vehicular edge networks, the hidden resource state at timeslot-t (h_t) represents the aggregated features of the local/edge resource availability, traffic patterns, and congestion queue conditions within tier-1. The calculation of h_t involves a linear transformation of the current input state x_t and the previous hidden resource state h_{t-1} (temporal dynamics of the vehicular environment). Equation (4) represents the output o_t of the transformation in a single hidden unit, which is the weighted sum of the tier-1 states that represents the aggregated input to the hidden state. U denotes the size of the input vector x_t, and V is the size of the hidden state vector h_{t-1}. $W_{hx_u} \cdot x_{t_u}$ signifies the impact of the tier-1 x_t on the hidden resource state, and $W_{hh_v} \cdot h_{t-1_v}$ captures the temporal dependencies from h_{t-1}. The bias b_h is a constant vector added to the weighted sum that provides an additional degree of freedom and control over the h_t calculation. Equation (5) shows that the hidden resource state h_t is then passed through the hyperbolic tangent ($tanh$) activation function. h_t encapsulates significant features of current states in tier-1 by considering both instantaneous conditions and experience replays.

$$o_t = \sum_{u=1}^{U} W_{hx_u} \cdot x_{t_u} + \sum_{v=1}^{V} W_{hh_v} \cdot h_{t-1_v} + b_h \tag{4}$$

$$h_t = \tanh(o_t) \tag{5}$$

The main DQN component, a primary RDQ3N agent, leverages h_t along with other relevant states in tier-2 to assist action selection in the second phase. The convergence of RNN and DQN components enables the proposed RDQ3N agent to learn and adapt the slicing policies with optimal actions based on the temporal expansion of hidden resource states in the vehicular edge and core environments. After obtaining h_t, transformation is computed for output QoS_t following Equations (6) and (7). For a single output unit in predicting QoS for vehicular services, we expand the output transformation z_t, and later use activation functions to set a fix label class of V2X-QoS criticality, namely, QoS_{t_u} of the input requested service demands. P denotes the size of the output vector QoS_t.

$$z_t = \sum_{v=1}^{V} W_{Qh_v} \cdot h_{t_v} + b_Q \tag{6}$$

$$QoS_{t_u} = \frac{e^{z_{t_u}}}{\sum_{p=1}^{P} e^{z_{t_p}}} \tag{7}$$

In the second phase, the hidden state observation and prediction (h_t and QoS_t) serve as features to the main DQN component of the RDQ3N-assisted intelligent slice controller. RDQ3N executes the main DQN class to obtain the optimal policy π^* on the slicing management. Algorithm 1 presents the integration of the main DQN to RDQ3N agent on the built vehicular environment. We build a virtual representation based on the predicted and hidden state resources in tier-1 and tier-2, following current resource allocations for each slice instance. We transform the observed physical states in tier-1 into a functional state representation that encodes the resource availability and current slice allocations. In the main DQN, we determine the early (hyper) parameters on both online and target DNNs. In pseudocode line 4, RDQ3N iterates through training steps. The state observations are randomized from batches of RNN outputs of the requested service, including e_t and x_t. The RDQ3N agent sets the slice identification of e_t, whether it belongs to L_1, L_2, or L_3. The resource allocation to the identified slice is obtained and re-configured (CPU, memory, and bandwidth), depending on the current policy π and partial ε-greedy. The performance might be ineffective during the training phase; however, after obtaining π^*, the performance will be optimized accordingly. RDQ3N also schedules the tasks or steering for the selected slice, associated to intelligent slice controller with ***SliceController()*** function. After applying (1) $L_i(e_t)$ on service request clustering, (2) $\rho_t(m|e_t)$ on tier-2 allocation, and (3) $\varphi_t(s|e_t)$ on tier-1 scheduling, the RDQ3N agent monitors the physical infrastructure response and formulates s_{t+1} and $r_t(.|r_t^{TP}, r_t^{LC}, r_t^{RP})$. Experience replay B is a compute-intensive input for RDQ3N, which requires random mini-batch sampling to perform the ***OptimizeOnline()*** function. Then, we use the sampling batch one-by-one to train predicted-Q as $Q_{pre}(s_t, a_t)$ and improve the online network parameters θ_{on}. We give the general target-Q formulation as $Q_{tar}(s_t, a_t)$, and we aim to identify the temporal difference (TD) error δ and minimize the loss of the networks. We exchange the weights θ_{tar} for periodic improvement using θ_{on}. The hyperparameter τ determines the balance between early and updated values.

Algorithm 1. Training RDQ3N for QoS-Driven Slicing Management

Initialize the main DQN with early online and target DNNs
Initialize all state observation $\sim s_t(.|h_t \ \& \ QoS_t)$
Initialize experience replay B
For each step **do**:
  Sample s_t of vehicular edge-core systems and NFV-based infrastructure
  For each timeslot t **do**:
    Detect service request $e_t, x_t \rightarrow h_t, QoS_t$
    Select $L_i(e_t), \rho_t(m|e_t), \ \varphi_t(s|e_t)$ based on current π and partial ε-greedy
    SliceController(): deploy a_t to SDN steering and NFV-enabled infrastructure
    Formulate s_{t+1} and $r_t(.|r_t^{TP}, r_t^{LC}, r_t^{RP})$
    $B \leftarrow (s_t, a_t, r_t, s_{t+1})$ // store the batch to experience replay
    Sample mini-batch from B and perform ***OptimizeOnline()***:
      $Q_{tar}(s_t, a_t) = r_t + \gamma \max_{a'} Q_{tar}(s_{t+1}, a')$ // general target-Q
      Use online DNN: $Q_{pre}(s_t, a_t)$ // early-Q
      $\delta = Q_{tar}(s_t, a_t) - Q_{pre}(s_t, a_t)$ // TD error
      $Loss = \frac{1}{N}\sum_{k=1}^{N} \delta_k^2$ // loss function
      Minimize Loss: Backpropagation
      $\theta_{on} \leftarrow \theta_{on} - \alpha \nabla \theta_{on} LOSS$ // update weights on online DNN
    Update target DNNs:
      $\theta_{tar} \leftarrow \tau\theta_{online} + (1-\tau)\theta_{tar}$ // update weights on online DNN
    Obtain output reactive QoS metrics $QoS_t(.|L_t + T_t + P_t + U_t)$
    // compared to main RNN output
    Update reward efficiency and ***OptimizeTarget()*** for π^*
    Alter environment to s_{t+1} for next-timeslot
  End for
End for

The reward function r_t emphasizes actions that enhance QoS and resource utilization while penalizing suboptimal adjustments to the experience replay. We assess the stability and effectiveness of learned policies under changing service demands throughout multiple timeslots. Each QoS_t of slicing V2X services is measured by the threshold of the identified slices. Depending on the satisfaction criteria, we update epsilon (ε)-greedy. RDQ3N keeps training until policy π^* is obtained with minimized loss and satisfied QoS expectations.

When obtaining optimal action from RDQ3N, the functionality of the intelligent slice controller consists of flow rule steering, property configuration, and self-organizing adaptation. From the main RNN component of RDQ3N, we continuously monitor the V2X incoming service requests and generate QoS predictions. Action-assisted steering leads to re-routing traffic for specific slices, adjusting resource properties in MEC-equipped MBS, and modifying scheduling parameters in offloading queues of SBS. The intelligent slice controller translates RDQ3N recommendations into configuration updates for the SDN controller and NFV orchestrator, which update central and local flow entry rules. The intelligent slice controller also continuously monitors the performance of the slices following the implemented configuration, which is evaluated as rewards. Further software execution steps of the proposed framework are given in Figure 2, Section 5.2.

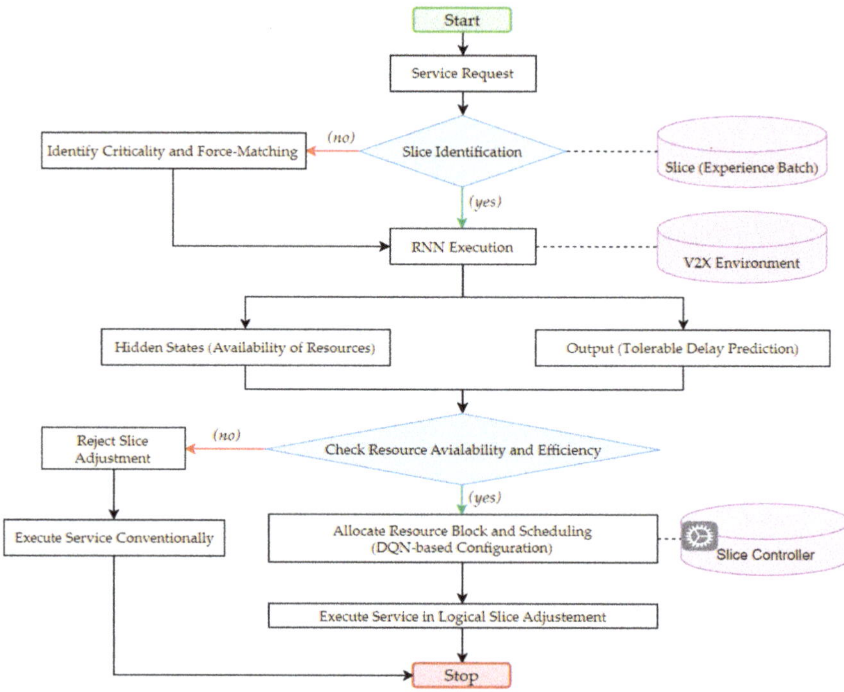

Figure 2. Overview flowchart of the proposed RDQ3N experiment.

5. Experiment and Results

5.1. Environment Setup

The (hyper) parameters of the simulation are given in Table 4. There are three major entities to discuss in this subsection, as follows:

- ***Integration infrastructure.*** To effectively evaluate the performance of the intelligent slice controller, a comprehensive simulation setup was designed on Mininet-based host core topology and PyTorch-based RDQ3N training. PyTorch was selected as the framework due to its flexibility and dynamic computation graph, while Gymnasium provided support functions for developing the representational V2X environment. Mininet was employed for SDN (RYU controller) and NFV simulation to offer a scalable flow entry installation and customizable resource virtualization.
- ***RDQ3N execution.*** The RNN configuration, with a hidden size of 256, two layers, and an input size of 10, was chosen to handle features in this V2X environment. A dropout rate of 0.2 helped prevent overfitting in our process. The DQN configuration, with a learning rate of 0.001, a discount factor of 0.99, and an exploration rate starting at 0.9, ensured stable training and balanced immediate and future rewards. A large replay buffer size (1,000,000) and batch size (128) facilitated effective experience replay. Target DNN update frequency of 1000 steps provided stability, and an initial exploration phase of 5000 steps helped gather the V2X experiences. The environment parameters, with a small action space (three actions) and a state space of 12 parameters, simplified learning complexity and also provided enough observations for the RDQ3N agent.

Longer training steps for the DQN (100,000) compared to the RNN (10,000) accounted for the criteria of Q convergence. ReLU, Sigmoid, and Tanh activation functions offered non-linearity for diverse action selections and prediction classes throughout the framework. The training steps match the transmission flows generated by the Mininet emulator for 10 h.

- ***Service flow and request operations.*** The slice configuration is three instances, and there are 120 vehicle nodes, 10 tier-1 (SBS), 2 tier-2 nodes (MBS), and service requests at a rate of 1000 per second, aimed at modeling a realistic V2X environment. This comprehensive setup aimed to strike a balance between the real-world environment, computational efficiency, and our proposed QoS-driven model in vehicular communications.

Table 4. Primary (hyper) parameter setup on PyTorch-based RDQ3N and Mininet environments.

Purpose	Specification
RNN and DQN hosts	Python (PyTorch, Gymnasium)
SDN and NFV core hosts	Mininet
SDN controller	RYU
Interfaces between RDQ3N and controller hosts	RESTful APIs
Hidden size	256
RNN number of layers	2
RNN input size	10
RNN dropout	0.2
Learning rate	0.001
Discount factor	0.99
Exploration rate	0.9 (exponential decay: 1000)
Minimum epsilon	0.05
Replay buffer size	1,000,000
Batch size	128
Target DNN update frequency	1000 steps
Target DNN update rate	0.005
Initial exploration steps	5000
Action space	3
State space	12
RNN training steps	10,000
DQN training steps	100,000
Mininet (service flow creation) times	10 h (5 h for training, 5 h for validation)
Action functions	ReLU, Sigmoid, Tanh
Number of slice instances	3
Number of vehicle nodes	120
Tier-1 and tier-2 nodes	(10), (2)
Service requests per second	1000

5.2. Proposed and Reference Schemes

Figure 2 presents the overview flowchart of the proposed RDQ3N framework associated with the V2X environment and slice controller. The process of the proposed scheme can be described as follows.

- ***Initial service request***: The execution phase begins with the reception of a service request from a node-v or vehicular application. The processing request encapsulates the specific requirements of the application.
- ***Slice identification***: The next step involves identifying the appropriate network slice for the received service request, which is achieved through a two-way approach:
    - Slice container (experience batch) or new RNN execution: The system maintains a slice container, also known as an experience batch, which stores slice identifiers and early QoS predictions. The experience batch is leveraged to efficiently identify the most suitable slice for the current service request (less computing response).
- ***Criticality identification and force-matching***: In cases where the slice cannot be identified for the service request, a criticality identification process is triggered. Our system assesses the urgency of the service request and force-matches it with the closest slice instance (class 1, 2, or 3) to ensure real-time service delivery.
- ***RNN-assisted proactivity***: When the slice is identified from the experience batch or force-matching, the main RNN from RDQ3N is employed to predict QoS_t and gather the hidden resource availability states h_t. The main RNN component interacts with the V2X environment (tier-1) to gather real-time input state x_t.
- ***QoS prediction validation***: The predicted QoS output is compared against the maximum tolerable delay associated with the pre-determined slice. If the predicted QoS does not meet the delay threshold, the predicted feature is synchronized to the RDQ3N, the main DQN agent. The feedback mechanism enables the main DQN component in RDQ3N to refine its policy for future slice identifications.
- ***Resource allocation and scheduling***: Once the slice is identified, the execution phase proceeds with resource allocation and scheduling.
    - Resource availability gathering: The main RNN component in RDQ3N, in conjunction with the V2X environment (tier-1), gathers input and output features to maintain hidden states reflecting resource availability and QoS predictions. This real-time resource monitoring enables a major weight in the placement process.
    - Resource availability utilization: The intelligent controller checks the availability and efficiency of the resources associated with the selected slice. If the resources are sufficient and efficiently utilized, the main DQN component in RDQ3N proceeds with current actions $L_i(e_t)$, $\rho_t(m|e_t)$, and $\varphi_t(s|e_t)$.
    - Resource unavailability handling: If the required resources (for serving the current requests) are not available at the specified timeslot, the slice instantiation and adjustment are rejected. In this case, the framework routes the service request to be executed conventionally without logical slice isolation.

For the reference schemes, we selected three policy settings: (1) resource-oriented slicing management using genetic algorithm (RS-GA), (2) priority-based slicing management using weighted fair queuing (PS-WFQ), and (3) multi-agent on tier-2 and tier-3 (MA-SM) to coordinate the slicing management.

5.3. Evaluation Metrics

In this subsection, the results and discussion of the evaluation metrics of the proposed and reference schemes are given. We consider the following metrics as the evaluation of our study to determine the strengths and weaknesses of this framework following the simulation configuration described in Section 5.2. Each metric is explained as follows:

- The loss (train, validation) and accuracy (train, validation) in executing the main RNN associated with tier-1 vehicular edge environment input are captured and illustrated in Figure 3. The training of the RDQ3N (RNN) for predicting early QoS and hidden states based on resource availability using Mininet-generated flow features demonstrates significant effects on slicing management. Initially, the main RNN components obtained high training and validation losses with low accuracy within the first 100 steps.

However, as training progresses, both training and validation losses decrease substantially, and accuracy reaches 98.71% at step [170/10,000]. The model maintains high accuracy and relatively low losses throughout the rest of the training process, with the exception of a slight increase in validation loss (early high peak of 0.1399) at step [2750/10,000]. The problem encountered with validation loss increment suggests a possible degree of overfitting or the need for further regularization. Despite this, the validation accuracy remains high, at an average of 97.18%, indicating that the model is still capable of making reliable QoS predictions. We obtained this result to showcase the applicability of the RDQ3N model to improve NS control in vehicular communications. However, the slight increase in validation loss at the later stage of training suggests that the model may not generalize well to new, namely unseen vehicular communication scenarios. Therefore, we integrated DQN into RDQ3N to tackle the generalization by evaluating the actions taken with three sub-rewards concurrently. With future unsatisfied rewards, RDQ3N activates the actions on hyperparameter tuning, batch normalization, regularization, and cross-validation for better assessment of the model and generalizability.

- The main reward, given in Equation (3), consists of three potential QoS-driven subcomponents: (1) throughput r_t^{TP}, (2) latency r_t^{LC}, and (3) resource efficiency r_t^{RP}. Figure 4 illustrates each sub-reward within 100,000 steps of RDQ3N (DQN component). The optimal main reward is a total value of 100, which balances each sub-reward from ranges of 20 to 40, depending on the incoming service requests in that particular timeslot. However, throughout our simulation execution (Mininet-generated flows), the congestion states varied, and the reward performance fluctuated to reflect the randomness of the V2X environment. Our proposed QoS-driven settings prioritized latency performance (slices 1 and 2) higher than throughput-maximized slice 3 (infotainment streaming). Therefore, the better results of the sub-reward of latency reflect our expectations compared to the other two sub-rewards. Resource efficiency remains challenging to overcome at every high-peak congestion state, which illustrates a low fluctuation value range between 0 to the highest peak of 17.87 and at the end of 10.09. However, RDQ3N remains effective in showcasing satisfied service request processing, acceptance, and successful ratios compared to reference schemes.
- The average delay per slice, total throughput, acceptance ratios, and successful ratios of the V2X service requests are plotted in Figure 5 in parts (a), (b), (c), and (d), respectively. Within 10 h of service flow generations at a constant rate of 1000 requests per second, our proposed RDQ3N-assisted intelligent slice controller obtained an average delay of 14.24 ms for safety-critical slice (L_1), which is 4.99 ms, 0.99 ms, and 11.31 ms lower than the average delay of MA-SM, PS-WFQ, and RS-GA, respectively. For the other 2 slice instances (L_2 and L_3), the improvements from our scheme achieved (14 ms, 1.61 ms), (15.9 ms, 12.35 ms), and (32.1 ms, 11.37 ms) better than MA-SM, PS-WFQ, and RS-GA, respectively. In Figure 5b, the total throughput represents the efficiency of bandwidth resource allocation and scheduling for each instance. By comparing reference schemes, we can showcase that the RDQ3N improvements are highly notable, with 2255.88 Mb/s, 1671.02 Mb/s, and 3050.12 Mb/s on slices 1, 2, and 3, respectively. PS-WFQ prioritized slice 1 as the most significant serving resource (2254.12 Mb/s); however, PS-WFQ's throughput on slice 3 had the lowest outcome (2921.09 Mb/s) among all the management schemes due to the bias toward less priority. Our scheme aimed to balance (1) priority in constrained states and (2) high transmission rates in non-congested states; therefore, the throughput outcome for all three slices is well preserved. The throughputs of MA-SM and RS-GA are (2200.14 Mb/s, 1662.88 Mb/s, and 3021.12 Mb/s) and (2191.04 Mb/s, 1650.23 Mb/s, 3013.91 Mb/s), on each slice instance, respectively. The improved throughput of RDQ3N is critical for massive machine-type communications, ultra-dense requests, and resource-constrained environments. Figure 5c illustrates the acceptance ratios of initial service requests, which depend on how each scheme optimizes the resource availability, QoS efficiency, and slice instantiation process. Within

the first 4 h flow generation, RDQ3N had an acceptance ratio below 99.00%; however, from the fifth-hour simulation, RDQ3N gained experience batch, environment awareness, and Q convergence that led to better performances from 99.01% to 99.79%. From this point to the end of simulation time, RDQ3N maintained a satisfied acceptance rate on average of 99.46%, which is 0.7016%, 2.6891%, and 1.2491% higher than MA-SM, PS-WFQ, and RS-GA, respectively. From the fifth-hour timeframe, MA-SM, PS-WFQ, and RS-GA achieved only 98.76%, 96.77%, and 98.21% acceptance ratios, respectively. The acceptance ratios illustrate how the slice controller is able to proactively plan the steering path and resource orchestration placement; however, only successful (completion) ratios can truly determine the reliability metric of the management scheme. Figure 5d showcases the success percentage of all accepted service requests. Table 5 gives the successful ratios of each slice between the proposed and reference schemes.

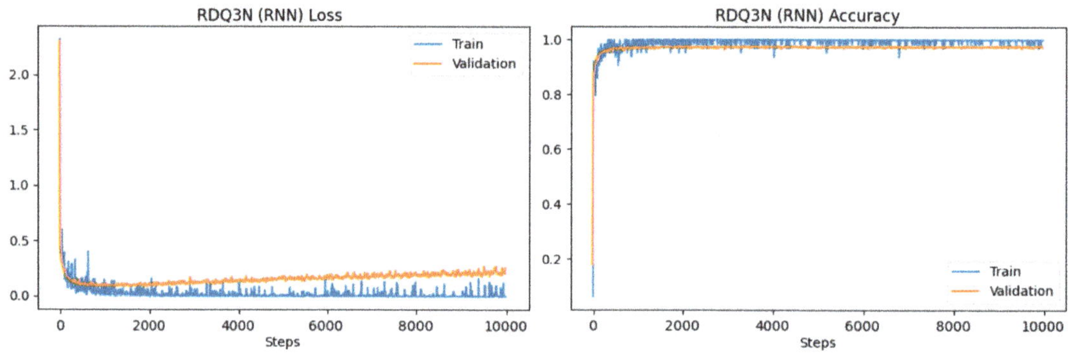

Figure 3. Loss and accuracy on RDQ3N (RNN component) over Mininet-generated flows.

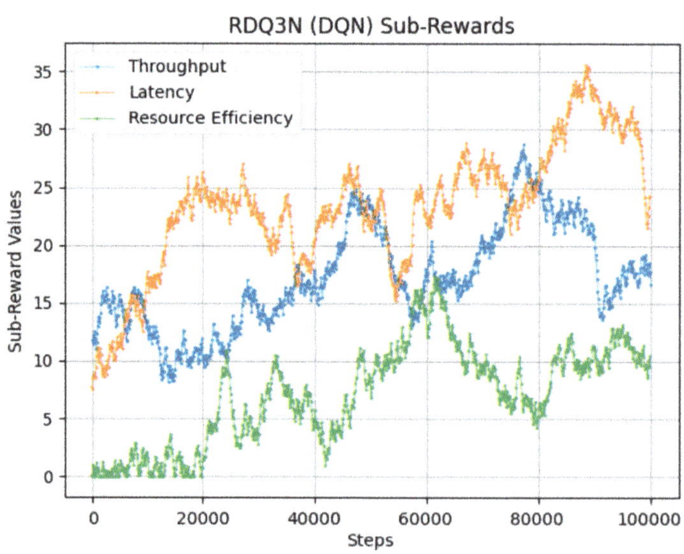

Figure 4. Three sub-rewards of RDQ3N (DQN component) over a virtual representation environment (with tier-2 states and RNN-based hidden state resource availability).

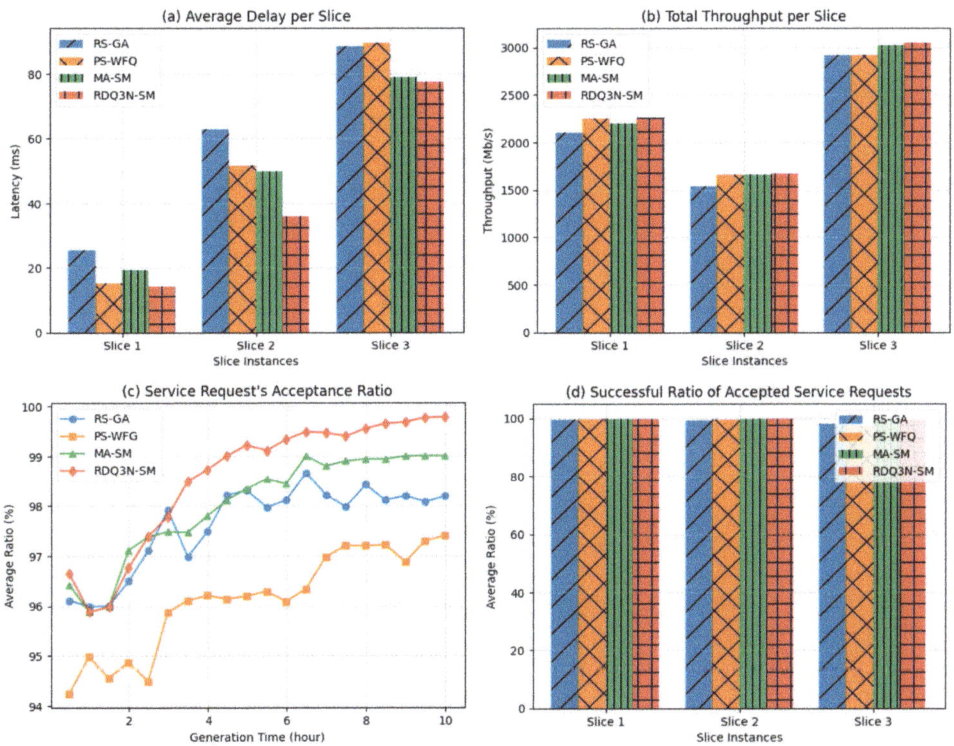

Figure 5. Results on (**a**) average delay per slice, (**b**) total throughput per slice, (**c**) acceptance ratios, and (**d**) successful (completion) ratios between proposed and reference schemes.

Table 5. Successful (completion) ratios of all accepted service requests in each slice.

Condition	RS-GA	PS-WFG	MA-SM	RDQ3N-SM
Slice 1	99.56%	99.92%	99.89%	100.00%
Slice 2	99.45%	99.59%	99.81%	99.95%
Slice 3	98.23%	97.63%	99.79%	99.91%

RDQ3N has demonstrated significant improvements in terms of latency, reliability, and resource efficiency. By leveraging the main RNN and DQN components, RDQ3N effectively predicts early QoS and hidden states based on resource availability, which enables proactive resource orchestration and slice instantiation. The approach simulation led to reduced average delay, improved throughput, and higher acceptance ratios, which enhanced the reliability and performance of V2X services ranging from high safety-critical to non-safety services. The evaluation metrics showcased the key contributions of the RDQ3N scheme in the QoS-driven mechanism to ensure that critical V2X services receive the necessary bandwidth/computing resources while balancing the needs of all slice instances to maximize overall QoS performances.

6. Conclusions and Future Studies

The deployment of RDQ3N in a three-tier vehicular edge-core architecture demonstrated its effectiveness in SDN- and NFV-enabled systems for E2E V2X slicing management. Our experiments showed that RDQ3N enhanced reliability metrics, latency, and resource efficiency for serving massive vehicular services, ranging within safety-critical,

ITS-supported, and infotainment services. The QoS-driven models are formulated with the objective of evaluating core RDQ3N rewards to emphasize the primary goal of the controller and orchestrator settings. The proactivity in predicting QoS and resource availability from the main RNN component made the proposed scheme a compelling solution for achieving diverse V2X QoS requirements. Furthermore, the main DQN component corrected RNN errors through adaptation in a virtual representational environment and optimized actions with the functional capability to allocate and schedule resources in real-time cross-tier orchestration. The intelligent slice controller leveraged the potential outputs and recommendations from RDQ3N and optimized the physical infrastructure (V2X environment) with SDN-based steering and NFV-enabled MEC.

Our future studies will discuss scalability challenges through standardization and interoperability that support proof-of-concept in the intelligent slicing orchestration of vehicular communications systems. We will consider end-to-end integration, hierarchical orchestration, and continuous learning. Thus, we expand the implementation on real-world data that can achieve greater scalability, efficiently manage more service instances, and meet the dynamic demands of a large-scale vehicular network.

Author Contributions: Conceptualization, P.T. and S.K.; methodology, P.T. and S.R.; software, P.T. and S.R.; validation, I.S. and S.R.; formal analysis, I.S. and S.R.; investigation, S.K.; resources, S.K.; data curation, P.T.; writing—original draft preparation, P.T.; writing—review and editing, I.S., S.R. and S.K.; visualization, I.S.; supervision, S.K.; project administration, S.K.; funding acquisition, S.K. All authors have read and agreed to the published version of the manuscript.

Funding: This work was supported by an Institute of Information and Communications Technology Planning and Evaluation (IITP) grant funded by the Korea government (MSIT) (No. RS-2022-00167197, Development of Intelligent 5G/6G Infrastructure Technology for the Smart City), in part by the National Research Foundation of Korea (NRF), Ministry of Education, through the Basic Science Research Program under Grant NRF-2020R1I1A3066543, in part by BK21 FOUR (Fostering Outstanding Universities for Research) under Grant 5199990914048, and in part by the Soonchunhyang University Research Fund.

Data Availability Statement: The data presented in this study are available on request from the corresponding author.

Conflicts of Interest: The authors declare no conflicts of interest.

References

1. Zeadally, S.; Javed, M.A.; Hamida, E.B. Vehicular Communications for ITS: Standardization and Challenges. *IEEE Commun. Stand. Mag.* **2020**, *4*, 11–17. [CrossRef]
2. Arena, F.; Pau, G. An Overview of Vehicular Communications. *Future Internet* **2019**, *11*, 27. [CrossRef]
3. Wang, X.; Li, X.; Jing, T.; Yu, H. An Integrated Dependability Guarantee Provisioning for Cluster-Based IoV Networks with Slicing. *IET Intell. Transp. Syst.* **2023**, *17*, 1752–1768. [CrossRef]
4. Sanchez-Iborra, R.; Santa, J.; Gallego-Madrid, J.; Covaci, S.; Skarmeta, A.F. Empowering the Internet of Vehicles with Multi-RAT 5G Network Slicing. *Sensors* **2019**, *19*, 3107. [CrossRef]
5. Wijethilaka, S.; Liyanage, M. Survey on Network Slicing for Internet of Things Realization in 5G Networks. *IEEE Commun. Surv. Tutor.* **2021**, *23*, 1. [CrossRef]
6. Afolabi, I.; Taleb, T.; Samdanis, K.; Ksentini, A.; Flinck, H. Network Slicing and Softwarization: A Survey on Principles, Enabling Technologies, and Solutions. *IEEE Commun. Surv. Tutor.* **2018**, *20*, 2429–2453. [CrossRef]
7. Su, R.; Zhang, D.; Venkatesan, R.; Gong, Z.; Li, C.; Ding, F.; Jiang, F.; Zhu, Z. Resource Allocation for Network Slicing in 5G Telecommunication Networks: A Survey of Principles and Models. *IEEE Netw.* **2019**, *33*, 172–179. [CrossRef]
8. Debbabi, F.; Jmal, R.; Fourati, L.C.; Ksentini, A. Algorithmics and Modeling Aspects of Network Slicing in 5G and Beyonds Network: Survey. *IEEE Access* **2020**, *8*, 162748–162762. [CrossRef]
9. Wang, X.; Mao, S.; Gong, M.X. An Overview of 3GPP Cellular Vehicle-To-Everything Standards. *GetMobile: Mob. Comput. Commun.* **2017**, *21*, 19–25. [CrossRef]
10. Mohammed, M.H.; Agrawal, D.; Dixit, P.; Bhattacharyya, B. Deep Learning for Network Slicing and Self-Healing in 5G Systems. In Proceedings of the 2021 IEEE International Conference on Signal Processing, Information, Communication & Systems (SPICSCON), Dhaka, Banglades, 3–4 December 2021.
11. Liu, Q.; Choi, N.; Han, T. Deep Reinforcement Learning for End-To-End Network Slicing: Challenges and Solutions. *IEEE Netw.* **2022**, *37*, 222–228. [CrossRef]

12. Sarra, B.; Saleh, M.; Saad, H. Ensuring QoS and Efficiency of Vehicular Networks by SDVN-IoV. In Proceedings of the 2020 International Conference on Advanced Aspects of Software Engineering (ICAASE), Constantine, Algeria, 28–30 November 2022.
13. Wu, Q.; Shi, S.; Wan, Z.; Fan, Q.; Fan, P.; Zhang, C. Towards V2I Age-Aware Fairness Access: A DQN Based Intelligent Vehicular Node Training and Test Method. *Chin. J. Electron.* **2023**, *32*, 1230–1244.
14. Wu, Z.; Lu, Z.; Hung, P.C.K.; Huang, S.-C.; Tong, Y.; Wang, Z. QaMeC: A QoS-Driven IoVs Application Optimizing Deployment Scheme in Multimedia Edge Clouds. *Future Gener. Comput. Syst.* **2019**, *92*, 17–28. [CrossRef]
15. Aldhyani, T.H.H.; Alrasheedi, M.; Alqarni, A.A.; Alzahrani, M.Y.; Bamhdi, A.M. Intelligent Hybrid Model to Enhance Time Series Models for Predicting Network Traffic. *IEEE Access* **2020**, *8*, 130431–130451. [CrossRef]
16. Hu, Y.; Zhu, L.; Zhang, J.; Cai, Z.; Han, J. Migration and Energy Aware Network Traffic Prediction Method Based on LSTM in NFV Environment. *KSII Trans. Internet Inf. Syst.* **2023**, *17*, 896–915.
17. Tam, P.; Kang, S.; Ros, S.; Kim, S. Enhancing QoS with LSTM-Based Prediction for Congestion-Aware Aggregation Scheduling in Edge Federated Learning. *Electronics* **2023**, *12*, 3615. [CrossRef]
18. Park, J.; Son, J.G.; Kim, D. Resource Metric Refining Module for AIOps Learning Data in Kubernetes Microservice. *KSII Trans. Internet Inf. Syst.* **2023**, *17*, 1545–1559.
19. Song, I.; Tam, P.; Kang, S.; Ros, S.; Kim, S. DRL-Based Backbone SDN Control Methods in UAV-Assisted Networks for Computational Resource Efficiency. *Electronics* **2023**, *12*, 2984. [CrossRef]
20. *ONF White Paper*; Software-Defined Networking: The New Norm for Networks; Open Networking Foundation: Palo Alto, CA, USA, 2012.
21. *ETSI GS NFV-INF 001*; Network Functions Virtualisation (NFV); Infrastructure Overview. ETSI: Sophia Antipolis, France, 2015.
22. Sabella, D.; Vaillant, A.; Kuure, P.; Rauschenbach, U.; Giust, F. Mobile-Edge Computing Architecture: The Role of MEC in the Internet of Things. *IEEE Consum. Electron. Mag.* **2016**, *5*, 84–91. [CrossRef]
23. Chowdhury, N.M.M.K.; Boutaba, R. A Survey of Network Virtualization. *Comput. Netw.* **2010**, *54*, 862–876. [CrossRef]
24. Ren, Y.; Guo, A.; Song, C. Multi-Slice Joint Task Offloading and Resource Allocation Scheme for Massive MIMO Enabled Network. *KSII Trans. Internet Inf. Syst.* **2023**, *17*, 794–815.
25. Song, I. A Study on 5G Service Methods by Using BOCR Model and ANP. *KSII Trans. Internet Inf. Syst.* **2022**, *16*, 3299–3312.
26. Li, X.; Rao, J.; Zhang, H.; Callard, A. Network Slicing with Elastic SFC. In Proceedings of the 2017 IEEE 86th Vehicular Technology Conference (VTC-Fall), Toronto, ON, Canada, 24–27 September 2017.
27. Ros, S.; Eang, C.; Tam, P.; Kim, S. ML/SDN-Based MEC Resource Management for QoS Assurances. In Proceedings of the Advances in Computer Science and Ubiquitous Computing (CUTECSA 2022), Vientiane, Laos, 19–21 December 2022; Park, J.S., Yang, L.T., Pan, Y., Park, J.H., Eds.; Lecture Notes in Electrical Engineering. Springer: Singapore, 2023; Volume 1028.
28. *ETSI GS NFV-EVE 005*; Network Function Virtualisation (NFV); Ecosystem; Report on SDN Usage in NFV Architectural Framework. ETSI: Sophia Antipolis, France, 2015.
29. Wu, W.; Zhou, C.; Li, M.; Wu, H.; Zhou, H.; Zhang, N.; Shen, X.S.; Zhuang, W. AI-Native Network Slicing for 6G Networks. *IEEE Wirel. Commun.* **2022**, *29*, 96–103. [CrossRef]
30. Togou, M.A.; Bi, T.; Dev, K.; McDonnell, K.; Milenovic, A.; Tewari, H.; Muntean, G.-M. DBNS: A Distributed Blockchain-Enabled Network Slicing Framework for 5G Networks. *IEEE Commun. Mag.* **2020**, *58*, 90–96. [CrossRef]
31. Messaoud, S.; Bradai, A.; Ben Ahmed, O.; Quang, P.; Atri, M.; Hossain, M.S. Deep Federated Q-Learning-Based Network Slicing for Industrial IoT. *IEEE Trans. Ind. Inform.* **2020**, *17*, 1. [CrossRef]
32. Li, M.; Gao, J.; Zhou, C.; Shen, X.S.; Zhuang, W. Slicing-Based Artificial Intelligence Service Provisioning on the Network Edge: Balancing AI Service Performance and Resource Consumption of Data Management. *IEEE Veh. Technol. Mag.* **2021**, *16*, 16–26. [CrossRef]
33. Zhou, W.; Islam, A.; Chang, K. Real-Time RL-Based 5G Network Slicing Design and Traffic Model Distribution: Implementation for V2X and EMBB Services. *KSII Trans. Internet Inf. Syst.* **2023**, *17*, 2573–2589.
34. Phyu, H.; Naboulsi, D.; Stanica, R. Machine Learning in Network Slicing–A Survey. *IEEE Access* **2023**, *11*, 39123–39153. [CrossRef]
35. Liang, T.; Chen, M.; Yin, Y.; Zhou, L.; Ying, H. Recurrent Neural Network Based Collaborative Filtering for QoS Prediction in IoV. *IEEE Trans. Intell. Transp. Syst.* **2022**, *23*, 2400–2410. [CrossRef]
36. Tam, P.; Song, I.; Kang, S.; Ros, S.; Kim, S. Graph Neural Networks for Intelligent Modelling in Network Management and Orchestration: A Survey on Communications. *Electronics* **2022**, *11*, 3371. [CrossRef]
37. Hurtado Sánchez, J.A.; Casilimas, K.; Caicedo Rendon, O.M. Deep Reinforcement Learning for Resource Management on Network Slicing: A Survey. *Sensors* **2022**, *22*, 3031. [CrossRef]
38. Cui, Y.; He, S.; Wang, R.; He, P.; Wang, D.; Huang, X. Multi-Agent Reinforcement Learning for Slicing Resource Allocation in Vehicular Networks. *IEEE Trans. Intell. Transp. Syst.* **2023**, 1–12. [CrossRef]
39. Kamal, M.A.; Raza, H.W.; Alam, M.M.; Su'ud, M.M.; Sajak, A. binti A.B. Resource Allocation Schemes for 5G Network: A Systematic Review. *Sensors* **2021**, *21*, 6588. [CrossRef] [PubMed]
40. Pratik, T.; Anurag, T.; Kulkarni, A. GREENSKY: A Fair Energy-Aware Optimization Model for UAVs in Next-Generation Wireless Networks. *Green Energy Intell. Transp.* **2023**, *3*, 100130.
41. Shen, H.; Wang, T.; Heng, Y.; Bai, G. Joint Resource Slicing and Vehicle Association for Drone-Assisted Vehicular Networks. *Drones* **2023**, *7*, 534. [CrossRef]

42. Mlika, Z.; Cherkaoui, S. Network Slicing with MEC and Deep Reinforcement Learning for the Internet of Vehicles. *IEEE Netw.* **2021**, *35*, 132–138. [CrossRef]
43. Wu, Y.; Dai, H.-N.; Wang, H.; Xiong, Z.; Guo, S. A Survey of Intelligent Network Slicing Management for Industrial IoT: Integrated Approaches for Smart Transportation, Smart Energy, and Smart Factory. *IEEE Commun. Surv. Tutor.* **2022**, *24*, 1175–1211. [CrossRef]
44. Li, Y.; Zhang, G.; Ren, C. Network Slicing for IoV: QoS-Oriented Radio Resource Optimization. In Proceedings of the 2022 IEEE 6th Information Technology and Mechatronics Engineering Conference (ITOEC), Chongqing, China, 16–26 October 2022.
45. Khan, H.; Luoto, P.; Samarakoon, S.; Bennis, M.; Matti, L. Network Slicing for Vehicular Communication. *Trans. Emerg. Telecommun. Technol.* **2021**, *32*, e3652. [CrossRef]
46. Smida, K.; Tounsi, H.; Frikha, M.; Song, Y.-Q. FENS: Fog-Enabled Network Slicing in SDN/NFV-Based IoV. *Wirel. Pers. Commun.* **2022**, *128*, 2175–2202. [CrossRef]

Disclaimer/Publisher's Note: The statements, opinions and data contained in all publications are solely those of the individual author(s) and contributor(s) and not of MDPI and/or the editor(s). MDPI and/or the editor(s) disclaim responsibility for any injury to people or property resulting from any ideas, methods, instructions or products referred to in the content.

electronics

MDPI

Article

Vulnerable Road User Safety Using Mobile Phones with Vehicle-to-VRU Communication

Sukru Yaren Gelbal, Bilin Aksun-Guvenc and Levent Guvenc *

Automated Driving Laboratory, Ohio State University, Columbus, OH 43212, USA; gelbal.1@osu.edu (S.Y.G.); aksunguvenc.1@osu.edu (B.A.-G.)
* Correspondence: guvenc.1@osu.edu

Abstract: Pedestrians, bicyclists, and scooterists are Vulnerable Road Users (VRUs) in traffic accidents. The number of fatalities and injuries in traffic accidents involving vulnerable road users has been steadily increasing in the last two decades in the U.S., even though road vehicles now have perception sensors like cameras to detect risk and issue collision warnings or apply emergency braking. Perception sensors like cameras are highly affected by lighting and weather conditions. Cameras, radar, and lidar cannot detect vulnerable road users in partially occluded and occluded situations. This paper proposes the use of Vehicle-to-VRU communication to inform nearby vehicles of VRUs on trajectories with a potential collision risk. An Android smartphone app with low-energy Bluetooth (BLE) advertising is developed and used for this communication. The same app is also used to collect motion data of VRUs for training. VRU motion data are smoothed using a Kalman filter, and an LSTM neural network is used for future motion prediction. This information is used in an algorithm comparing Time-To-collision-Zone (TTZ) for the vehicle and VRU, and issues driver warnings with different severity levels. The warning severity level is based on the analysis of real data from a smart intersection for close vehicle and VRU interactions. The resulting driver warning system is demonstrated using proof-of-concept experiments. The method can easily be extended to a VRU collision-mitigation system.

Keywords: vulnerable road user safety; vehicle-to-VRU communication; pedestrian collision warning

1. Introduction

A significant number of traffic accidents occur each year and result in a large number of injuries and deaths [1]. To reduce these incidents, researchers have introduced new and better active and passive protection measures each year to improve passenger and other road user safety [2–4]. In traffic, pedestrians are among the most vulnerable road users (VRUs) [5–7], and pedestrian safety is a priority for the U.S. Federal Highway Administration's (FHWA) Office of Safety [8]. The City of New York, for example, is testing new technologies to improve the safety of pedestrians [9]. According to National Highway Traffic Safety Administration (NHTSA) statistics, more than 6000 pedestrians are killed in traffic accidents every year in the U.S. Although pedestrians are the most common VRU type, other VRUs, such as bicyclists [10] and scooter users [11], are also very susceptible to high-collision-risk scenarios and crashes.

To improve the safety of VRUs as they interact with autonomous and non-autonomous vehicles, an important tool that can be utilized is vehicle-to-everything (V2X) communication [12–15]. This can be achieved on an individual basis, where each road user carries a message-broadcasting device, and on an infrastructure basis, where an intersection is designed as a smart intersection that detects VRUs as well as vehicles, and publishes information for them as if they are all connected [16,17], or through the over-the-air sharing of relevant Connected Vehicle (CV) data in centralized servers. Although an automated vehicle can obtain more accurate information about the VRUs using its own sensors, there are conditions in which this is not reliable or possible, such as bad weather (heavy fog, rain) or a no-line-of-sight (NLOS) situation in which the target of interest is blocked by an obstacle or another vehicle [18]. When the information is transferred via wireless communication, these types of situations can be remedied. Since Bluetooth advertising mode and Bluetooth 5.0 technology are now more widely available in current mobile phones [19,20], it is now a feasible communication method between VRUs and other road users [21]. VRUs can carry their mobile phones in their hands or in their pockets, which affects the range of communication. Current mobile phones possess a large number of sensors that can be used for tracking or predicting the motion and behavior of pedestrians, bicyclists, or scooters [22,23]. Using, for example, the Android Bluetooth Low Energy (BLE) application programming interface (API) [24], it is possible to develop an Android app that can be used for this purpose. It is possible to achieve different benefits when using this approach, depending on the different modes of Bluetooth communication [25].

Pedestrian safety systems need knowledge of the possible future path of nearby pedestrians to determine the possibility of a near-future collision risk [26–28]. Even with the correct information transfer from a pedestrian's phone using Bluetooth or other communication means, predicting the path of a VRU can be tricky, especially for pedestrians [29]. Different types of methods, such as grid-based ones [30], constant velocity models [31], and neural networks [32], have been utilized in the literature to predict pedestrian paths. Generative Adversarial Networks (GANs) [33] and Long Short-Term Memory (LSTM) networks [34] have been used and observed in the literature to provide reliable results for prediction. The memory-based feedback nature of the LSTM network helps with the motion patterns of pedestrian movements, especially at intersections. The prediction method can be combined with the data obtained from sensors or wireless communication. One problem posed here could be the low-frequency update rate of normal GPS sensors of 1 Hz if a pedestrian is the source of information. However, the GPS, acceleration, and gyro sensors available on a phone can be utilized with a Kalman Filter [35] for higher-frequency location-tracking. Utilizing the sensors in a pedestrian's mobile phone is very beneficial when combined with Bluetooth advertisement capabilities. This approach helps ensure information transfer between road users and, most importantly, allows us to implement a safety application that can be installed conveniently on mobile phones and can be used by both the driver of a vehicle and a VRU.

This paper focuses on VRU safety using V2P communication. The methods and results presented in the paper are applicable to all VRUs, but the experimental results reported here are for pedestrians. The contributions of the paper are as follows:

- The development of an Android app for collecting mobile phone sensor data of VRUs;
- Using the collected data to predict the future motion of a VRU using Kalman filtering and LSTM-based motion prediction;
- Converting the future motion and localization data of the VRU to a Personal Safety Message (PSM);
- Broadcasting the PSM using Bluetooth advertising or over-the-air using internet connectivity to nearby vehicles;
- Further developing and using the same Android app to receive PSM data in a vehicle;
- Developing and implementing programs in a Bluetooth board to similarly broadcast and receive PSM data if needed;
- Developing and implementing a pedestrian collision warning system using the PSM data;
- Analyzing available vehicle and pedestrian interactions in a smart intersection to calibrate the pedestrian collision warning system;
- Experimentally developing and demonstrating a full pedestrian collision avoidance system.

In this paper, V2P communication is used to improve the safety of VRUs, especially pedestrians, using a driver warning system. The VRU mobile phone sensors are used to predict future motion to better assess the risk of collision with nearby vehicles and issue warning messages. These warnings can also be used to begin the automated slowing down and braking of the vehicle. The organization of the rest of the paper is as follows. The V2P app's development and implementation are presented in Section 2. V2P communication using the developed app is investigated in Section 3. The V2P-communication-based pedestrian safety system is developed in Section 4. Simulation and experimental results are presented and discussed in Section 5. The paper ends with conclusions and recommendations in Section 6.

2. V2P App Development and Implementation

To the determine collision risk and execute the necessary safety precautions, pedestrian motion information needs to be obtained by the vehicle. By transferring this over a wireless link, we also ensure that, even if there is NLOS and perception sensors are not able to detect the pedestrian, the vehicle would still be able to detect the pedestrian and track their location, speed and direction. Therefore, the main enablers of this implementation are its relatively wide availability and the Bluetooth 5.0 capabilities of recent Android phone models. However, it should be noted that the algorithms and methods developed in this paper for vehicle and pedestrian interaction analyses, driver alerts and warnings, pedestrian motion and intent prediction, and collision risk determination are independent the communication method that is used.

Bluetooth communication technology was experimented with to analyze the practical abilities of its implementation in smartphones. First, communication experiments were carried out on the nRF52840-DK Bluetooth development board [36]. Afterwards, an Android application was coded using the BLE module in Android API 28. This software has the ability to encode, transmit, receive, and decode messages via Bluetooth 5.0 through extended advertisement. PSM from the Society of Automotive Engineers (SAE) J2735 standard [37] was used to establish the pedestrian information message structure. Experiments were first carried out using the extended advertisement mode to achieve a long-range and large data-carrying capacity, and very good results were obtained. The mode of communication was later switched to normal advertisement, as this is available in more smartphones. As well as using the location, speed, acceleration and heading fields to calculate the collision possibility and issue a warning, PSM also has fields such as user type, device use state, cross request, cluster size, and attachment. These fields will be useful for improving the algorithm in the future for pedestrian group clustering, pedestrian behavior prediction, and more accurate collision predictions for a VRU with pets, carts, or a wheelchair. The standard is designed to allow for this message to be published at 10 Hz, which is a reasonable frequency for our calculations. The pedestrian transmits the PSM, and the vehicle receives it via the driver warning implementation. The PSM communication between the VRU and vehicle(s) is illustrated in Figure 1. The PSM-populated fields are the location (latitude and longitude), heading, speed, and acceleration, as shown in Figure 2.

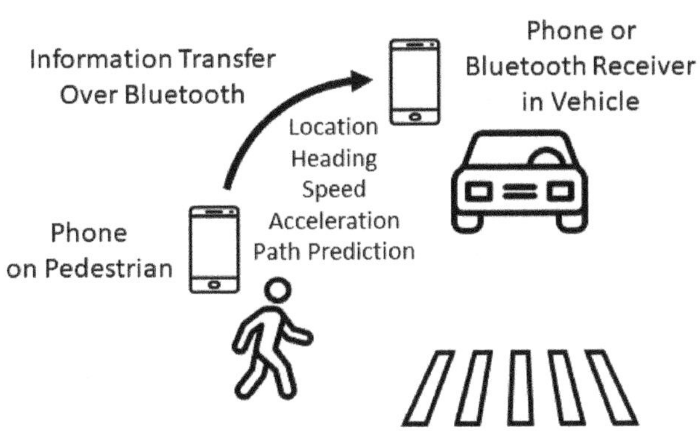

Figure 1. Illustration of V2P communication over Bluetooth.

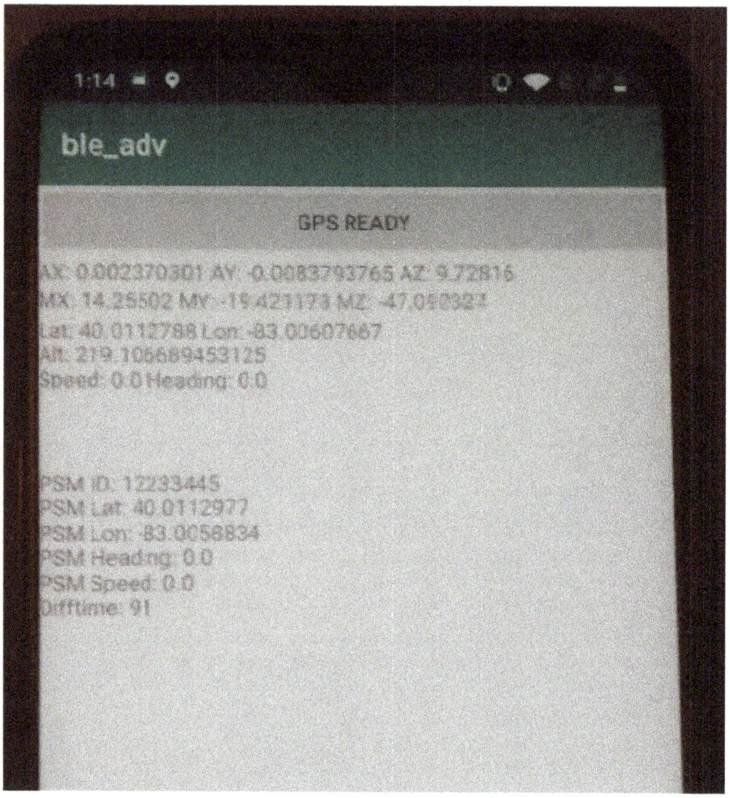

Figure 2. PSM over Bluetooth V2P app running on a smartphone.

An Android phone can be used in the vehicle to obtain the pedestrian's information, handle the warning calculations, and show a warning and the recommended speed profile (if desired) to the driver. If the vehicle is autonomous, a Bluetooth receiver board such as the nRF52840-DK can be used to obtain the messages and transfer the information to computing unit(s) for comfortable braking and stopping, as well as for collision avoidance. This board has a serial interface. This was combined with a W5500 Ethernet shield [38] to transfer the information received via Bluetooth through the Ethernet connection to achieve better results. A picture of the combined BLE-receiving board can be seen in Figure 3.

Figure 3. The nRF52840-DK board with a mounted W5500 Ethernet shield.

3. V2P Communication Experiments

After the implementation of V2P into the smartphone, some initial testing for determining communication range was conducted. This testing involved both nRF52840-DK Bluetooth boards and Android smartphones in obstructed and non-obstructed cases along with normal and extended Bluetooth advertisement modes. The results obtained are presented in Table 1. The Bluetooth board has the longest communication range, of more than 250 m, in the extended advertising mode. This corresponds to the first row of the table. The obstruction of the smartphone by the pedestrian body results in the lowest communication range in normal Bluetooth advertising mode when communicating with another smartphone in the vehicle. This corresponds to the last row of the table. Data transfer over WebSocket [39] functionality was also implemented in the application. However, this was changed later to internal storage during the experiments because the WebSocket connection required constant internet access. Moreover, due to GPS sensor updates being paused when the phone screen is locked, GPS callback was tied to a background process to ensure regular updates. The process flow diagram of the application is shown in Figure 4.

Table 1. Bluetooth communication range experiments.

Device	Condition	Distance in Normal Advertisement (m)	Distance in Extended Advertisement (m)
nRF52840-DK	No obstructions	191	253
Android	No obstructions	78	114
Android	Intentionally obstructed by pedestrian body	27	55

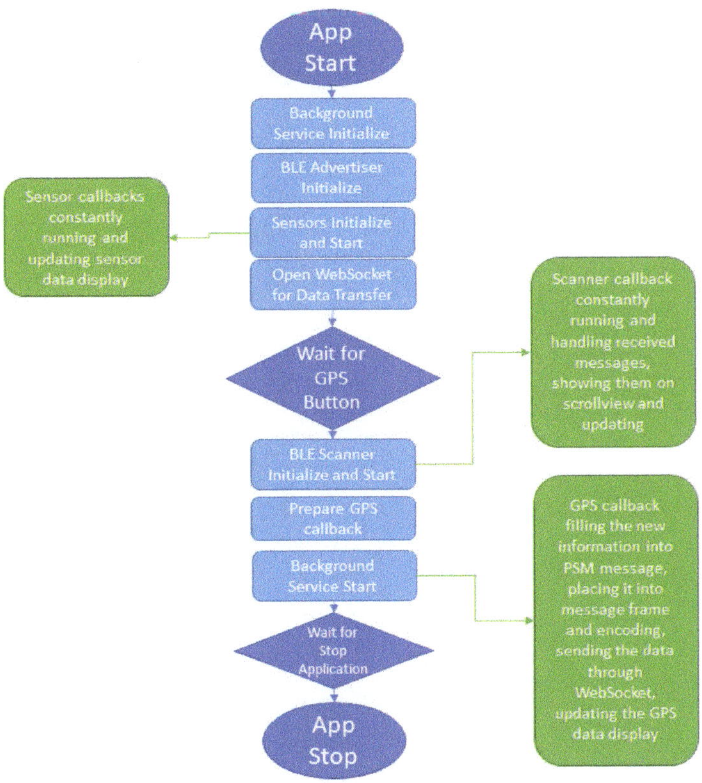

Figure 4. PSM over Bluetooth V2P app process flow diagram.

Along with the V2P broadcasting mobile application discussed above, a data collection application was developed and implemented in the Android phone. This application was developed separately from the previous app to support older model Android phones which do not have Bluetooth 5.0. The application, running on an Android phone, is displayed in Figure 5. This application receives data from several sensors built in the Android phone and records them locally with timestamps in the form of csv files. The sensors used are acceleration, gyro, step counter, GPS, orientation, proximity, and light sensor. A feature for uploading the data into a remote server was also implemented. Data plots after initial testing of the app are shown in Figure 6.

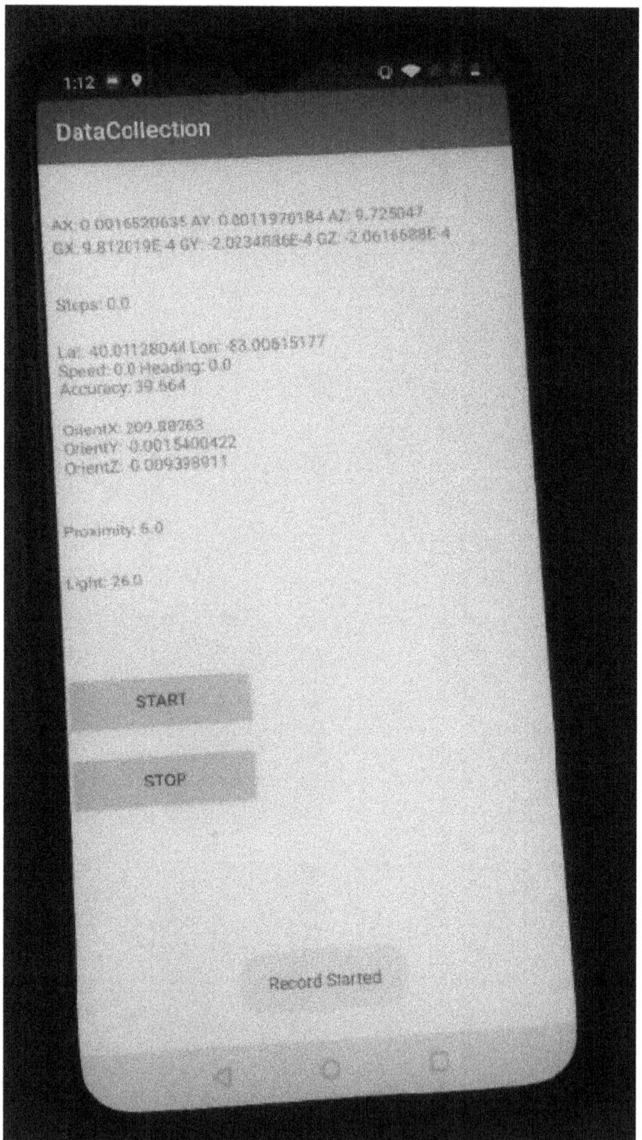

Figure 5. Data collection app running on a smartphone.

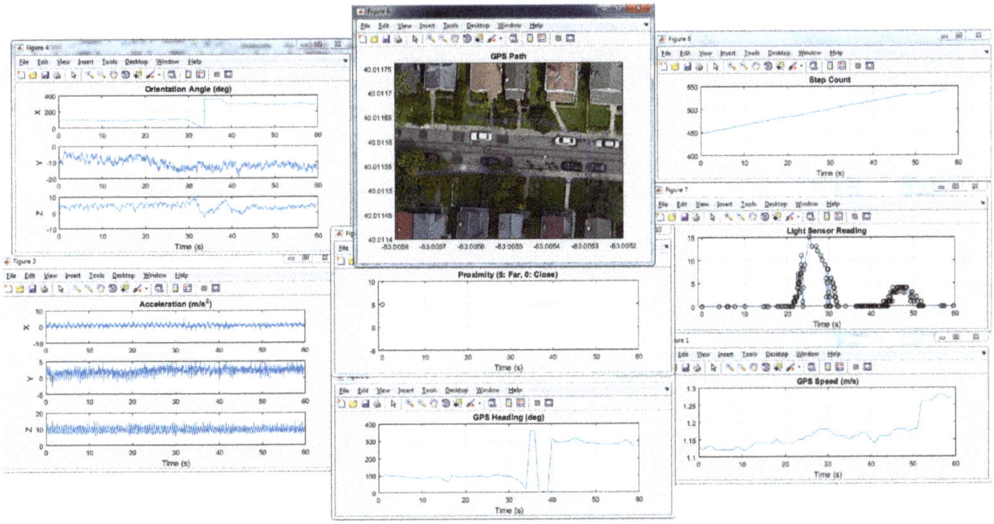

Figure 6. Collected data plotted in MATLAB. Blue circles show pedestrian position in middle top plot. Black circles show light sensor reading in right middle plot. Other blue lines show mobile phone sensor data.

4. V2P-Communication-Based Pedestrian Safety Warnings

The two applications that were previously discussed regarding V2P communication and data recording were later combined into a single application built for recent Android phones with Bluetooth 5.0. The user interface was also improved. Then, a custom J2735 library was prepared to encode and decode PSM messages. Several quality-of-life changes were implemented later. These include manual or automatic user selection, manual or random PSM ID, easier file access for recorded data, a debug mode for developers, and a user mode for end-users. The new app screen is shown in Figure 7. PSM decoding, encoding, broadcasting and data collection were combined in this final application. The user interface was simplified and improved. Icons were added to warn the driver in case of a collision risk. The warning design, parameters and the user interface are discussed in more detail later in the paper.

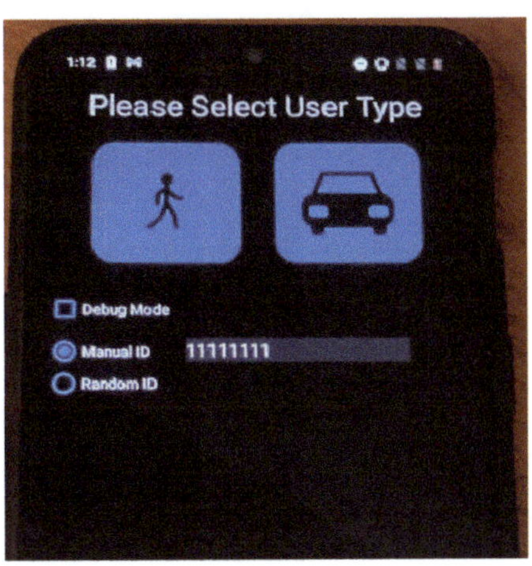

Figure 7. Pedestrian safety app running on a smartphone.

4.1. Safety Approach

The overall architecture of the V2P-communication-based pedestrian safety approach of this paper is illustrated in Figure 8. The vehicle is approaching a crosswalk that a pedestrian is intending to cross in Figure 8. The elements that make up this architecture are explained. The first element in the approach is pedestrian path prediction and tracking. This is utilized to predict the pedestrian's location in the near future with LSTM neural networks, as well as to track his/her motion at a higher sampling rate utilizing the Kalman filter. This is used to make safety warnings and action decisions. These include how to warn the driver or how to adjust the speed of the vehicle if it is autonomous.

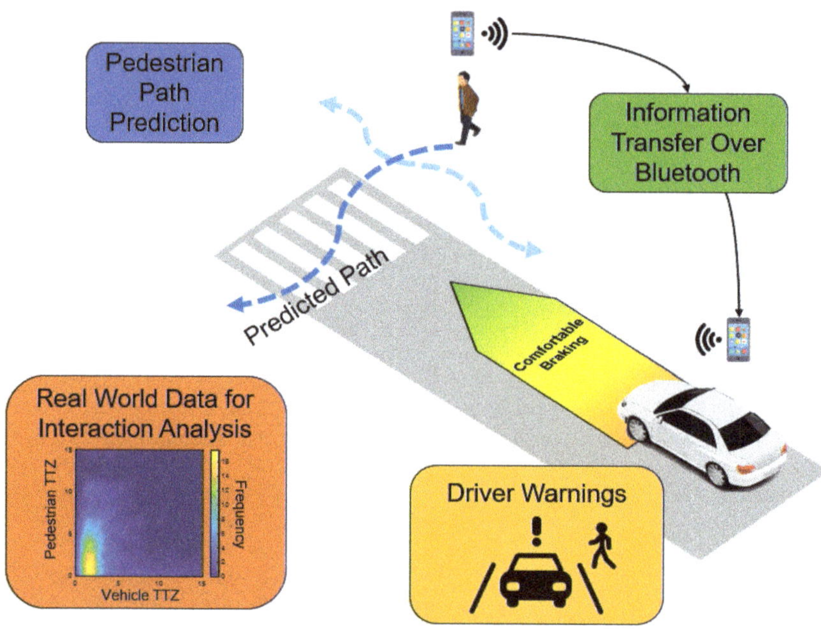

Figure 8. Architecture of the overall safety approach.

The second element in the architecture, communication, or information transfer focuses on wireless communication to provide information about the pedestrian, especially in cases involving reduced visibility or NLOS, to the nearby vehicle. Mobile phone Bluetooth or extended Bluetooth are used in this paper due to the enhanced support they provide when carrying larger chunks of information over longer distances. On top of that, with the dual-frequency GPS technology that is recently starting to be implemented in mobile devices, localization accuracies of a few decimeters can be achieved [40]. Most importantly, mobile phones are very common and can be utilized in non-autonomous vehicles, as well to warn the driver. Because of this, the means of wireless communication, as a way to discuss and demonstrate the VRU safety approach, is chosen as Bluetooth in this paper. However, it should be noted that the VRU collision warning and avoidance braking methods of this paper do not depend on the communication method used.

The next two elements of the architecture, real-world data analysis and driver warnings, are intertwined. The warning system was designed to allow for non-autonomous vehicles to utilize this safety system by showing the driver warnings, of varying degrees of severity, of a collision risk situation. Real-world data analysis plays a very important role in determining the severity of these warnings. An analysis of real interactions between vehicles and pedestrians at an intersection was used to achieve a naturalistic slowing down and stopping of the vehicle by providing the driver with the correct warnings at the right times.

4.2. Pedestrian Path-Tracking and Prediction

The path-tracking and prediction system is designed to be a combination of Kalman filtering and neural-network-based path prediction. While neural-network-based prediction predicts motion information multiple seconds into the future, Kalman filtering increases the frequency of location to above the 1 Hz GPS update rate in mobile phones. The diagram of the neural network and the Kalman filter path-tracking and prediction algorithm is shown in Figure 9. The mobile phone GPS sensor is utilized at its maximum rate of 1 Hz to obtain location and velocity information, which are then fed into the LSTM neural network that predicts the future location of the pedestrian. Although a single neural network can be used to predict multiple locations of the pedestrian path, multiple neural networks were trained separately to ensure a better prediction accuracy, along with a smaller network size. Therefore, the algorithm uses multiple small neural networks to predict future location within a timeframe of a few seconds. Naturally, the further in the future the algorithm tries to predict, the less accurate it becomes. Along with the neural network prediction, the Kalman filter uses GPS information at 1 Hz, as well as other higher update rate sensors such as orientation and acceleration, to track the pedestrian location at higher sampling than 1 Hz.

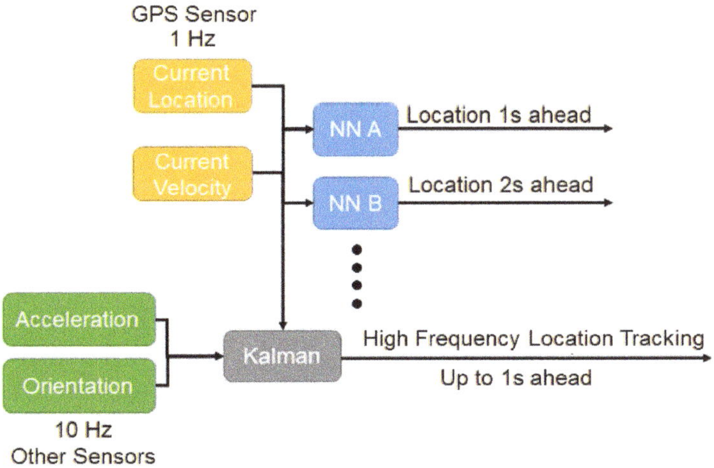

Figure 9. Pedestrian-tracking and path prediction using LSTM neural networks and a Kalman filter.

The LSTM type recurrent neural network was used because of its memory, since pedestrian movement patterns highly benefit from both short- and long-term memory, especially at an intersection. LSTM was also tested in other research in the field [34] and shown to perform better for pedestrian path prediction compared to other options. These networks were trained with both real-world data collected from mobile phones and synthetically generated pedestrian data from the Vissim environment. Some of these samples are presented in Figure 10. The trained neural network can be transferred into mobile devices using the TensorFlow Lite library for runtime operation [41].

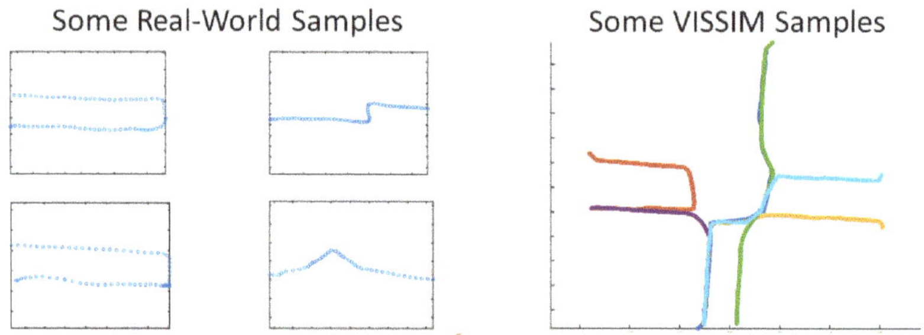

Figure 10. Some of the samples used to train the LSTM neural networks.

4.3. Pedestrian Behavior Prediction

Android phones have numerous sensors that provide high-frequency data. Some of these sensors were utilized to predict pedestrian behavior. An LSTM-based neural network structure was created, with these sensor readings used as multiple inputs. The first structuret that was experimented with had a single output with values ranging between 0 and 1. Different intervals within this range were trained as different behaviors. This structure, along with the inputs and outputs, is illustrated in Figure 11.

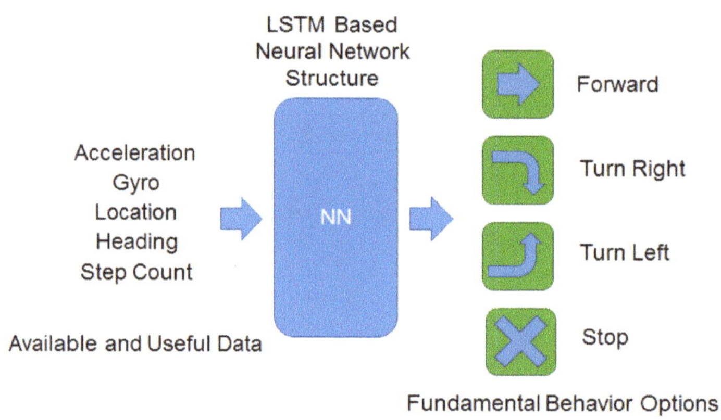

Figure 11. Pedestrian behavior prediction's initial structure.

To train the neural network of Figure 11, real-world data were collected with the data collection application that was explained earlier. Afterwards, the data were manually labeled by utilizing a custom data-labeling tool, developed in MATLAB version 2022 specifically for this purpose. Using the sensor information visualization features of the labeling tool, the labeling process was significantly shortened. The network was trained using the labeled data and then tested. The first designed structure, as mentioned before, had a single output with different value intervals indicating different behavior. With repeated experiments, the sensors that resulted in unnecessary noise or inconsistent measurements and that provided minimal benefit to the detection, such as acceleration and step count sensors, were removed from the input for better accuracy. Figure 12 shows the testing results of the trained neural network structure.

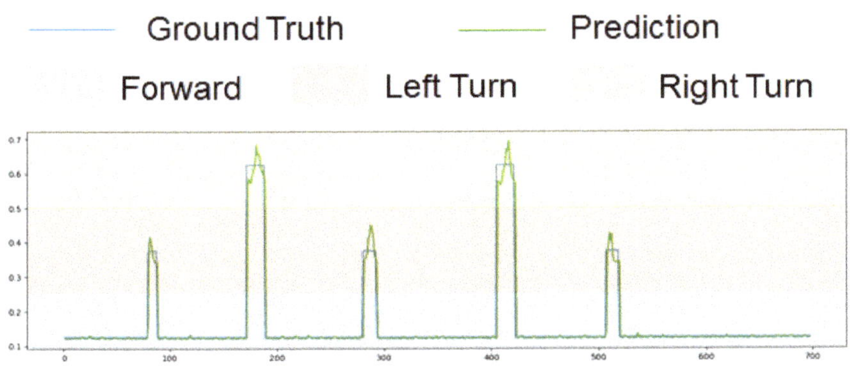

Figure 12. Single output prediction results.

As indicated in Figure 12, different colored areas indicate different behavior. When the output is between 0.5 and 0.75, the pedestrian is making a right turn. If the output is between 0.5 and 0.25, the pedestrian is making a left turn. Finally, an output between 0 and 0.25 means the pedestrian is moving forward. The graph in Figure 12 shows that the prediction results (green line) are very close to the ground truth (blue line), and always fall within the range of correct behavior. The second structure that was experimented with was created by modifying the first structure to incorporate multiple outputs to increase learning capacity, scalability, and compatibility with the literature. These outputs were then assigned to different modes of behavior, and interpreted as confidence levels. The network was trained with same data as before, although these data were now labeled for multiple outputs. The test results shown in Figure 13 have three different outputs, as opposed to one output in the previous figure. Each output has a confidence level between 0 and 100. The predicted behavior has the highest confidence level value. It can be seen that the prediction (orange line) follows the ground truth (blue line) closely. Although there is some noise in certain sections, the predictions were correct in almost all the test data.

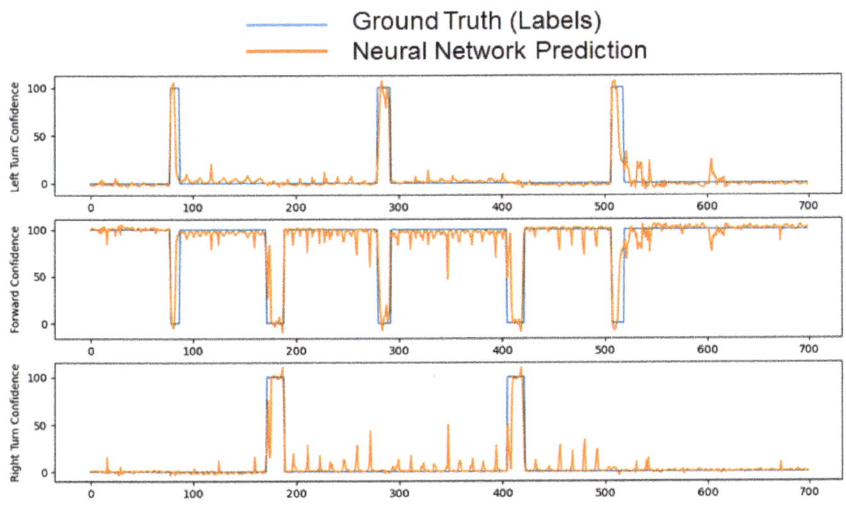

Figure 13. Multi-output prediction results.

4.4. Pedestrian–Vehicle Interaction

The base scenario geometry for pedestrian–vehicle interaction assumes that the vehicle and the pedestrian are moving straight along paths that are perpendicular to each other. This is similar to one of the configurations in a recent National Highway Traffic Safety Administration (NHTSA) report on the causes of accidents [8]. The base scenario geometry and variables are illustrated in Figure 14. The collision risk zone is defined at the crosswalk as an area of potential conflict, where vehicle speed is v_v, pedestrian speed v_p, and distances to the collision zone d_v, and d_p, for the vehicle and pedestrian, respectively, are defined as the significant variables. This configuration was created as a base scenario to be used throughout the paper for driver warning discussions. The scenario geometry can easily be modified to include left and right turns, as well with similar variables, where distance will be circular rather than linear. Using the variables mentioned above, Time to Zone (TTZ) can be calculated as:

$$TTZ_v = \frac{d_v}{v_v} \quad TTZ_p = \frac{d_p}{v_p}, \tag{1}$$

where TTZ_v, and TTZ_p are vehicle and pedestrian time to zones, respectively. The calculations for pedestrians can be combined with the tracking and prediction algorithm to calculate a more accurate TTZ_p, especially if the pedestrian is about to turn onto the crosswalk. Depending on how many seconds into the future the location is predicted, TTZ_p can be completely replaced or extrapolated using a constant-velocity or other, similar motion model [31].

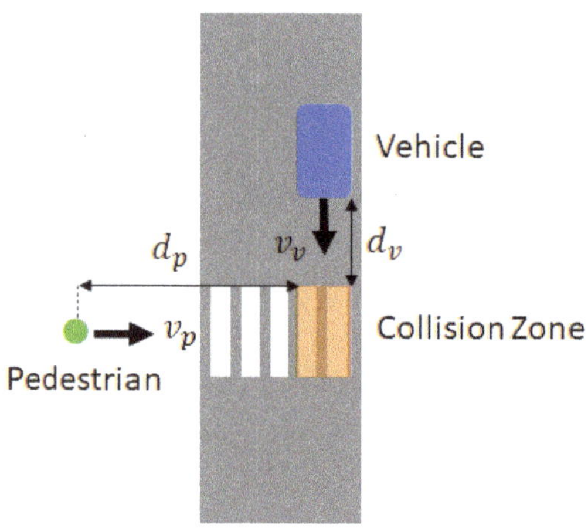

Figure 14. Geometrical configuration and variables for base scenario.

After the TTZ values are obtained, we can state the high collision risk condition as

$$TTZ_p - t_s < TTZ_v < TTZ_p + t_s, \tag{2}$$

or

$$|TTZ_v - TTZ_p| < t_s. \tag{3}$$

t_s is the safety margin time, and is either constant or a dynamic value based on GPS accuracy and driver preferences. If the conditions in Equations (2) or (3) are satisfied, there is a possible collision in the future. This condition is further investigated in the next subsection where this correspondence is visualized, and parameters are defined and determined using real-world interaction data. *TTZ* can be calculated and used to determine a possible collision risk in the future and its severity. For each second from the start of the interaction (depicted as interaction time), there will be TTZ_v and TTZ_p values. An interaction diagram, such as that presented in Figure 15, can be created using the correspondence of *TTZ* values for the pedestrian and the vehicle.

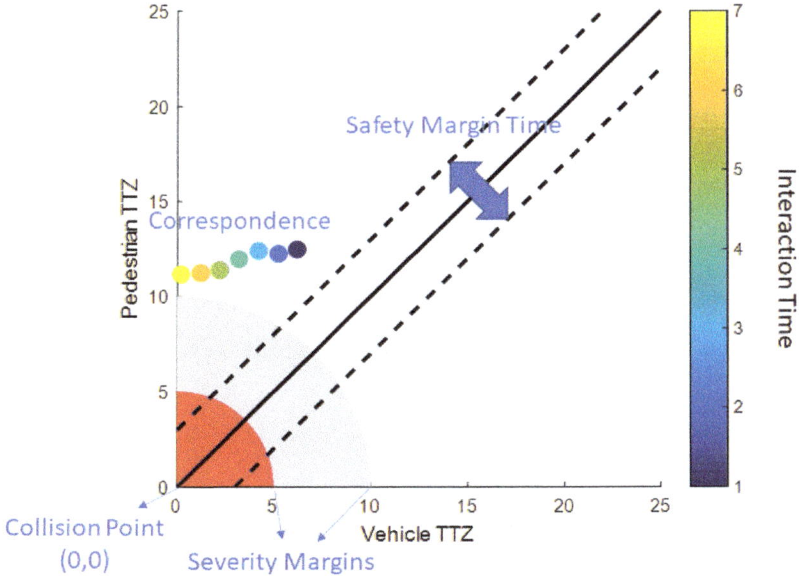

Figure 15. Pedestrian (VRU) and vehicle interaction diagram.

TTZ values for the vehicle and pedestrian are shown as small colored circles in the example in Figure 15, where the color of the circle determines the interaction time, with the legend provided on the right-hand-side vertical axis. Since both *TTZ* values are equal to zero, i.e., an interaction at the origin means collision, the closer the correspondence for the ordered pair (TTZ_v, TTZ_p) is to the origin, the more severe the collision risk. Severity margins can be defined as circular areas around the origin indicating different levels of severity. Defining the severity margins is highly important for non-autonomous vehicles, since the severity of the situation should be properly conveyed to the human drivers to allow for them to react correctly in a timely manner. Therefore, values for these margins were used as the main design parameters for the warning calculation. Another important parameter is the safety margin time, which represents the difference between the *TTZ* values for the vehicle and the pedestrian, like t_s, defined in the previous subsection. If the *TTZ* values are reasonably different from each other, either the vehicle or the pedestrian would cross the zone first, without any risk of collision. Therefore, the warning system only considers the correspondences that occur within this safety margin time region in Figure 15.

To better visualize how the interaction graph would look with corresponding vehicle and pedestrian behavior, Figure 16 was created as an illustration. A vehicle approaching the crosswalk at the same time as a pedestrian is illustrated in Figure 16. Steps 1, 2, and 3 represent the vehicle and pedestrian locations as the vehicle is slowing down. The vehicle *TTZ* is shown to be moving towards zero at first, but because the vehicle is slowing down, the values start to increase and the corresponding points move beyond the critical part of the diagram as the vehicle stops. The final image at the right side of Figure 16 shows the overall *TTZ* correspondence trajectory.

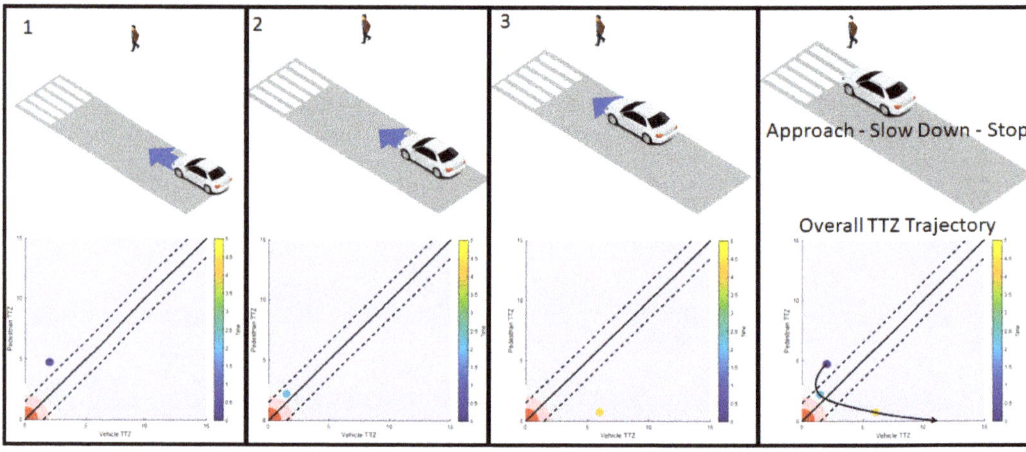

Figure 16. Illustration of vehicle and pedestrian behavior with successive interaction diagrams.

4.5. Real World Data Processing

To gain insight from real-world interactions, data from a real-world intersection at Marysville, Ohio, U.S., were collected and processed. This intersection is a smart intersection where four cameras were set up [16] to detect vehicles and pedestrians and send this information to the roadside unit (RSU) installed at the intersection. This RSU broadcasted basic safety messages (BSM) for each pedestrian as well as for each vehicle, along with their respective location and speed information, as if they were carrying a dedicated short-range communication (DSRC) broadcasting device. Although these data are very large and would not be practical to physically listen to and record, the City of Marysville made them available for public use through the Smart Columbus Operating System (SCOS) database [40]. The data were downloaded and processed to extract the vehicle and pedestrian interactions that fit the base scenario. This data contained eight months of BSM messages; some months contained more interactions than others.

Extracted interactions were processed to calculate *TTZ* for interacting vehicles and pedestrians. Then, interaction graphs were created showing all these interactions combined. To understand the similarities within all the interaction data, a heatmap graph was also created. These graphs can be seen in Figure 17. The interaction graph on the left side of Figure 17 shows each different interaction with a different color. Since the number of interactions is very high, it is difficult to interpret this, and a vehicle–pedestrian interaction density heatmap is used for better readability on the right-hand side of Figure 17. The yellow area shows the densest points in the graph. When the vehicle is slowing down, vehicle *TTZ* stays at around similar values for a while as pedestrian *TTZ* steadily decreases. The dense area in the heatmap graph corresponds to the area in which vehicles are slowing down when they see a pedestrian close to the intersection. Using this information, captured from real-world data, allows for us to design our warning system to mimic the natural behavior of drivers when they encounter nearby pedestrians.

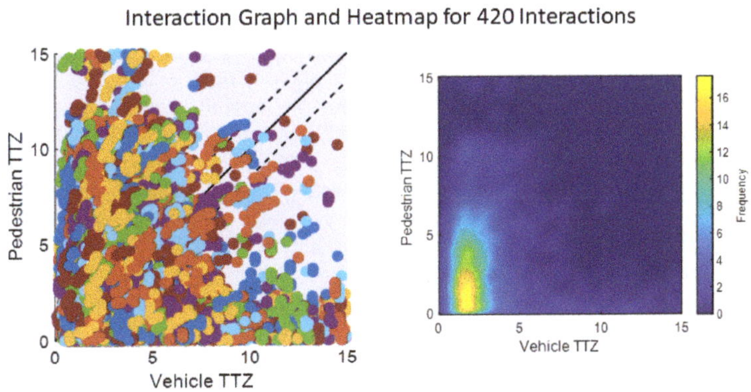

Figure 17. Nearby vehicle and pedestrian interaction graph (**left**) and heatmap (**right**) based on real-world data.

Interactions were divided according to the sunset time for each respective month, and are plotted again in Figure 18 for a comparison of night and daytime interactions. As seen in this figure, drivers at this intersection behaved more cautiously at nighttime, resulting in an approximately 0.5 s shift in the dense area in the heatmap. This information can also be incorporated into safety algorithms, either as a preference or to provide a more comfortable and cautious experience during the evening and night.

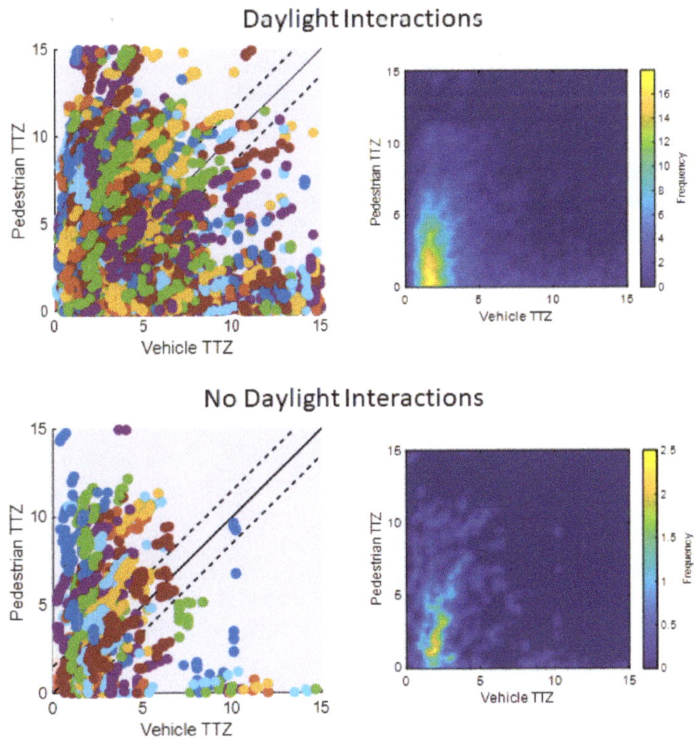

Figure 18. Comparison of daylight and no-daylight interactions.

Using the design parameters and real-world behavior information discussed in the previous section, driver warnings were designed for a safe stop. The warning system was designed with different severity values to convey the level of danger to the driver. This will help the driver decide how to slow down and stop. The main criteria used to determine the severity of the warning are the severity margins discussed in the previous section. With the current approach, which uses the *TTZ* as the main criteria to determine the current severity, we incorporate speed along with the remaining distance, as well as the *TTZ* correspondence for the pedestrian and vehicle. This prevents us from warning the driver unnecessarily in cases where either the pedestrian or the vehicle will be able to pass the crosswalk without the need to stop. Three margins were defined for the driver warning—Inform, Warn, and Emergency—with increasing degree of severity. The color-coding and explanations of these margins are shown in Figure 19.

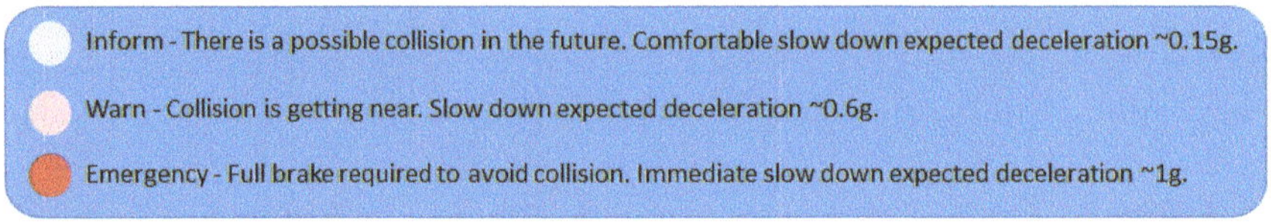

Figure 19. Warning severities and meanings.

Initially, there are no warnings, even though there may be the possibility of a collision in the distant future. As the risk of a future collision becomes higher, and as the value of TTZ_v reduces below a pre-determined threshold value, the warning system enters the Inform state. The Inform state corresponds to the lightest severity, where a future collision could easily be avoided by a comfortable speed reduction. As the risk of future collision increases, and as the TTZ_v value decreases below another pre-determined threshold, the system enters the Warn state, which means that the vehicle should start slowing down as soon as possible, using a relatively uncomfortable amount of deceleration to stop safely. These two margins can be chosen as constant values, and can later be modified adaptively or manually by the driver. To determine the value of these margins, the processed real-world behavior data that were discussed above were used. A histogram graph of the data was used to determine a base value ranging from 1 to 3.3 s for the Inform margin. After determining the base value for this first margin, a Vissim simulation environment with the base scenario was constructed to further test and adjust this value in simulations. An attention lag was also added to incorporate the driver reaction time. The test was conducted with a very comfortable deceleration to obtain a value that results in a conservative driver warning. The general flow of the simulations is illustrated in Figure 20.

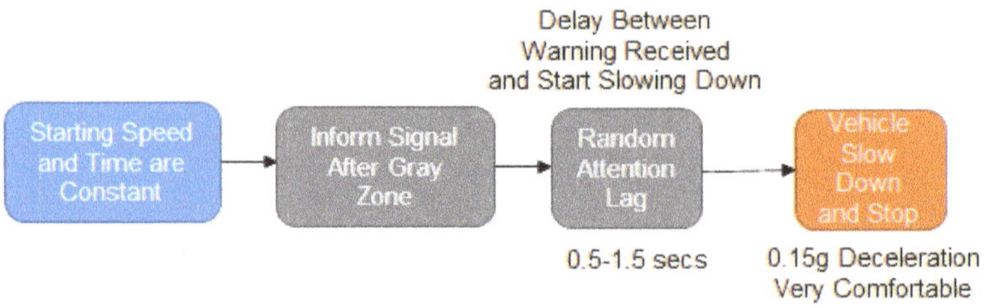

Figure 20. Flow of Vissim simulation to adjust the Inform margin.

As stated above, the simulations were conducted at a constant start speed and time where the only changing values were the random attention lag, and the Inform margin value. Interaction diagrams for multiple simulations, run across different Inform margin start values, are shown in Figure 21. When the diagram data were shifted by 1 s (diagram on left), there appear to be certain cases when the vehicle cannot stop in time with comfortable braking and needs to resort to emergency braking. When the data were shifted by 1.5 s (graph in the middle of Figure 21), although there were no accidents and vehicles were able to slow down and stop comfortably, an unexpected increase in reaction time still resulted in some unwanted situations. Therefore, the value shifted by 2 s (diagram on the right side of Figure 21) was selected, as this is the most conservative value, while still being close to the base value. This adjustment brought the Inform margin into the range of from 3 to 5.3 s.

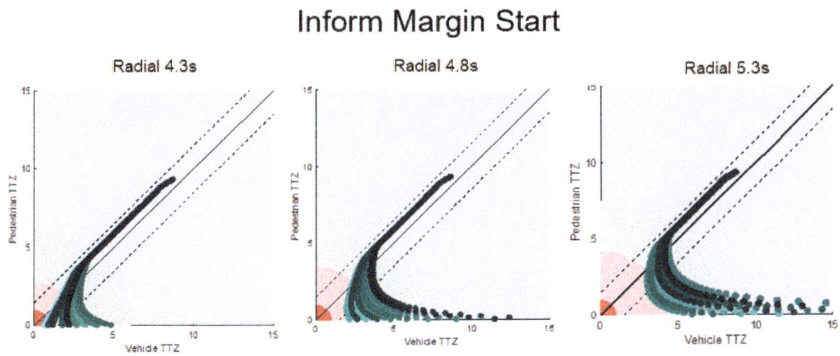

Figure 21. Simulation results for adjusted Inform margin (gray area) values. The color-coded severity areas are for the default values assigned before the Vissim simulation study. The gray area for the inform margin is adjusted afterwards so that the most conservative case on the right subplot is between 3 s and 5.3 s.

After the Inform margin is determined with both lower and upper threshold values, the Warn margin (pink area) does not need a range determination, since this becomes the area between the dynamically calculated Emergency margin and the constant Inform margin. The third and final severity margin is the red Emergency margin. In this case, the vehicle must perform an emergency braking to stop safely. The threshold value that is used will be dynamic. It will depend on the vehicle speed and distance. It should be computed at each step to issue a warning to the driver, if necessary. The minimum amount of deceleration a_{min} that is needed to stop the vehicle at a given distance from the conflict zone at the given vehicle speed can be calculated as:

$$a_{\min} = \frac{v_v^2}{2d_v} \quad (4)$$

When the minimum deceleration is below the braking deceleration limit, the driver is required to apply full braking to stop the vehicle. This will result in the red Emergency severity zone.

5. Simulation and Experimental Results and Discussion

This section discusses the evaluation and testing of the VRU safety warning system that was developed. The evaluations include pedestrian path prediction evaluation, Vissim simulations, and real-world testing.

5.1. Kalman Filter and LSTM Pedestrian Path Prediction Testing

As discussed previously, a system with LSTM neural networks in combination with Kalman filtering was developed for pedestrian path prediction and tracking. The test results and discussion are presented here, starting with the Kalman filter. The Kalman filter was first implemented in MATLAB and tested using real-world data recorded from an Android mobile device. Android location data from GPS, and acceleration data from its accelerometer sensor, were integrated together to achieve higher-resolution tracking. Since the mobile device GPS updates at 1 Hz, the use of the Kalman filter is highly beneficial for the calculations and for broadcasting messages, since the SAE J2735 standard messages are published at 10 Hz. Initial testing was conducted offline by replaying the recorded data through the algorithm. Some of the test results are plotted in Figure 22. GPS points are represented by blue circles, while Kalman filter updates are represented by red cross signs. As seen in Figure 22, the filter can successfully track the pedestrian at a higher frequency with fewer errors. Error calculations were also carried out to determine the accuracy of the algorithm, and the results were found to be satisfactory. The values are presented in Table 2 as the Average Displacement Error (ADE) in meters. After initial testing, the Kalman filtering algorithm was implemented in the Android device and was tested in a real-world setting. The results are plotted on top of a satellite image and are shown in Figure 23. As seen in the figure, the algorithm can provide higher-resolution location information compared to the Android GPS sensor. More importantly, it prevents sudden jumps in location, which can be seen in the top right corner of the plots.

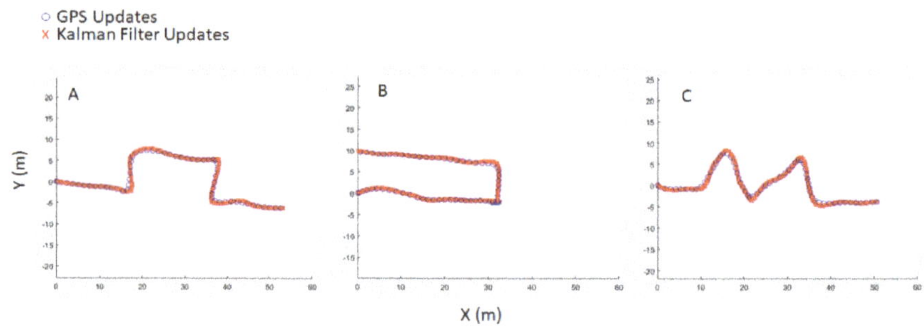

Figure 22. Kalman Filter testing with Android mobile phone data. A, B, and C correspond to different experiments.

Table 2. Tracking error for Kalman filter testing.

Test	ADE (m)
A	0.2866
B	0.2112
C	0.3109

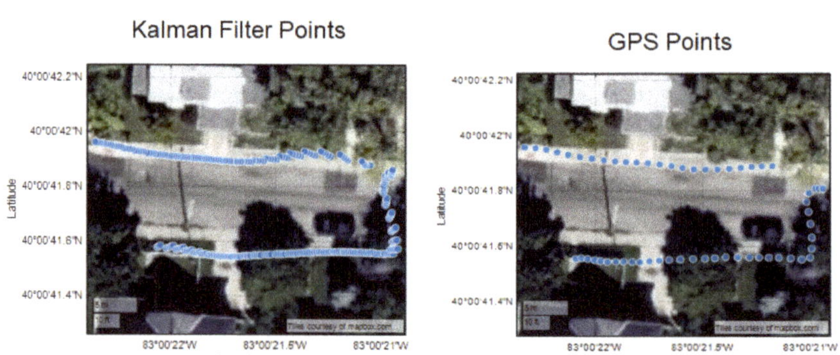

Figure 23. Kalman filter real-world testing with Android mobile device.

The results in Figure 23 show that future motion prediction has problems when the pedestrian makes sudden turns at a relatively high speed. A constant turn or another motion model may be more useful in those cases than a constant-velocity model during Kalman filtering. Errors in phone-GPS-based location also cause problems and will influence the *TTZ* computations, which may result in an inadequate warning. Tuning the

TTZ based warnings based on real data will help improve this. It is also expected that the localization accuracy of mobile phones will keep improving, meaning that this will be a smaller problem in the future. The possibility of these errors needs to be considered in real-world deployments.

The LSTM neural network for pedestrian future path prediction was trained and tested offline using data collected with an Android mobile device. Pedestrian location and speed data were collected and fed to the network to test 1 s future prediction first. The validation data were not included in the training dataset. The results for some of the tests are plotted in Figures 24 and 25. Collected pedestrian path test data are plotted as a blue line, while the neural network predictions for each step are plotted on top of these data as an orange line. The predictions of 1 s into the future are very accurate for both test cases: A and B. Along with the path prediction plots, the calculated prediction error is presented in Table 3. The error is smaller than that obtained with the Kalman filter.

Table 3. Prediction error for LSTM.

Test	ADE (m)
A	0.1132
B	0.1556

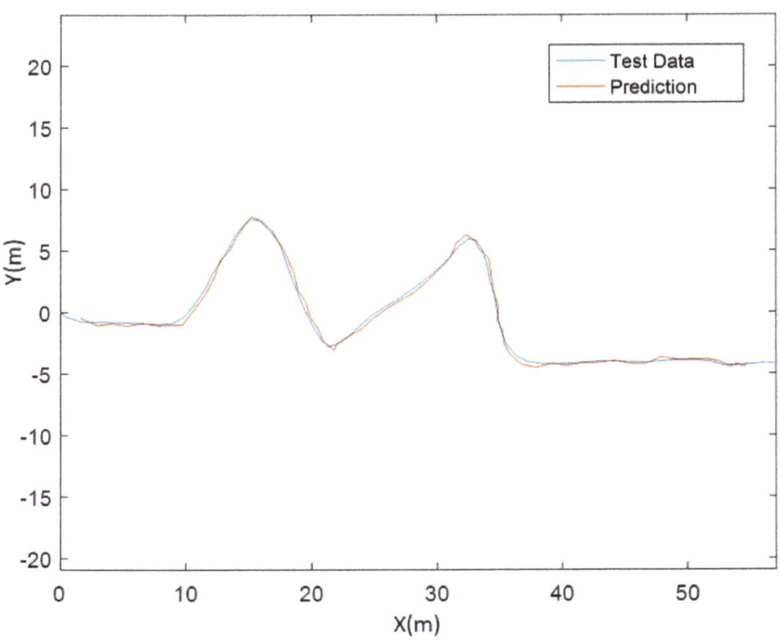

Figure 24. LSTM pedestrian path prediction test A.

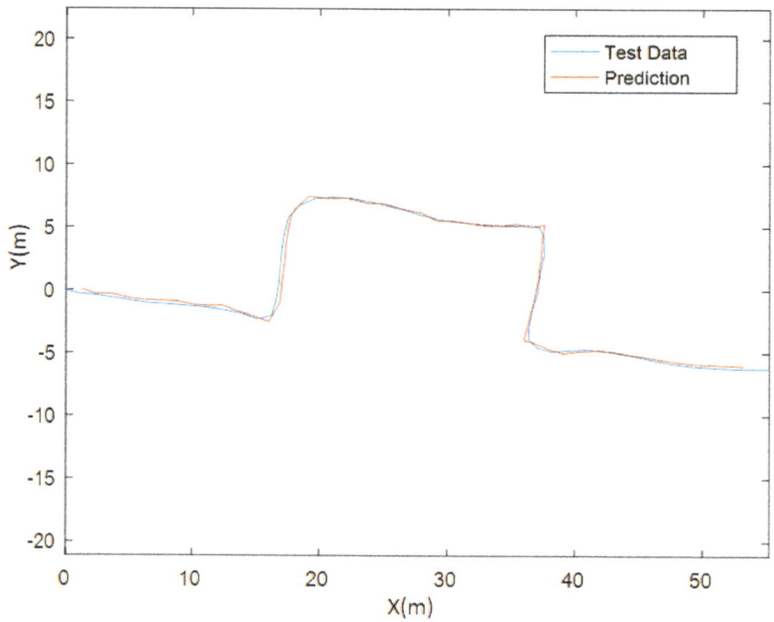

Figure 25. LSTM pedestrian path prediction test B.

5.2. Vissim Simulation Testing

A Vissim simulation environment was constructed with the base scenario geometry discussed above. As the vehicle speed is controllable in the simulation, several aspects of the driver warning design during slowing down and stopping were tested. First, speed information from real-world data was extracted and fed into the simulation to test the safe stop algorithm. The main idea of this simulation test was to see if the interactions obtained from the simulation were like the real-world interactions. Some differences were expected, since the design of the timings was more conservative, and deceleration was set to a very comfortable, constant, 0.15 g value in the simulations. Finally, the simulation structure was created with the flow illustrated in Figure 26. Vehicles start with a random speed profile selected from a random real-world vehicle in the data, where the Inform state is issued at 5.3 s. This is followed by a random attention lag. At the end, the vehicle starts to slow down and stops with 0.15 g deceleration.

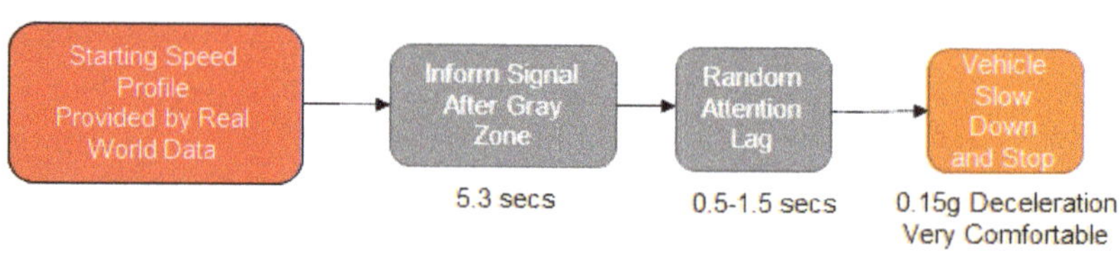

Figure 26. Simulation flow for Vissim interaction testing.

Real-world speed data, along with measurement noise, helped to ensure the observed behavior was similar to the real-world data-processing. Extracted interactions from the simulation results are shown in Figure 27. One main feature is obvious when we look at the graph on the left compared to the previously discussed real-world data displayed in Figure 17. Vehicle *TTZ* values do not come as close to zero as those in the graph, obtained from real-world data. This behavior indicates more conservative behavior. On the right side, in the heatmap graph, the area of dense points is like the real-world counterpart, while they are also farther from the origin point than the real data, indicating more conservative vehicle behavior. The more conservative but similar behavior observed in these simulations shows that the fundamental approach is correct. This method is useful for capturing the interactions, and can be modified to obtain a better fit later, if desired.

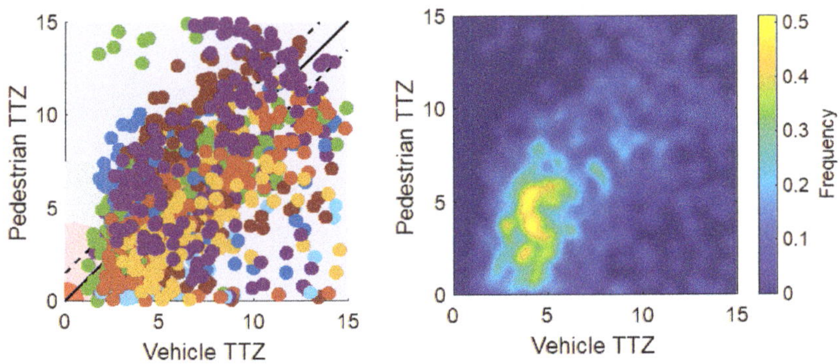

Figure 27. Vissim pedestrian-vehicle interaction testing.

The Vissim simulations and real-world data are qualitatively compared in Figures 17 and 27. The results were observed to be similar. Quantitatively, the vehicle–pedestrian interactions never entered the red emergency zone in Vissim, while this happened many times in the experimental data. The yellow area of high density in the real data showed that vehicles kept about 2 s of *TTZ* when a pedestrian was at the intersection. In contrast, Vissim vehicles kept a *TTZ* of about 4.5 s in the same situation, showing the conservativeness of the Vissim simulations. This should be adjusted in future work. It is also recommended to use real data in Vissim simulations in the future for the better replication of real-world experiments.

All three states of Inform, Warn and Emergency were enabled in the simulation to test the design parameters at different reaction times, as shown in Figure 28. The three different driver reaction times of 0.75 s, 1.25 s, and 2.5 s were tested with a random starting speed and random deceleration within a determined range, to see how reaction times affect the safe approach speed and what the limits are. The results are shown in Figure 29, where each graph shows multiple simulations with the vehicle starting at different speeds (y axis). The circles correspond to the severity state that caused the vehicle to stop. If the circle is gray, this means the vehicle comfortably stopped in the Inform state. If it is pink, stopping was a bit harder and the vehicle stopped in the Warn state. Lastly, if there is a red circle, this means the vehicle had to perform an emergency brake and stopped suddenly in order to not hit the pedestrian. With the current values for the design parameters, in terms of the maximum approach speed for safe stopping, without resorting to emergency braking, a 0.75 s reaction time results in approximately 100 km/h, a 1.25 s reaction time results in approximately 85 km/h, and a 2.5 s reaction time results in approximately 50 km/h when approaching the collision risk zone. Given that 1.25 s is a reasonable value for the average reaction time of a driver, the maximum approach speed for the current safe stop design is seen to be reasonable.

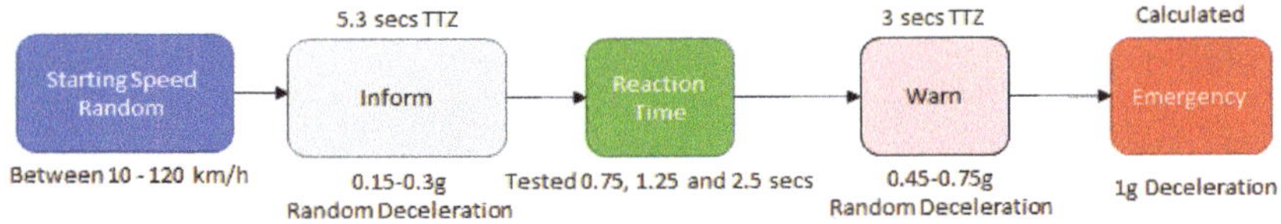

Figure 28. Simulation flow for Vissim driver reaction time testing.

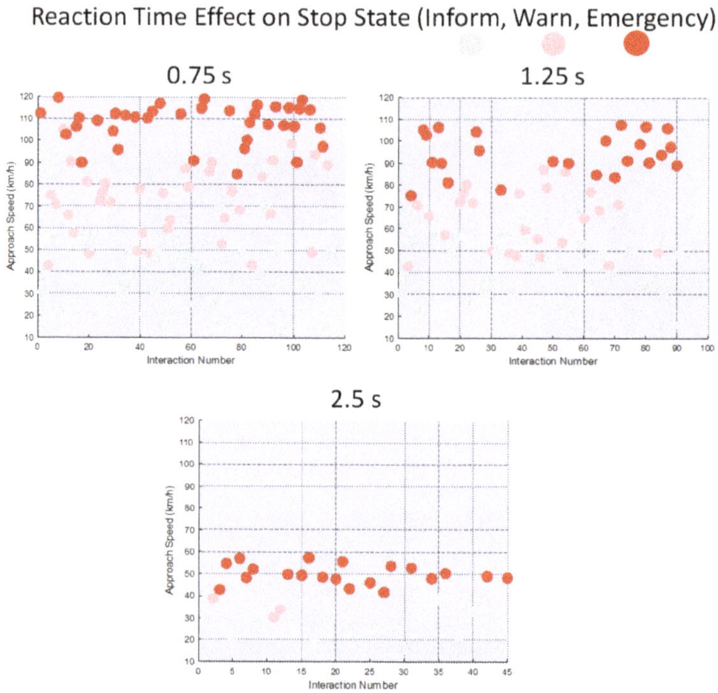

Figure 29. Simulation results for different driver reaction times.

5.3. Real-World Testing of Driver Warning System for Pedestrian Collision Avoidance

The mobile phone app was used to test the driver warnings in the real world. We created an NLOS situation at the uncontrolled intersection behind our lab building, illustrated in Figure 30, to demonstrate the efficacy of generating driver warnings for pedestrian collision avoidance. The blue arrow in Figure 30 shows the vehicle's motion towards the intersection while the green arrow shows the pedestrian's motion. The orange rectangle in Figure 30 shows the collision risk zone, which is used to calculate the *TTZ* values for the vehicle and the pedestrian. The vehicle had one phone and the pedestrian also had a similar phone, with both running the application during the testing. The app was used for communication between the phones, for driver warning computations, and for recording data. The forward-looking camera inside the vehicle was used to record the driver's view and the heads-up display (HUD), which was used to display the pedestrian safety warnings. Figure 31 shows different instances of the test, as recorded by the forward-looking camera.

The HUD is shown in Figure 31, with the phone screen facing upward and reflecting the driver warning generated by the app. The numbers from 1 to 4 correspond to different instances of the vehicle moving towards the intersection. The vehicle starts its forward motion while there is no warning, as shown in the first frame. The phone in the vehicle keeps receiving PSM over V2P, even during NLOS, as in frame 2, and a yellow-colored warning corresponding to the Inform state is displayed on the HUD. This indicates that there is a pedestrian approaching the intersection. The vehicle is closer to the intersection without slowing down in frame 3, when the warning becomes orange, corresponding to the Warn state. The vehicle stops in a safe manner as the crossing pedestrian is seen in frame 4. The HUD does not display a warning, as the vehicle has stopped, which successfully completes the real-world testing of the VRU safety application.

Figure 30. Real-world test scenario illustration.

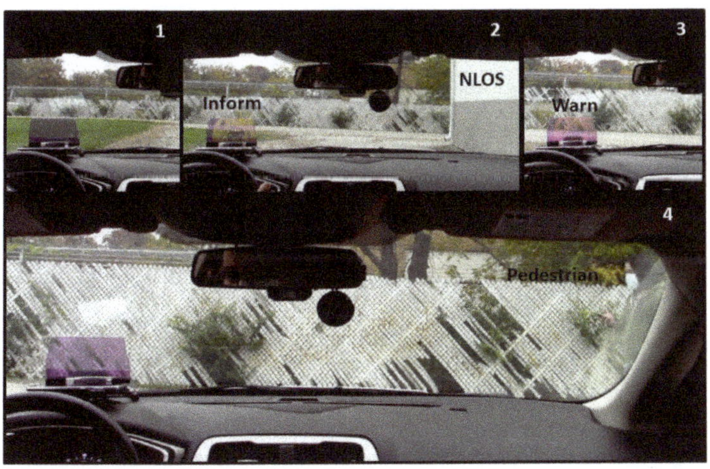

Figure 31. In vehicle snapshots recorded during testing. 1, 2, 3, and 4 show different snapshots as the vehicle approaches the intersection.

The data recorded by the phone app are summarized visually in Figure 32. The garage building forming the obstruction for NLOS is the gray rectangle in the middle of Figure 32. The pedestrian motion is shown with light blue circles. The vehicle positions are shown, with the color-coding corresponding to the warnings that were issued. Dark blue represents no warning. Yellow represents Inform and orange represents Warn. The vehicle starts with no warning while the pedestrian is in an NLOS position. The vehicle locations that are shown are its GPS measurements. The yellow warning is issued while the pedestrian is still behind the garage building. The orange warning is then issued as the vehicle moves closer to the collision zone without slowing down. The vehicle stops safely at the end, and warnings are disabled at the dark blue circle in Figure 32. The pedestrian continues their motion in a cross-path and passes in front of the stopped vehicle.

The experiment in Figure 32 had only one VRU, a pedestrian. Both mobile phones used, the one in the vehicle and the one held by the pedestrian, had some GPS localization errors, which were not large enough to affect the collision warning system in this experiment. VRUs carrying their mobile devices in their pockets have a shorter communication range, which would affect this NLOS experiment and reduce its performance. The vehicle would determine the presence of a dangerously moving pedestrian later in that case, and the driver would have to slow down faster. Multiple pedestrians and a combination of multiple pedestrians and bicyclists are also difficult situations as there will be more data. Currently, the VRU safety app calculates *TTZ*-based warnings simultaneously for each VRU, and the worst-case warning is used. Considering the combined dynamics of multiple VRUs may be useful in such cases to allow for the better prediction of future collision risk and reduce the computation time.

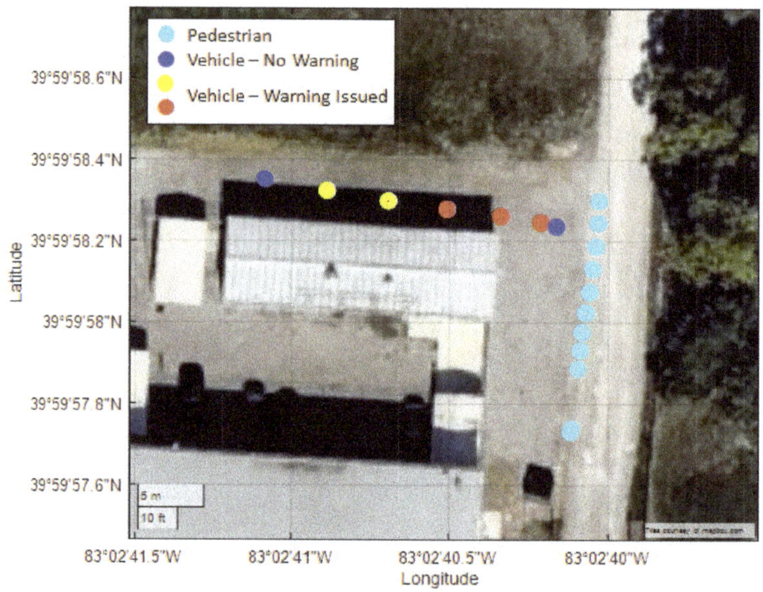

Figure 32. Visualization of communication and warning data during the testing.

The Human–Machine Interface (HMI) used in Figure 32 was finalized for our proof-of-concept testing after many iterations, using different choices to improve the attentiveness of the driver. It was finally decided to use a very simple flashing graphical warning with no text information. The experiment shown in Figure 32 and other similar experiments were repeated many times and the driver reaction times were seen to be adequate. Future pilots and larger-scale deployments will enable us to obtain a better understanding of driver reaction times.

6. Conclusions and Recommendations

A V2P-communication-based driver warning system for pedestrian safety was developed and evaluated in this paper. The paper also considered possible problems and developed methods to remedy these problems. These methods were evaluated in several ways, including realistic simulation environments and real-world testing. By utilizing mobile phone sensors and Bluetooth communication on the Android smartphones, a V2P safety application was developed to broadcast information for pedestrian motion while providing warnings to the human driver, when necessary, using the developed driver warning system. This application allows for easy access to a pedestrian safety system for public use. The developed application was also used for data collection from the sensors on the Android phone. With these sensors, the capabilities of the data collection application, and the virtual environments created in Vissim, neural network structures were trained and experimented for pedestrian path and behavior predictions. Kalman Filtering was also utilized for higher-frequency pedestrian motion-tracking.

One of the key contributions of the paper was the utilization of real-world data to create a baseline for driver safety warning system design parameters, with the goal of achieving more natural interaction behavior. In addition to this, some interesting observations were made while comparing several aspects of driver behavior, such as approach direction or the time of day. It was observed that drivers were more cautious while approaching an intersection with a pedestrian at times when sunlight is limited or not available. The data that were analyzed and used had a span of several months and included a large amount of BSM messages that were published for all vehicles and pedestrians approaching or crossing the street at the chosen smart intersection, as well as signal phase and timing (SPaT) messages.

Among the current limitations that need to be further explored in detail is how to treat crowds of pedestrians (VRUs). Although the way in which the application is currently designed to work will not create problems in terms of providing a warning when there are multiple pedestrians, problems may arise due to the highly increased processing load and congestion in the Bluetooth communication band. Another limitation of the system that should be mentioned is the GPS accuracy of the Android phones. Although the accuracy was satisfactory for the provision of reliable data and warnings in many data collection and warning system experiments, this might not always be the case. The dynamics of the vehicle [41] on a road may also affect GPS accuracy. However, the Kalman Filter in the current implementation helps with this issue, especially for irregular jumps in the GPS location. One of the expected improvements is dual-frequency GPS. As dual-frequency GPS smartphones become more common, the GPS accuracy issue will be reduced. On top of this, RTK GPS chips are getting smaller and cheaper over time, while providers are building stations to publish RTK correction messages over the internet for better accessibility. With the RTK GPS technology becoming more widely available, the GPS accuracy for small devices like smartphones will be significantly improved. This will help to successfully implement the V2P-communication-based pedestrian safety system presented here.

As the work reported here is a proof-of-concept, only Android phones with Bluetooth connectivity were considered. The authors experimented with the different phones available to them, ranging from relatively old and simple ones to more recent ones with more enhanced capabilities. The Android application was developed and checked to make sure that it worked on different versions of Android and for the different mobile phones that were available to the authors. For real-world deployments with many users, an iOS implementation of the app is also necessary, considering the large number of iPhone users. There are also many VRUs who do not have a mobile phone or who use basic mobile phones, on which the app cannot be installed. It is recommended to use low-cost wearable devices that only act as a beacon broadcasting the presence of the VRU in such cases. Nearby vehicles should also use their perception sensors to detect and track VRUs in cases where the VRU does not have a mobile phone, or where the app cannot be installed on their mobile phone.

Communication delays between the VRU and vehicle mobile phones (communication devices) are also an important factor that needs to be considered. Here, the communication occurred at a 10 Hz rate, and the communication delay in the experiments reported here and the other tests that the authors conducted showed that this was not a serious problem for the Vehicle-to-VRU-communication-based VRU safety approach presented in this paper. All PSM messages are time-stamped such that the communication delay value is known when the message is received on the vehicle side. The use of a Kalman filter and the neural-network-based prediction of the future motion of the VRU and vehicle also help in alleviating the effect of the communication delay.

While this paper focused on a driver warning system, a collision avoidance system with collision-free trajectory computations and trajectory-tracking control [3] are also needed and will be the focus of future work. The parameter-space-based robust control methods [42,43] can be used for robust trajectory-tracking control for a collision mitigation system to reject disturbances in the presence of model uncertainty. Future work can also focus on the fused use of V2P communication and AV perception sensors [44], and the use of socially acceptable VRU collision avoidance in low-speed campus-like deployments [45].

Future work should focus on improving the applicability of this Vehicle-to-VRU-communication-based VRU safety system. An iOS implementation of the app should also be developed for iPhone users. Pilot and larger-scale deployments and further data analyses are necessary to improve the app for practical use and user acceptance. It will then be possible to make the app freely available to users. Android Auto [46] and Apple CarPlay [47] allow for mobile phones to be connected to the displays of current vehicles. It will be useful to use this connectivity to display VRU safety warnings using the vehicle display, instead of using an HUD. It would be best to further improve this method for automated braking, and braking and steering, to establish a more comprehensive VRU safety system in the future. The addition of communication devices that can communicate with both the vehicles and the VRUs at intersections will be helpful to improve the range of communication and reduce latency problems due to multiple simultaneous actors. In addition to Bluetooth and direct and over-the-cloud internet connectivity, ultra-wide band (UWB) connectivity can also be used [48].

Author Contributions: Conceptualization, S.Y.G., B.A.-G. and L.G.; methodology, S.Y.G., B.A.-G. and L.G.; software, S.Y.G.; validation, S.Y.G.; formal analysis, S.Y.G.; investigation, S.Y.G., B.A.-G. and L.G.; resources, B.A.-G. and L.G.; data curation, S.Y.G.; writing—original draft preparation, S.Y.G.; writing—review and editing, B.A.-G. and L.G.; visualization, S.Y.G.; supervision, B.A.-G. and L.G.; project administration, B.A.-G. and L.G.; funding acquisition, B.A.-G. and L.G. All authors have read and agreed to the published version of the manuscript.

Funding: This research was partially funded by the Ford Alliance program (OSU0076).

Data Availability Statement: Data are contained within the article.

Acknowledgments: The authors thank the Automated Driving Lab at the Ohio State University. The authors thank Gopichandra Surnilla and Jay Parikh for insightful discussions on V2P communication and VRU safety. The authors thank the Smart Columbus project of the City of Columbus for making the intersection data in Marysville available through their Smart Columbus Operating System and the Public Service Department of the City of Marysville for their help in interpreting the data.

Conflicts of Interest: The authors declare no conflicts of interest.

References

1. *Early Estimate of Motor Vehicle Traffic Fatalities in 2018*; NHTSA: Washington, DC, USA, 2019.
2. Guvenc, L.; Guvenc, B.A.; Emirler, M.T. Connected and Autonomous Vehicles. In *Internet of Things and Data Analytics Handbook*; John Wiley & Sons, Ltd.: Hoboken, NJ, USA, 2017; pp. 581–595, ISBN 978-1-119-17360-1.
3. Guvenc, L.; Aksun-Guvenc, B.; Zhu, S.; Gelbal, S.Y. *Autonomous Road Vehicle Path Planning and Tracking Control*; Book Series on Control Systems Theory and Application; Wiley/IEEE Press: New York, NY, USA, 2021; ISBN 978-1-119-74794-9.
4. Gelbal, S.Y.; Aksun-Guvenc, B.; Guvenc, L. SmartShuttle: A Unified, Scalable and Replicable Approach to Connected and Automated Driving in a Smart City. In Proceedings of the Science of Smart City Operations and Platforms Engineering in Partnership with Global City Teams Challenge (SCOPE-GCTC) Workshop, Pittsburgh, PA, USA, 18–21 April 2017.
5. Bagheri, M.; Siekkinen, M.; Nurminen, J.K. Cloud-Based Pedestrian Road-Safety with Situation-Adaptive Energy-Efficient Communication. *IEEE Intell. Transp. Syst. Mag.* **2016**, *8*, 45–62. [CrossRef]
6. WHO—Advice for the Public. Available online: https://www.who.int/emergencies/diseases/novel-coronavirus-2019/advice-for-public#:~:text=Maintain%20at%20least%20a%201,The%20further%20away,%20the%20better (accessed on 1 November 2020).
7. Gandhi, T.; Trivedi, M.M. Pedestrian Protection Systems: Issues, Survey, and Challenges. *IEEE Trans. Intell. Transp. Syst.* **2007**, *8*, 413–430. [CrossRef]
8. Swanson, E.; Yanagisawa, M.; Najm, W.G.; Foderaro, F.; Azeredo, P. *Crash Avoidance Needs and Countermeasure Profiles for Safety Applications Based on Light-Vehicle-to-Pedestrian Communications*; National Highway Traffic Safety Administration (NTHSA): Washington, DC, USA, 2016.
9. Hickey, K. NYC Looks to Tech to Improve Pedestrian Safety. 2016. Available online: https://gcn.com/articles/2016/10/19/nyc-ped-sig.aspx (accessed on 10 December 2021).
10. Dozza, M.; Werneke, J. Introducing naturalistic cycling data: What factors influence bicyclists' safety in the real world? *Transp. Res. Part F Traffic Psychol. Behav.* **2014**, *24*, 83–91. [CrossRef]

11. Ma, Q.; Yang, H.; Mayhue, A.; Sun, Y.; Huang, Z.; Ma, Y. E-Scooter safety: The riding risk analysis based on mobile sensing data. *Accid. Anal. Prev.* **2021**, *151*, 105954. [CrossRef]
12. NTHSA. *Vehicle-To-Vehicle Communication Technology for Light Vehicles*; NHTSA: Washington, DC, USA, 2019.
13. Bae, J.-K.; Park, M.-C.; Yang, E.-J.; Seo, D.-W. Implementation and Performance Evaluation for DSRC-Based Vehicular Communication System. *IEEE Access* **2021**, *9*, 6878–6887. [CrossRef]
14. Gyawali, S.; Xu, S.; Qian, Y.; Hu, R.Q. Challenges and Solutions for Cellular Based V2X Communications. *IEEE Commun. Surv. Tutor.* **2021**, *23*, 222–255.
15. Moradi-Pari, E.; Tian, D.; Bahramgiri, M.; Rajab, S.; Bai, S. DSRC Versus LTE-V2X: Empirical Performance Analysis of Direct Vehicular Communication Technologies. *IEEE Trans. Intell. Transp. Syst.* **2023**, *24*, 4889–4903. [CrossRef]
16. Honda. Honda Demonstrates New "Smart Intersection" Technology that Enables Vehicles to Virtually See Through and around Buildings. 2018. Available online: https://hondanews.com/en-US/releases/honda-demonstrates-new-smart-intersection-technology-that-enables-vehicles-to-virtually-see-through-and-around-buildings (accessed on 10 December 2021).
17. Moradi-Pari, E.; Tian, D.; Mahjoub, H.N.; Bai, S. The Smart Intersection: A Solution to Early-Stage Vehicle-to-Everything Deployment. *IEEE Intell. Transp. Syst. Mag.* **2022**, *14*, 88–102. [CrossRef]
18. Merdrignac, P.; Shagdar, O.; Nashashibi, F. Fusion of Perception and V2P Communication Systems for the Safety of Vulnerable Road Users. *IEEE Trans. Intell. Transp. Syst.* **2017**, *18*, 1740–1751. [CrossRef]
19. Gentner, C.; Günther, D.; Kindt, P.H. Identifying the BLE Advertising Channel for Reliable Distance Estimation on Smartphones. *IEEE Access* **2022**, *10*, 9563–9575. [CrossRef]
20. Tsai, Y.-R.; Chen, Y.-C. Opportunistic Connectionless Undirected Information Dissemination Based on Bluetooth Low Energy Advertising Technology on Smartphones. *IEEE Access* **2021**, *9*, 155851–155860. [CrossRef]
21. Wu, M.; Ma, B.; Liu, Z.; Xiu, L.; Zhang, L. BLE-horn: A smartphone-based bluetooth low energy vehicle-to-pedestrian safety system. In Proceedings of the 2017 9th International Conference on Wireless Communications and Signal Processing (WCSP), Nanjing, China, 11–13 October 2017; pp. 1–6. [CrossRef]
22. Wang, Y.; Chen, Y.J.; Yang, J.; Gruteser, M.; Martin, R.P.; Liu, H.; Liu, L.; Karatas, C. Determining Driver Phone Use by Exploiting Smartphone Integrated Sensors. *IEEE Trans. Mob. Comput.* **2016**, *15*, 1965–1981. [CrossRef]
23. Ramah, S.-E.; Bouhoute, A.; Boubouh, K.; Berrada, I. One Step Further Towards Real-Time Driving Maneuver Recognition Using Phone Sensors. *IEEE Trans. Intell. Transp. Syst.* **2021**, *22*, 6599–6611. [CrossRef]
24. Google. Android Bluetooth Low Energy Documentation. Available online: https://developer.android.com/guide/topics/connectivity/bluetooth/ble-overview (accessed on 10 December 2021).
25. Warriach, E.U.; Witte, S. Approach for performance investigation of different Bluetooth modules and communication modes. In Proceedings of the 2008 4th International Conference on Emerging Technologies, Rawalpindi, Pakistan, 18–19 October 2008; pp. 167–171. [CrossRef]
26. Zhang, Y.; Wang, X.; Zhuo, K.; Jiao, W.; Yang, W. Research on pedestrian vehicle collision warning based on path prediction. In Proceedings of the 2023 7th International Conference on Transportation Information and Safety (ICTIS), Xi'an, China, 4–6 August 2023; pp. 1267–1272. [CrossRef]
27. Xiog, G.; Yang, T.; Li, M.; Zhang, Y.; Song, W.; Gong, J. A Novel V2X-based Pedestrian Collision Avoidance System and the Effects Analysis of Communication Delay and Packet Loss on Its application. In Proceedings of the 2018 IEEE International Conference on Vehicular Electronics and Safety (ICVES), Madrid, Spain, 12–14 September 2018; pp. 1–6. [CrossRef]
28. Ararat, O.; Guvenc, B.A. Development of a Collision Avoidance Algorithm Using Elastic Band Theory. *IFAC Proc. Vol.* **2008**, *41*, 8520–8525. [CrossRef]
29. Mínguez, R.Q.; Alonso, I.P.; Fernández-Llorca, D.; Sotelo, M.Á. Pedestrian Path, Pose, and Intention Prediction Through Gaussian Process Dynamical Models and Pedestrian Activity Recognition. *IEEE Trans. Intell. Transp. Syst.* **2019**, *20*, 1803–1814. [CrossRef]
30. Min, S.; Lee, J.-Y.; Jung, K.-D. Real-time Path Prediction and Grid-based Path Modeling Method. *Int. J. Appl. Eng. Res.* **2017**, *12*, 9997–10001.
31. Scholler, C.; Aravantinos, V.; Lay, F.; Knoll, A. What the Constant Velocity Model Can Teach Us About Pedestrian Motion Prediction. *IEEE Robot. Autom. Lett.* **2020**, *5*, 1696–1703. [CrossRef]
32. Alahi, A.; Goel, K.; Ramanathan, V.; Robicquet, A.; Fei-Fei, L.; Savarese, S. Social LSTM: Human Trajectory Prediction in Crowded Spaces. In Proceedings of the IEEE Conference on Computer Vision and Pattern Recognition (CVPR), Las Vegas, NV, USA, 27–30 June 2016.
33. Gupta, A.; Johnson, J.; Fei-Fei, L.; Savarese, S.; Alahi, A. Social GAN: Socially Acceptable Trajectories with Generative Adversarial Networks. In Proceedings of the IEEE/CVF Conference on Computer Vision and Pattern Recognition, Salt Lake City, UT, USA, 18–23 June 2018.
34. Xue, H.; Huynh, D.; Reynolds, M. Location-Velocity Attention for Pedestrian Trajectory Prediction. In Proceedings of the IEEE Winter Conference on Applications of Computer Vision (WACV), Waikoloa, HI, USA, 7–11 January 2019.
35. Sorenson, H.W. *Kalman Filtering: Theory and Application*; IEEE Press: Piscataway, NJ, USA, 1985.
36. nrf52840-DK Bluetooth Development Board. Nordic Semiconductor. Available online: https://www.nordicsemi.com/Products/Development-hardware/nrf52840-dk (accessed on 10 December 2021).
37. SAE. *J2735 Dedicated Short Range Communications (DSRC) Message Set Dictionary*; SAE: Warrendale, PA, USA, 2016.

38. WIZnet W5500 Ethernet Shield. Available online: https://www.wiznet.io/product-item/w5500-ethernet-shield/ (accessed on 10 December 2021).
39. Kulshrestha, T.; Niyogi, R.; Patel, D. A fast and scalable crowd sensing based trajectory tracking system. In Proceedings of the 2017 Tenth International Conference on Contemporary Computing (IC3), Noida, India, 10–12 August 2017; pp. 1–6. [CrossRef]
40. SCOS Database. Available online: https://www.smartcolumbusos.com/ (accessed on 10 December 2021).
41. Ozcan, D.; Sonmez, U.; Guvenc, L. Optimisation of the Nonlinear Suspension Characteristics of a Light Commercial Vehicle. *Int. J. Veh. Technol.* **2013**, *2013*, 562424.
42. Guvenc, L.; Aksun-Guvenc, B.; Demirel, B.; Emirler, M. *Control of Mechatronic Systems*; IET: London, UK, 2017.
43. Necipoglu, S.; Cebeci, S.A.; Has, Y.E.; Guvenc, L.; Basdogan, Ç. A Robust Repetitive Controller for Fast AFM Imaging. *IEEE Trans. Nanotechnol.* **2011**, *10*, 1074–1082. [CrossRef]
44. Baek, M.; Mun, J.; Kim, W.; Choi, D.; Yim, J.; Lee, S. Driving Environment Perception Based on the Fusion of Vehicular Wireless Communications and Automotive Remote Sensors. *Sensors* **2021**, *21*, 1860. [CrossRef]
45. Shiomi, M.; Zanlungo, F.; Hayashi, K.; Kanda, T. Towards a Socially Acceptable Collision Avoidance for a Mobile Robot Navigating Among Pedestrians Using a Pedestrian Model. *Int. J. Soc. Robot.* **2014**, *6*, 443–455. [CrossRef]
46. Android Auto. Available online: https://www.android.com/auto/ (accessed on 23 December 2023).
47. Apple, Car Keys and CarPlay, A Smarter Ride from Start to Finish. Available online: https://www.apple.com/ios/carplay/ (accessed on 23 December 2023).
48. Kuutti, S.; Fallah, S.; Katsaros, K.; Dianati, M.; Mccullough, F.; Mouzakitis, A. A Survey of the State-of-the-Art Localization Techniques and Their Potentials for Autonomous Vehicle Applications. *IEEE Internet Things J.* **2018**, *5*, 829–846. [CrossRef]

Disclaimer/Publisher's Note: The statements, opinions and data contained in all publications are solely those of the individual author(s) and contributor(s) and not of MDPI and/or the editor(s). MDPI and/or the editor(s) disclaim responsibility for any injury to people or property resulting from any ideas, methods, instructions or products referred to in the content.

electronics

MDPI

Article

Efficient Route Planning Using Temporal Reliance of Link Quality for Highway IoV Traffic Environment

Ritesh Yaduwanshi [1], Sushil Kumar [1], Arvind Kumar [2], Omprakash Kaiwartya [3,*], Deepti [2], Mohammad Aljaidi [4] and Jaime Lloret [5]

1 School of Computer & Systems Sciences, Jawaharlal Nehru University, New Delhi 110067, India
2 School of Computing Science & Engineering, Galgotias University, Greater Noida 203201, India
3 Department of Computer Science, Nottingham Trent University, Clifton Lane, Nottingham NG11 8NS, UK
4 Department of Computer Science, Faculty of Information Technology, Zarqa University, Zarqa 13110, Jordan
5 Department of Communications, Universitat Politecnica de Valencia, Camino Vera s/n, 46022 Valencia, Spain
* Correspondence: omprakash.kaiwartya@ntu.ac.uk

Abstract: Intermittently connected vehicular networks, terrain of the highway, and high mobility of the vehicles are the main critical constraints of highway IoV (Internet of Vehicles) traffic environment. These cause GPS outage problem and the existence of short-lived wireless mobile links that reduce the performance of designed routing approaches. Nevertheless, geographic routing has attracted a lot of attention from researchers as a potential means of accurate and efficient information delivery. Various distance-based routing protocols have been proposed in the literature, with an emphasis on restricting the forwarding area to the next forwarding vehicle. Many of these protocols have issues with significant one-hop link disconnection, long end-to-end delays, and low throughput even at normal vehicle speeds in high-vehicular density environments due to frequently interrupted wireless links. In this paper, an efficient geocast routing (EGR) approach for highway IoV–traffic environment considering the shadowing fading condition is proposed. In EGR, a geometrical localization for GPS outage problem and a temporal link quality estimation model considering underlying vehicular movement have been proposed. Geocast routing to select a next forwarding vehicle from forward region by utilizing temporal link quality is proposed for four different scenarios. To evaluate the effectiveness and scalability of EGR, a comparative performance evaluation based on simulations has been performed. It is clear from the analysis of the results that EGR performs better than state-of-the-art approaches in highway traffic environment in terms of handling the problem of wireless communication link breakage and throughput, as well as ensuring the faster delivery of the messages.

Keywords: link quality; link lifetime; connected vehicle; routing; highway traffic

1. Introduction

The vehicular ad hoc network (VANET), which connects vehicles to pedestrians, parking areas, roadside equipment, and city infrastructure, is being augmented by the notion of the Internet of Vehicles (IoV) [1]. IoV incorporates many significant Internet of Things elements to deliver a wide range of new services that are beneficial to people and society. IoV vehicles are great things that work together to share information and communicate with one another via the Internet. Data dissemination, traffic management, cost administrations, and location-based services are all important IoV applications. The provision of connected vehicles to prevent accidents, lower vehicle assemblage when partnering accidents, and offer alarms related to street and crossing point status is one of the most crucial applications of IoV. Underlying VANETs are now essential for offering entertainment and safety applications due to the co-evolution of standards like IEEE 802.11p and IEEE 1609 that facilitate Wireless Access in a Vehicular Environment (WAVE) [2]. The major characteristics that set it apart from other mobile ad hoc network are its high vehicle speed, dynamic mobility, and random topology.

The most notable characteristic of VANETs is the high node mobility, which underlies a number of VANET-specific characteristics that demand the provision of appropriate solutions [3,4]. All mobile networks have the inherent property of temporal connectivity caused by node mobility, which is much more apparent in the case of vehicular communications. Due to frequent communication link breakdowns between vehicles and poor performance, this poses serious issues. The topology continuously changes in a number of ways. In more detail, these issues make designing routing for VANETs very complex [5]. The changing aspects involved suggest that the actual connectivity abilities change quickly, so gathered routing data quickly become outdated and the formed communication routes quickly become invalid. Route reconstruction requires a huge amount of network resources, and the consequent interruption of information flow results in significant delays. Link quality has been accepted as a crucial performance component for VANETs by recent study. The frequency of route reconstructions will increase if the established routes are transient. There is no assurance that the new route will not break as soon as possible after it is established, leading to a series of subsequent route reconstructions. Authors addressed this issue in [6], suggested that VANETs have a relatively small network diameter due to the fact that many routes break down before they can be used. Authors in [7] suggest using the dissemination of GPS information to determine the desired link life span and to illustrate substantial performance improvement when the routing method tentatively involves link quality into consideration. This is important because it allows for the construction of long-lived routes and mitigates the occurrence of frequent communication interruptions.

In recent times, location-oriented vehicular communication, also referred to as geocasting, has attracted a lot of attention due to the fact that it may be applied to a wide variety of ITS applications [6]. The location of vehicles as determined by GPS has been put to use in a number of different applications, including the measurement of the prediction of collisions at intersections [7], vehicle density on roadways [8], and the determination of the location of roadside units [9]. When using GPS to localize a vehicle, one must implicitly assume the non-disruptive availability of the GPS signal of acceptable quality for information dissemination in VANETs [10]. Due to increasing urbanization, including high-rise structures, multi-level flyovers, and highway bridges, GPS outages are becoming a significant challenge [11]. If a vehicle experiences a GPS outage in a specific highway traffic environment, how can other vehicles in the same environment determine its GPS location? As a result, if the sending vehicle experiences GPS outage, its neighboring vehicles will also experience GPS outage [12]. To address this problem, we present a geometrical localization technique in this paper.

Geographic forwarding protocols choose the shortest path, which could result in a greater packet error rate due to the poor quality of a link. The unequal distribution of vehicles on the highways makes it more difficult to select a route; for example, the shortest path in terms of geographic distance may experience more frequent path disconnections even though it is the shortest path. It is possible for there to be frequent communication disconnections between vehicles due to the dynamic and fast changing topology of vehicular networks [13]. There are a number of geographic routing protocols available that are based on GPS location or geography and quality of one link connection [13–17]. Junction-based Geographic Distance Routing (J-GEDIR) [13] using greedy forwarding with a bias toward the node closest to the destination at a junction is suggested. Peripheral-node-based Geographic Distance Routing (P-GEDIR) [14] is proposed to select the next forwarding vehicle from a forwarding region. An important part of the P-GEDIR is the idea of a "peripheral node," or a node far from the sender that is selected by the next forwarding node. The next forwarding vehicle in the Voronoi region approaching the destination is chosen using Voronoi-diagram-based Geographic Distance Routing (V-GEDIR) [15]. In P-GEDIR, the border nodes are chosen as the next forwarding vehicle, and the coverage area is partitioned vertically toward the destination. These protocols [13–15] result in a minimal number of next forwarding vehicles being selected since they limit the size of the forwarding zone. They begin to work poorly if there are many vehicles in the congested reduced forwarding area. In addition, these protocols choose a next forwarding vehicle at the border to reduce the overall hop count. But because of their greater speed, transmission problems are more likely to occur in border vehicles. The quality of a hop connection link is significantly diminished by nearby impediments in the highway traffic setting, such as buildings and trees. All the above mentioned protocols disregarded this factor. To the best of our knowledge, a routing technique has not been developed considering specific connection links that proactively adapt to a topology that is continually changing due to vehicle mobility; instead, it has only used physical-layer-related information to provide criteria for the selection of the next forwarding vehicle.

In this context, efficient geocast routing using the temporal dependence of link quality for highway IoV–traffic environment is proposed. More specifically, a highway traffic mobility model to find locations of the vehicles where GPS outage problem occurred and a temporal link quality estimation model are used to assist the EGR to select the next forwarding vehicle from the forwarding region. An analytical framework that measures the probability of precisely selecting the longest-lasting link among a set of specified links between a sending vehicle and its neighboring vehicles is presented. The main contribution of the paper is listed as follows:

(1) Firstly, geometrical localization for GPS outage is presented to know the GPS locations of the vehicles.
(2) Secondly, a temporal link quality estimation model is proposed to capture the tendency of future link quality and residual link lifetime for the underlying structure for the vehicular traffic mobility model for highways.
(3) Thirdly, a new kind of geocast routing is proposed using geometrical localization, temporal transmission probability, and residual link lifetime under shadowing fading condition for a highway traffic environment.
(4) Finally, to show the benefit of the proposed routing, a comparison is made with state-of-the-art works in terms of one-hop link disconnection, throughput, and delay.

The remainder of this paper is organized in the following manner. The most recent geocast routing methods used in VANETs are reviewed in Section 2. The proposed EGR approach for a highway traffic environment is presented in Section 3. Section 4 presents the simulation's results. The conclusion and future work are explained in Section 5.

2. Related Work

This section provides a review of geocast routing protocols. Link quality is considered one of the key factors in geocast routing that has a substantial impact on the performance of these routing approaches. We categorize the existing geocast routing approaches into two subsections based on link quality.

2.1. GPS-Location-Based Geocast Routing without Link Quality

Based on the geographical information of indented vehicles for packet dispersal, a hybrid flooding technique has been proposed [18]. In dynamic vehicular settings, with the exception of those with a higher node density, a hybrid technique that was created by combining the position-aware technique with the counter-based technique improves message dependability. Two forwarding algorithms are used in CTRC-based geographic routing: area sending for hop-to-multihop information transmission and line forwarding for hop-to-hop information transmission [19]. The transmission range is decreased with range forwarding. Because of this, the impact of dynamic mobility patterns and vehicle node speeds on the ratio of data delivery is considerably diminished. Additionally, the hop count is greatly increased. It has been proposed that border node most forward within a radius can decrease end-to-end delay in VANETs [20]. The ideal route with the highest chance of connecting to the network is selected using connectivity-aware routing (CAR) [21], which is derived by a probabilistic model employing information from statistical traffic statistics. However, connectivity may be impacted by the node density calculation's error. In [22], the accuracy of CAR was significantly enhanced. Using real-time node density data produced by an on-the-fly density collection approach, the best link connection for the route is identified in this study. These studies mentioned above overlook the analysis of connectivity while taking junction nodes into account. Cache is utilized in Cache-Agent-based Geocast (CAG) routing in [23] to solve the packet reject issue. A connectivity assertion mechanism was put forth in CAG to guarantee that the cached packet would not be deleted. It is woefully inadequate to only take into account node density when evaluating connectivity in geocast routing. Because, despite the high node density in urban VANETs, communication can be disrupted in some circumstances due to the presence of traffic signals. Junction-based Geographic Distance Routing (J-GEDIR) [14] and peripheral-node-based Geographic Distance Routing (P-GEDIR) [15] are the two most popular and recent geographic routing protocols. The next hop cars are chosen by J-GEDIR in junction nodes within the transmission range. P-GEDIR chooses all border nodes as next-hop vehicles from the half-circular section moving toward the destination after vertically dividing the circular coverage zone into two half-circular portions. By reducing the forwarding region's size, these protocols can choose fewer next-hop vehicles.

All of the aforementioned geocast routing approaches considered either flooding techniques or vehicle density for routing purposes. The flooding in multi-hop routing contribute to an increase in network overheads. In addition, these routing approaches often experienced high path disconnection, and their efficiency began to degrade with increasing vehicle speed, making them unsuitable for high traffic environments. In highway, the quality of the link is severely affected by the obstacles of the surrounding, which can include buildings, trees, and other natural features. These protocols and its alternative neglected to take this factor into account.

2.2. GPS-Location-Based Geocast Routing with Link Quality

The impact of radio channel propagation, such as noise, fading, shadowing, etc. on one-hop link connection is one of the crucial issues to consider while studying geocast routing in VANETs. An approach for geocasting [24] in VANETs that uses multiple metrics to choose the next-hop vehicle has been proposed by the authors. The suggested technique makes use of metrics that are based on links and vehicle density to search a reliable next forwarding vehicle for message dissemination in the geocast region. Intermittent Geocast Routing (IGR) [25] in urban VDTNs with storage-carry-forward mode. The IGR operates in two modes: intersection mode and road segment mode. In the intersection mode, the network connectivity characteristic is studied to determine the probability of connectedness for any road segment. In the following hop, the maximum connected road stretch is chosen. Then, on the selected road segment, the vehicular nodes estimate the effective connection interval based on the real road and vehicle circumstances.

To enhance the geographical routing strategy over VANET, a prediction-based geographical routing (PGR) approach is suggested using link quality measure. It reconsidered the forwarding and repair procedures of the routing approach given in [26]. R. Kasana et al. [27] performed an analysis of the precision of the location information of vehicles, which is dependent on the effectiveness and scalability of geographical routing. Through the use of GPS or other positioning technologies, the vehicle is able to control its own location. The geographical routing takes into consideration the precise location information implicitly. However, taking into account the GPS's accuracy limitations and the fact that signals might be blocked by the environments along roadways, this assumption is not reasonable. The erroneous location information is the root cause of the performance issues that occur as a result of using geographical routing methods. Specifically, Oliveira et al. [28]. demonstrate that inaccuracies in location during geographic routing (in wireless sensor networks) can lead to poor performance or even total failures. In geographic routing, it is necessary for each node to have information regarding its present position, which can be obtained by the use of GPS or another type of localization technique (Qureshi et al. [29]). In wireless sensor networks, one of the most crucial tasks to complete is the installation of a localization system. As a result of the geographic correlation of the sensed data, location information is frequently used to name the obtained data, address nodes, and regions and also to improve the performance of a wide variety of geographic algorithms. GPS-location-based Link quality and Degree of connectivity based Geographic Distance Routing (SLD-GEDIR) [17] is proposed, selecting the most dependable forwarding vehicle by taking into consideration one-hop link quality.

Nevertheless, the aforementioned routing approaches do not take into account dynamic vehicles movement patterns while considering one-hop link quality as one of the routing metrics, and these routing approaches are not suitable under a vehicular network scenario with intermittent connectivity under sparse vehicular networks and where GPS outages occur. However, in a VANETs scenario, the network delay and overhead increase as these protocols use more routing metrics, which in turn comes with higher process complexities.

3. The Proposed EGR Approach for Highway Traffic Environment

In this section, first, we present geometrical localization for GPS outage and the temporal locations of the moving vehicles that are facing GPS outage problem and its environment on a highway. Second, the residual link lifetime estimation model to capture the tendency of future link quality and predict the future lifetime of the link between vehicles is presented. Last, the proposed geocast routing, both geometrical localization and link quality for a highway traffic environment, is presented.

3.1. Highway Traffic Environment

Consider a highways environment where each road intersects the others at a right angle; this intersection is known as a junction point (see Figure 1). There are road side units (RSUs) to assist the transmission of the messages at every intersection. To ascertain the geographical properties of moving vehicles, GPS location service equipment is integrated into every vehicle. Each vehicle should have an onboard navigation system to pinpoint the precise location of the closest traffic intersections. Dedicated short-range communication (DSRC) is utilized to optimize the communicational pattern. We further suppose that the network area consists of a large number of virtual junctions. Due to the lack of next-hop vehicles that can send the packets to their destinations, the sender vehicles hold the forwarded packets in their caches. The buildings and trees are located in the side of the highway, and the highways pass through different types of terrain. The shadowing fading is considered that hinders the quality of links between the vehicles due to the presence of trees and buildings on the sides of the highway and the terrain types of the highway. In addition, increasing urbanization leads to GPS outage problem.

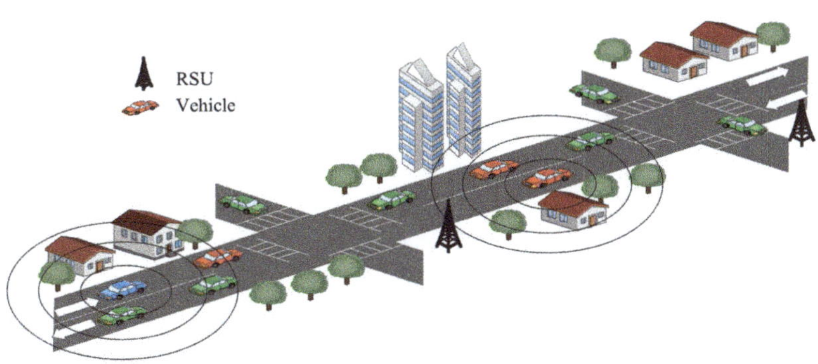

Figure 1. A highways traffic Environment.

3.2. Geometrical Localization for GPS Outage

The majority of the GPS-assisted localization algorithms assume the availability of a sufficient number of location-aware neighboring vehicles for the purpose of routing. In a realistic highway traffic environment, the assumption, which include GPS-capable surrounding vehicles is not feasible for routing purposes. When a vehicle's GPS signal is blocked in certain highway traffic environments, how are its neighbors in those environments supposed to know where they are relative to their own GPS signals? To solve this problem, we present geometrical localization to compute the locations of moving vehicles that are facing a GPS outage problem.

Every vehicle is capable of measuring speed, direction, and time with a degree of precision that is adequate. In the event that the GPS in a vehicle fails to work, an EGR system will allow the vehicle to estimate its location using geometrical localization. We consider two vehicles N_i and N_j that are moving on a straight road segment (see Figure 2). The maximum transmission radius of a vehicle in a perfect environment is assumed to be R. Its values for different types of vehicles is not equal. Let (x_i, y_i) and (x_j, y_j) represent the position of the vehicles N_i and N_j at time t respectively. In addition, let v_i and v_j represent the velocity of the vehicles and the direction of movement for both the vehicles N_i and N_j are θ_i and θ_j respectively. After a short while, if the vehicle experiences a GPS outage issue on the same road segment, the unknown position vectors of the vehicles N_i and N_j at time t' are (x_i', y_i') and (x_j', y_j'), respectively, and the movement direction and velocities for both vehicles may change when needed. Using the geometrical rule, the unknown positions of the vehicles in case of GPS outage can calculated as follows. The position for both the vehicles at time t' can be expressed as

$$\left.\begin{array}{l} x_k' = x_k + v_k(t' - t)\cos\theta_k \\ y_k' = y_k + v_k(t' - t)\sin\theta_k \end{array}\right\} \tag{1}$$

where $k = \{i, j\}$. The temporal distance between both vehicles at time t' can be expressed

$$d(t') = \sqrt{(\Delta x + \Delta v \Delta t \Delta \cos\theta)^2 + (\Delta y + \Delta v \Delta t \Delta \sin\theta)^2} \tag{2}$$

where $\Delta x = (x_i - x_j)$, $\Delta v = (v_i - v_j)$, $\Delta\cos\theta = (\cos\theta_i - \cos\theta_j)$, $\Delta\sin\theta = (\sin\theta_i - \sin\theta_j)$, and $\Delta t = (t' - t)$.

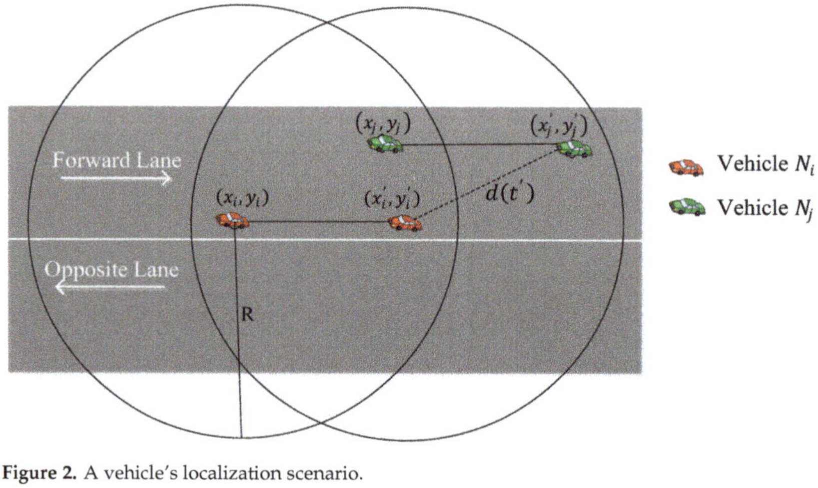

Figure 2. A vehicle's localization scenario.

3.3. Temporal Link Quality Estimation Model

In this section, we first calculate the link quality (temporal transmission probability) of each radio link between a sending vehicles and its neighboring vehicles that are moving in the same direction. Second, we estimate the remaining link lifetime of a specified link between a sending vehicle and its neighboring vehicles.

(*a*) *Temporal Transmission Probability*

We assume that there exists a sending vehicle N_i and a receiving vehicle N_j in the transmission range of the sender. Since the radio transmission range is affected by natural phenomena such as shadowing, multipath fading, noise, etc.; thus, the vehicle's radio transmission range is considered to be non-uniform in all directions. In the proposed model, we considered only the shadowing fading which is occurring due to the presence of some large transport vehicles or any other obstruction between vehicles N_i and N_j. Let vehicle N_i wish to send some packets to the neighboring vehicle N_j. The temporal received power $P_{j,R}(d(t'))$ at vehicle N_j at time t' using (2) can be expressed as

$$P_{j,R}\left(d(t')\right) = P_{i,T} - \overline{P}(d_0) - 10\eta log_{10}\left(\frac{\sqrt{(\Delta x + \Delta v \Delta t \cos\theta)^2 + (\Delta y + \Delta v \Delta t \sin\theta)^2}}{d_0}\right) + \chi_\sigma \quad (3)$$

where $P_{i,T}$ denotes the transmit power of the vehicle N_i; $\overline{P}(d_0)$ denotes the mean path loss at reference distance d_0, where the reference distance is generally assumed to be one; η is the path loss factor depending on the condition of the environment. The χ_σ denotes different levels of cutter in the received power due to shadowing fading, and it is assumed to be a Gaussian random variable with zero mean and σ^2 variance. The received threshold power of the vehicle N_j with the maximum transmission range R in a perfect environment can be expressed as

$$P_{j,R}\ (R) = P_{i,T} - \overline{P}(d_0) - 10\eta log_{10}\left(\frac{R}{d_0}\right) \quad (4)$$

The probability of successful transmission $\mathcal{P}_{i,j}(d(t'))$ between the sending vehicle N_i and the receiving vehicle N_j at time t' can be expressed as

$$\mathcal{P}_{i,j}(d(t')) = \mathcal{P}\left(P_{j,R}\ (d(t')) > P_{j,R}\ (R)\right) \quad (5)$$

$$= \mathcal{P}\left(\chi_\sigma > 10\eta log_{10}\left(\frac{\sqrt{(\Delta x + \Delta v \Delta t \Delta \cos\theta)^2 + (\Delta y + \Delta v \Delta t \Delta \sin\theta)^2}}{R}\right)\right) \quad (6)$$

where $\mathcal{P}(.)$ represents the probability function. The temporal transmission probability is given by

$$\mathcal{P}_{i,j}(d(t')) = \int_{\frac{10\eta log_{10}\left(\frac{\sqrt{(\Delta x+\Delta v\Delta t\Delta\cos\theta)^2+(\Delta y+\Delta v\Delta t\Delta\sin\theta)^2}}{R}\right)}{\sigma}}^{\infty} \frac{1}{\sqrt{2\pi}\,\sigma} e^{-\frac{x^2}{2\sigma^2}}\,dx \quad (7)$$

$$= Q\left(\frac{10\eta log_{10}\left(\frac{\sqrt{(\Delta x+\Delta v\Delta t\Delta\cos\theta)^2+(\Delta y+\Delta v\Delta t\Delta\sin\theta)^2}}{R}\right)}{\sigma}\right) \quad (8)$$

where $Q(u) = \int_u^\infty \frac{1}{\sqrt{2\pi}} e^{-\frac{t^2}{2}}\,dt$, which evaluates the tail probability of Gaussian distribution.

(*b*) *The Residual Link-Lifetime*

When using the distant vehicle for the purpose of routing, the measuring of the performance of the proposed routing does not achieve accuracy; thus, correctly finding a link between the sending vehicle and distant vehicles that will last longer is the purpose of this section. The residual link lifetime is the forthcoming time period in which the link remains active. The purpose of estimating the residual link lifetime during which an existing link between moving vehicles will continue to function properly under shadowing fading and to demonstrate how such a technique may be incorporated into a routing approach to achieve the required aim of the guaranteed delivery of the message with geocast routing for a highway traffic environment with links experiencing the adverse effect of shadowing fading. Let two vehicles N_i and N_j be connected at a time t'. The forthcoming temporal distance for both the vehicles remaining connected is $R - d(t')$. The uncertainty in the link-lifetime estimates can be introduced by temporal transmission probability $\mathcal{P}_{i,j}(d(t'))$, which represents variation in the link. The residual link-lifetime (T_{ll}) can be computed as

$$T_{ll} = \frac{R - d(t')}{|\Delta v|} \mathcal{P}_{i,j}(d(t')) \tag{9}$$

3.4. Geocast Routing

A description of the routing process of the proposed protocol is explained in this section. In a highway IoV traffic environment, the vehicles are intermittently connected and they show high-speed vehicle mobility. In addition, natural phenomena, such as shadowing fading, influence connectivity. These constraints make the routing a challenging task in a highway traffic environment. All of the individual links that make up the route for hop-by-hop message forwarding must remain accessible for the duration of the remaining link lifetime. Due to the special characteristics of the highway, we are forced to incorporate residual link lifetime and caching methodology into the proposed EGR.

Considering these critical issues, we present the route construction approach using a scenario depicted in Figure 3 in which a sending vehicle N_i estimates the temporal residual link-lifetime using Equation (9) for each vehicle N_j moving in a forward direction and are in the forward transmission region AOB (FTA) and transmission ranges of sending vehicle N_i. Herein, we define two useful functions *argmax*(U) and *argmin*(U) that are used in the following section.

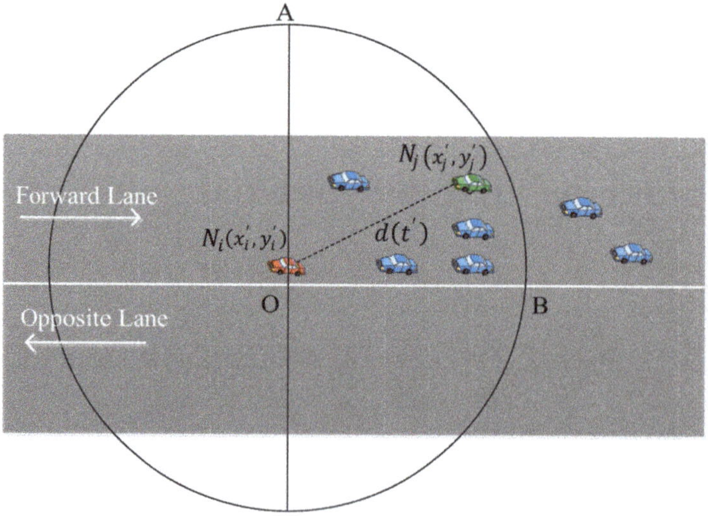

Figure 3. Selection of a vehicle for transmission.

Definition 1. *argmax(U) is the set of neighburing vehicles belonging in a FTA for which U attains the maximum value.*

Definition 2. *argmin(U) is the set of neighboring vehicles belonging in a FTA for which U attains the minimum value.*

Case 1: When vehicles present in FTA of forward lane

Let t_{tx}, t_{pr}, and t_{po} denote the transmission time of a packet on the link, the propagation time of the packet that travels via the link from the locations (x_i', y_i') and (x_j', y_j') of the vehicle N_i and vehicle N_j at time t', and the processing time of the packet at the receiving vehicle. Each packet is acknowledged to confirm its guaranteed delivery. The round trip time t_{rt}, in which a packet delivered successfully is the sum of transmission time, twice the propagation time, and the processing time. It can be given by

$$t_{rt} = t_{tx} + 2 \times t_{pr} + t_{po} \tag{10}$$

For routing purpose, the sending vehicle N_i chooses a vehicle N_j among neighboring vehicles that lay in its forward region that should satisfy the following condition:

1. The residual link lifetime T_{ll} of the radio link between both the vehicles is greater than the round trip time t_{rt}.
2. The temporal distance between the sending vehicle N_i and vehicle N_j should be high.

When the sending vehicle obtains temporal residual link-lifetime and temporal distance information about its neighboring vehicles, the sending vehicle N_i selects a neighboring vehicle with maximum $d(t')$ as the next forwarding vehicle such that

$$\left.\begin{array}{c} N_j = argmax\big(d(t') \times \mathcal{P}_{i,j}(d(t'))\big) \\ st.\ T_{ll} - t_{rt} > 0 \end{array}\right\} \tag{11}$$

where, herein, a vehicle with a maximum value of $d(t') \times \mathcal{P}_{i,j}(d(t'))$ is retuned by the *argmax() function*. Choosing a vehicle N_i and N_j using the above logic gives a radio link between moving vehicles that will continue to be useful for satisfactory packet transmission.

Case 2: When vehicles present in FTA in the opposite lane

Opposite-lane vehicles are also considered to participate in the routing process, as they can reduce the delay. This is done by a source vehicle that transmits packets to a relay vehicle that is part of its transmission range and which is moving in a backward direction relative to the sending vehicle. As shown in Figure 4, for routing purposes, the sending vehicle N_i did not find any neighboring vehicles in its forward transmission region AOB. In such a situation, it selects a vehicle N_{i+1} that belongs to the forward region BOC of the opposite lane and a vehicle N_{i+1} that has at least one neighboring vehicle N_j that is moving in a forward lane towards the destination. The round trip time in this case is equal to $2t_{rt}$.

For routing purpose, the sending vehicle N_i chooses a vehicle N_{i+1} among neighboring vehicles that are moving in the opposite lane and that should satisfy the following condition:

1. The product of the temporal distance and temporal transmission probability between the sending vehicle N_i and vehicle N_{i+1} should be high.
2. The vehicle N_{i+1} has at least one neighboring vehicle that is moving in a forward lane.

The sending vehicle N_i selects a vehicle N_j that is moving in a forward lane, using a neighboring vehicle N_{i+1} such that

$$\left.\begin{array}{l} N_{i+1} = argmax\big(d(t') \times \mathcal{P}_{i,j}(d(t'))\big) \\ N_j = argmin\big(d(t') \times \mathcal{P}_{i+1,j}(d(t'))\big) \\ \left(\frac{R-d(t')}{(v_i+v_{i+1})}\mathcal{P}_{i,i+1}(d(t')) + \frac{R-d(t')}{(v_{i+1}+v_j)}\mathcal{P}_{i+1,j}(d(t'))\right) - 2t_{rt} > 0 \end{array}\right\} \tag{12}$$

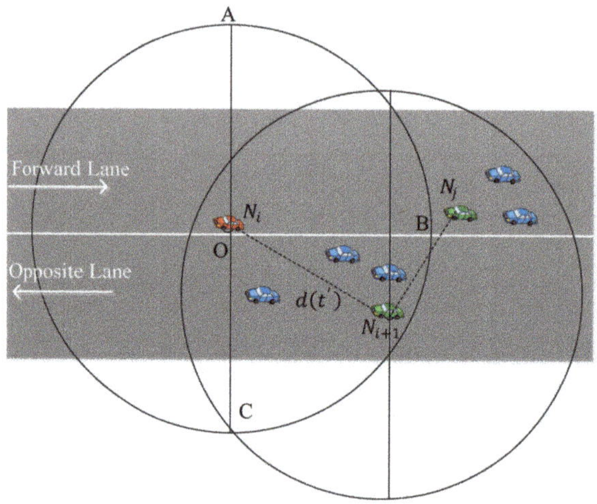

Figure 4. Selection of a vehicle in opposite lane.

Case 3: When no vehicles are present in forward lane or opposite lane

To keep the packet safe and its successful delivery to a suitable forwarding vehicle in the transmission range of the sender, each sending vehicle includes a short cache memory and searches a reliable forwarding vehicle that has the highest temporal distance between itsself and the sender and higher link quality under the same vehicular movement dynamics. If a sending vehicle N_i does not find a neighboring vehicle in the forward transmission region for both the forward lane and opposite lane (see Figure 5), in this case, sending vehicle N_i locally stores the packet in its cache until a neighboring vehicle is found in the forward direction. The cache runs in an event-driven manner rather than performing a routine check to see if the probable forwarding vehicle N_j for already cached packets is available. An event is generated as soon as a new vehicle N_j enters the transmission range of the current forwarding vehicle. The freshly generated event verifies whether or not there are any cached packets for it. This event-driven caching strategy reduces the amount of bandwidth that EGR needs but is not actually using.

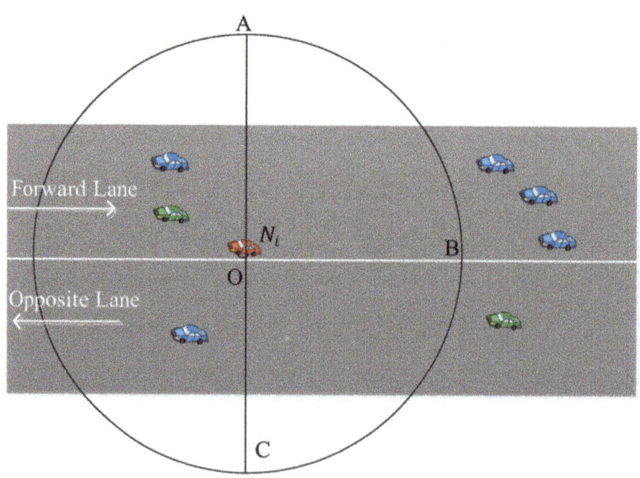

Figure 5. Re-caching by sending vehicle.

Case 4: When the sender vehicle deviates from a forward direction

When sending vehicle N_i wishes to turns either left or right from the message forward direction (see Figure 6), it tries to find a suitable neighboring vehicle in the forward transmission region in both the forward and opposite lanes. If a neighboring vehicle is found in its forward transmission area as shown in case 1 and case 2, it will send the packet to the neighboring vehicle. If no neighboring vehicle is present in the forward transmission area, the sending vehicle N_i chooses a neighboring vehicle that has an intention to travel in the message-forward direction, and the temporal distance between it and N_i is minimal. If no vehicles is found in the transmission range of N_i, then it chooses a road-side unit (RSU) to forward the message to a moving vehicle in a forward direction or caches the message until it finds a suitable vehicle moving in the required direction.

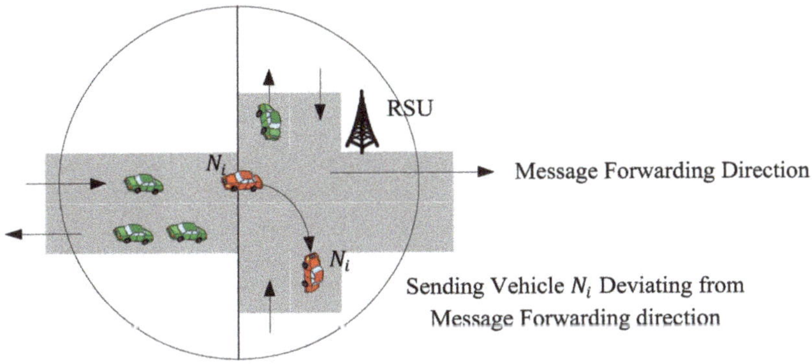

Figure 6. A sending vehicle deviation.

For routing purposes, the sending vehicle N_i chooses a vehicle N_j among the neighboring vehicles (in case no vehicle is found with case 1 and case 2) nearest to it and that should satisfy the following condition:

1. The temporal distance between N_i and N_j is minimal, and the temporal transmission probability of the link between vehicle N_i and vehicle N_j should be high.
2. Use a RSU in case no neighboring vehicle is present.

The sending vehicle N_i selects a neighboring vehicle with a minimum $d(t')$ as the next forwarding vehicle, such that

$$N_j = argmax(MIN(d(t')) \times \mathcal{P}_{i,j}(d(t'))) \tag{13}$$

where $MIN()$ function produces an element with a minimum value.

3.5. The Route Construction Algorithm

The proposed EGR routing algorithm is a temporal location-based geocast routing technique in which each sending vehicle N_i discovers the temporal locations of its neighboring vehicles utilizing the mathematical formulation given in Equation (1). This is also shown in step 1 of the proposed Algorithm 1

- *Neighbor Discovery*: Every vehicle N_i sends HELLO packets in its FTA including its coordinates. When a vehicle receives a HELLO packet, it responds with an "ECHO" packet containing positional data. Every vehicle creates a list of its neighbors after receiving these ECHO packets. Only those vehicles that are included in the relevant FTA will be included in the list.
- *Residual Link Lifetime Calculation*: The sender N_i estimates the residual link lifetime and temporal transmission probability that will be used to find the best forwarding vehicle N_j among the discovered neighboring vehicles, as shown in step 2.

- *Route Computation*: In step 3, the EGR searches for forwarding vehicle N_j using residual link lifetime and temporal transmission probability.

Algorithm 1. EGR Algorithm

Input: (x_i, y_i), $\left(x_j, y_j\right)$, v_i, v_j, θ_i and θ_j
Output: forwarding vehicle N_j

1. **Neighbor discovery:** Each vehicle $N_i \in N$ broadcasts HELLO packet to learn about its each neighboring vehicle N_j vehicle N_i computes its temporal location (x_i', y_i') and N_j computes its temporal location $\left(x_j', y_j'\right)$ using Equation (1) at time t'
2. **Residual link lifetime calculation: for** each vehicle N_j **do** N_i computes temporal distance $d(t')$ using Equation (2) N_i computes link quality $\mathcal{P}_{i,j}(d(t'))$ using Equation (8) N_i estimate residual link lifetime T_{ll} using Equation (9) **end for**
3. **Route construction: if** at least one neighbor exists in N_i's forward region of forward lane **and** no deviation **then** N_j= argmax$(d(t'))$, $st.\ T_{ll} - t_{rt} > 0$ Vehicle N_i sends packet to the vehicle N_j **else if** at least one neighbor exists in N_i's forward region of opposite lane **and** no deviation **then** choose N_j using Equation (12) Vehicle N_i sends packet to the vehicle N_j **else if** no vehicles present in forward lane and opposite lane **and** no deviation **then** Vehicle N_i caches the packet until a neighboring vehicle found its forward region **else if** (N_i wishes to deviate) $N_j = argmax\ \left(MIN(d(t')) \times \mathcal{P}_{i,j}(d(t'))\right)$ Vehicle N_i sends packet to the vehicle N_j **end if**

A flowchart (see Figure 7) of the proposed EGR for a highway traffic environment when shadowing fading is present is provided for better understanding the logic of it. The flowchart shows the main steps of EGR under fading channel conditions.

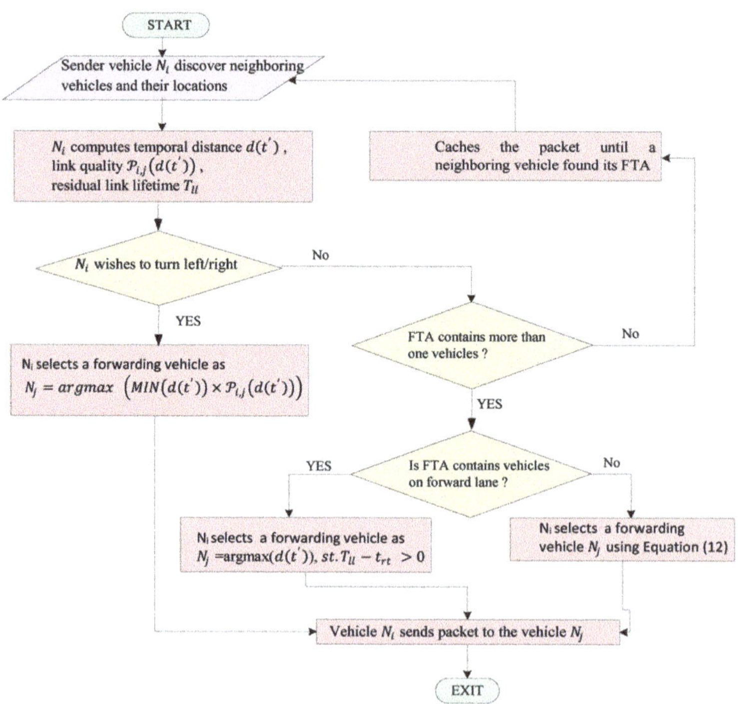

Figure 7. Flowchart of EGR.

4. Simulation and Performance Analysis

The results of the simulations performed assess how well the EGR technique performed in managing sparse vehicular networks in a highway traffic scenario for the high mobility of vehicles. Throughput, end-to-end delay, and one-hop link disconnections were evaluated in the simulation. The EGR algorithm's results are compared with the following contemporary routing approaches: SLD-GEDIR), J-GEDIR, and P-GEDIR [22].

4.1. Simulation Environment

The NS-2.34 network simulator has been used to simulate the EGR algorithm. An accurate mobility model and a highway traffic environment are produced using the "mobility model generator for vehicular networks" (MOVE) [30]. An open source, highly portable, microscopic, and continuous traffic modeling tool called simulation of urban mobility (SUMO) serves as the foundation for MOVE [31]. The majority of the key elements of the vehicular traffic environment, including the number of roads, the number of lanes, the quantity of intersections, the number of traffic lights in each intersection, the speed of vehicles, the probability that a vehicle will turn left or right at a specific intersection, etc., are set up and implemented using a road map editor and a vehicle movement editor. NS-2 has directly utilized the mobility trace produced by MOVE with SUMO's support. To evaluate EGR, two types of vehicle traffic environments—high-speed and sparsely connected vehicles—are taken into account. The three EGR modules, namely geometrical localization for GPS outage, link quality, and residual link lifetime, have been implemented [32–34]. This investigation focuses on determining if the three EGR modules reduce routing overheads, one-hop link disconnection, and increased throughput with increased vehicle velocity and vehicle density in a highway traffic environment. The simulation region consists of a set of two horizontal and two vertical highways that intersect, creating sixteen junction locations spaced equally apart. Every road has two lanes. Simulation parameters used for the experiment are shown in Table 1. After configuring the network and traffic flow with the parameter values that were taken into consideration, the simulations were run. Locations sending vehicles, destination vehicle, and geocasting region are arbitrarily chosen for each simulation run from one of two predetermined locations that are the same for all ten simulation runs. Each value used in the results analysis was obtained from an average of ten independent simulation runs.

Table 1. Simulation Parameters.

Parameters	Values	Parameters	Values
Total time	500 s	Ifqlen	50 packets
Simulation area	2000×2000 m^2	Channel type	Wireless
Vehicle speed	10–60 Km/h	Antenna model	Omni directional
Number of vehicles	500 vehicles	Propagation model	Shadowing
Source vehicles	20	MAC data rate	5 Mbps
Tranmission range	300 m	MAC protocol	IEEE 802.11p
Junction	8	CBR rate	6 Packets/s
Hello timeout	0.5 s	Packet type	UDP
Query period	2.5 s	Packet size	512 bytes
Traffic type	CBR	Geocast region size	500×500 m^2

4.2. Simulation Metrics

The performance of EGR and the related exiting routing approaches are compared using the following routing metrics.

- *End-to-End delay*: End-to-end delay is the amount of time for moving a packet from the source to the destination across a network. It is the sum of the delays experienced during transmission, propagation, and processing via each link from the source to the destination. Equation (14) represents the statistical formula that can be used to determine end-to-end delay (t_{ee}) in terms of milliseconds and is given by

$$t_{ee}(\text{ms}) = \frac{\sum_{i=1}^{10}\sum_{j=1}^{n_{PR}}(t_{ST}^{i,j} - t_{RT}^{i,j})}{n_{PR} \times 10} \tag{14}$$

where n_{PR} represents the number of packets delivered at the destination, $t_{ST}^{i,j}$ denotes the simulation time in the ith round during which the jth packet is sent, and $t_{RT}^{i,j}$ denotes the simulation time in the ith round during which the jth packet is delivered at the destination.

- *Throughput*: The throughput is the quantity of packets effectively sent to the destination in a unit simulation time. Bits per second are used to measure it. The limited number of forwarding possibilities available on the highway is due to the fact that there are alternative forms of transportation accessible, and intercity traffic is not extremely heavy. Equation (15), a statistical expression representing throughput (τ), is given by

$$\tau\ (\text{kbps}) = \frac{\sum_{i=1}^{10}(n_{PR} \times n_b)}{10} \tag{15}$$

where n_b represents the number of bits per packets.

- *One-hop Link Disconnection*: One-hop link disconnection is the number of unsuccessful attempts at packet transmission. The reliability of a link is demonstrated by this measure. It affirms the effectiveness of the forwarding vehicle selection process. Equation (16) represents the arithmetical formula that used to quantify one-hop link disconnection (l_d).

$$l_d(\%) = \left\{ \frac{\left(\sum_{i=1}^{10} \frac{n_{RD}}{n_{PS}}\right)}{10} \right\} \times 100 \tag{16}$$

where n_{RD} denotes the number of unsuccessful attempts to deliver a message from a sending vehicle to destination vehicles in the ith simulation run and n_{PS} denotes the number of packets delivered to the destination in the ith simulation run.

As presented in the four cases in Section 3.4, a single best forwarder recipient vehicle is selected considering the current traffic scenario, which might fall into any of the four case studies. We do agree that duplicate messages have been received by some recipient vehicles due to communication reestablishment. However, the duplicate messages are easily discardable considering the packet information, such as sequence number, hop count, and time-to-live timestamp.

4.3. Performance Analysis

The simulation results for analyzing the impact of GPS outage on EGR, effect of vehicle speed on the functionality of EGR, and the investigation of how vehicle density affects performance is presented in this section.

(a) Performance Analysis Under GPS Outage

In this section, we determine whether or not EGR is a viable and effective approach for information dissemination under GPS outage problems and how geometrical localization assists EGR to improve the its performance in terms of throughput, end-to-end delay, and one-hop link disconnection. For this simulation, the velocity of the vehicles are 30 km/h. GPS outage time is the length of time in which vehicles lost GPS signal.

Figure 8 shows the end-to-end delay of EGR versus the state-of-the-art geocast routing approaches. It is clear from the result that, as GPS outage time increases, the end-to-end delay for all the geocast routing approaches increases. The end-to-end delay for the EGR is stable and does not increase much with the increase in GPS outage time. However, it is increasing rapidly for J-GEDIR, P-GEDIR, and SLD-GEDIR. For example, GPS outage time is 50 ms; the end-to-end delay for EGR is 15 ms, and for J-GEDIR, P-GEDIR, and SLD-GEDIR, the delays are 50 ms, 65 ms, and 71 ms respectively. This is due to the fact that, with P-GEDIR boundary vehicles as forwarder, J-GEDIR chooses the vehicles from the furthest junctions. The boundary and farthest crossing vehicles are sometimes sensitive to link failure because of the frequent occurrence of GPS outage. All three, J-GEDIR, P-GEDIR, and SLD-GEDIR, do not employ any localization technique for GPS outage problems, whereas the EGR uses geometrical localization for the same problem. The proposed EGR gives an average of 70% lower end-to-end delay as compare to SLD-GEDIR, wherein the end-to-end delay for EGR is about 75% to 80% lower than that of J-GEDIR and P-GEDIR, respectively. Thus, when compared to state-of-the-art routing approaches, the proposed EGR performs better.

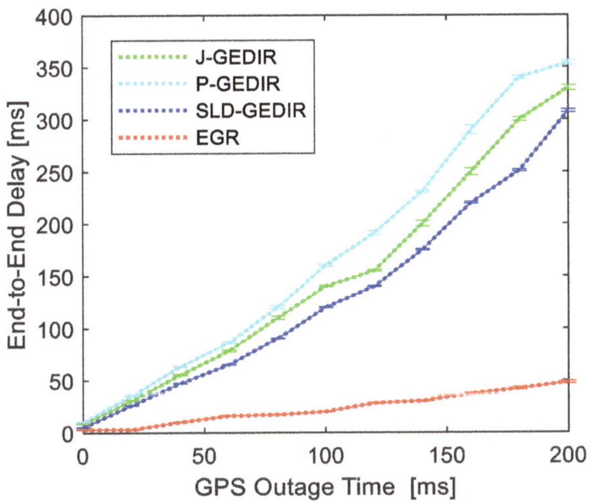

Figure 8. Effect of GPS outage on end-to-end delay.

Figure 9 shows the throughput of EGR versus state-of-the-art geocast routing approaches. It is clear from the result that, as GPS outage time increases, the throughput for all the geocast routing approaches decreases. For the EGR, it is stable and is not decreasing much with the increase in GPS outage time. However, it is decreasing rapidly for J-GEDIR, P-GEDIR, and SLD-GEDIR. For example, GPS outage time is 50 ms; the throughput for EGR is 240 Kbps, and, for J-GEDIR, P-GEDIR, and SLD-GEDIR, the throughput is 170 Kbps, 30 Kbps, and 31 Kbps, respectively. When GPS outage time is 150 ms, the throughput for EGR is 230 Kbps, and for J-GEDIR, P-GEDIR, and SLD-GEDIR, the throughput is 100 Kbps, 20 Kbps, and 17 Kbps, respectively. This is due to the fact that link failure occurs more frequently due to occurrence of GPS outage for all three, J-GEDIR, P-GEDIR, and SLD-GEDIR, and they do not employ any localization technique for the GPS outage problem, whereas the EGR uses geometrical localization to cope with the same problem. The proposed EGR gives an average of 72% better throughput compared to SLD-GEDIR, wherein the throughput for EGR is about 80% to 90% greater than that of J-GEDIR and P-GEDIR, respectively. The proposed EGR performs better compared to state-of-the-art routing approaches.

Figure 10 compares the results of one-hop link disconnection between EGR and contemporary routing approaches for different values of GPS outage time. It is evident that one-hop link disconnection for EGR is minimal and stable compared to other contemporary routing approaches for GPS outage times of 0 ms and 100 ms. It is because the forwarding vehicle selection process is completed within the residual link lifetime time However, as GPS outage increases to more than 100 ms, the one-hop link disconnections abruptly increase, since the forwarding vehicle selection process could not be finished in the residual link lifetime time in both EGR and SLD-GEDIR. In comparison to J-GEDIR, P-GEDIR, and SLD-GEDIR, the EGR performs better. It is due to the fact that the EGR uses the geometrical localization technique to cope with the GPS outage problem. For the proposed EGR, the average rate of one-hop link disconnection is about 15%, whereas it is about 40% for SLD-GEDIR, and it ranges from 60% to 80% for the rest of the J-GEDIR and P-GEDIR contemporary routing approaches. Thus, EGR performs better than the compared contemporary routing approaches.

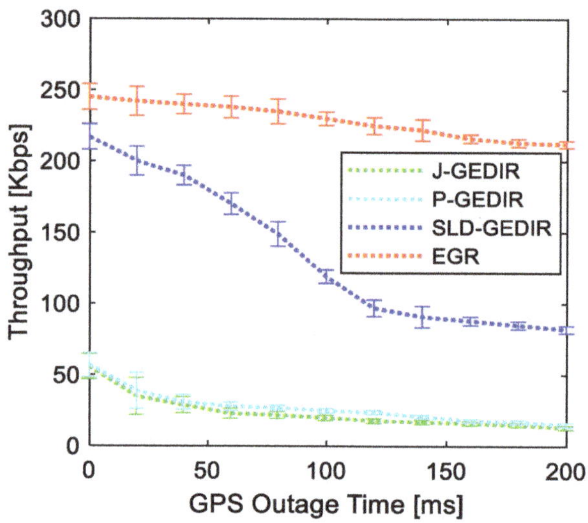

Figure 9. Effect of GPS outage on throughput.

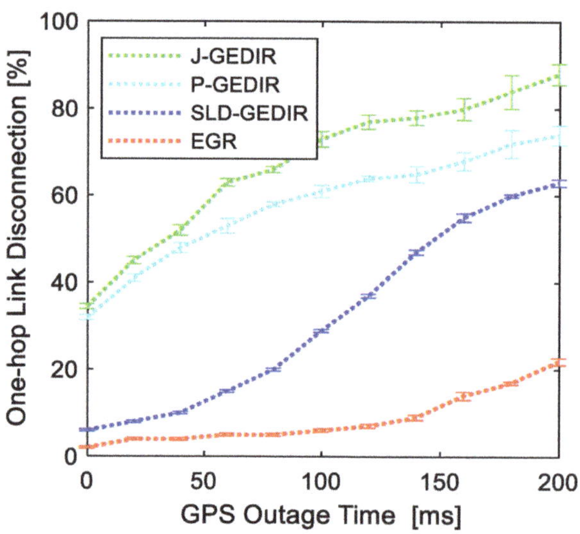

Figure 10. Effect of GPS outage on one-hop link disconnection.

(*b*) *Performance Analysis Under Vehicle Mobility*

The considered matrices have all been evaluated in the simulation for various vehicle velocity magnitudes. Vehicle velocity has been adjusted from 10 to 60 km/h in order to evaluate the performance of EGR. In this simulation, geometrical localization for GSP outage is used with EGR.

According to the result shown in Figure 11, the end-to-end delay for the EGR approach is approximately 4 ms for vehicle velocity between 10 and 40 km/h. Additionally, EGR considerably reduces end to end delay compared with the other routing approaches considered in this simulation. At a velocity greater than 40 km/h, the delay for both EGR and SLD-GEDIR is increasing with a higher rate, but the rate of increment is higher for SLD-GEDIR. This can be due to the forwarding vehicle selection procedure in EGR being done based on the most reliable link and also being made clear in favor of the most likely direction of the destination. At a velocity greater than 20 km/h, J-GEDIR and P-GEDIR performance begins to deteriorate. This is due to the fact that P-GEDIR chooses boundary vehicles as the forwarder, and J-GEDIR chooses the vehicles from the furthest junctions. The boundary and farthest crossing vehicles are sometimes sensitive to link failure because of the frequent movement of the traffic environment. The proposed EGR gives an average of 25% lower end-to-end delay compared to SLD-GEDIR, wherein the end-to-end delay for EGR is about 60% to 70% lower than that of J-GEDIR and P-GEDIR, respectively. Thus, when compared to state-of-the-art routing approaches, the proposed EGR performs better.

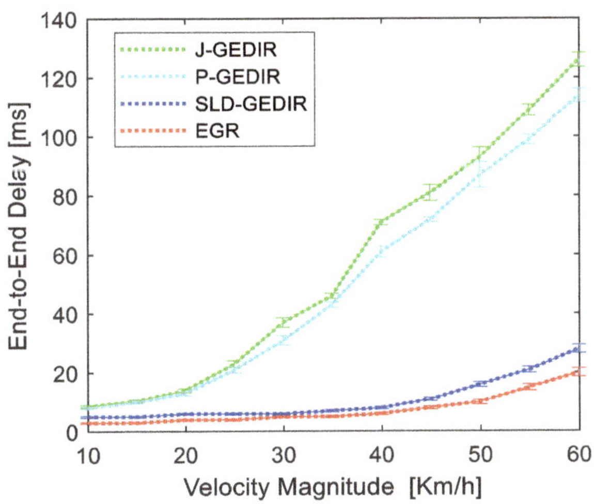

Figure 11. Effect of velocity on end-to-end delay.

In Figure 12, the findings demonstrate a throughput comparison between EGR and other procedures, taking into account variable velocity. The findings show that, when vehicle velocity increases, the throughput of EGR linearly decreases. The throughput for the existing routing approaches, however, dramatically drops. For example, the throughput values of J-GEDIR, P-GEDIR, SLD-GEDIR, and EGR at a vehicle velocity of 20 Km/h are 150 Kbps, 163 Kbps, 256 Kbps, and 265 Kbps, respectively; when the velocity increases to 40 Km/h, the throughput values decreased to 23 Kbps, 28 Kbps, 176 Kbps, and 231 Kbps respectively. This can be linked to the fact that, due to the high rate of link breakage with increased vehicle speed, modern protocols must frequently re-discover one-hop links. The link chosen by EGR is based on a temporal transmission probability and a residual link lifetime that increases its reliability. As a result, EGR consistently performs better than the existing routing approaches considered in this simulation.

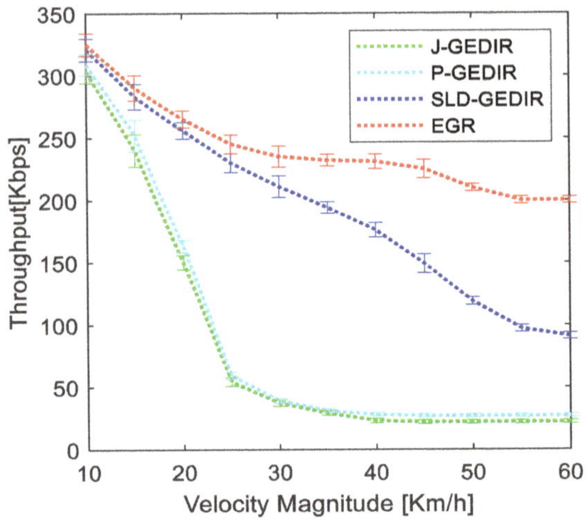

Figure 12. Effect of velocity on throughput.

Figure 13 compares the results of one-hop link disconnection between EGR and contemporary routing approaches. It is evident that one-hop link disconnection for EGR is minimal compared to other contemporary routing approaches for velocities between 10 and 60 km/h. It is because the most reliable link for the transmission of information between the sender vehicle and the farthest forwarding vehicle is picked. However, at speeds over 40 km/h, one-hop link disconnections abruptly increase, since the forwarding vehicle selection process could not be finished in the residual link lifetime time in both EGR and SLD-GEDIR. In comparison to J-GEDIR and P-GEDIR, both EGR and SLD-GEDIR perform the best in relation to the quantity of one-hop link disconnections. However, it is lower for EGR compared with SLD-GEDIR. Because the forwarding technique employed in J-GEDIR and P-GEDIR does not take velocity information into account in NHV selection, one-hop link disconnections in P-GEDIR and J-GEDIR are experienced rapidly above a velocity of 20 Km/h. Due to its higher number of hops between the source and destination regions, for the proposed EGR, the average rate of one-hop link disconnection is about 5%, whereas it is about 8% for SLD-GEDIR, and it ranges from 40% to 50% for the rest of the J-GEDIR and P-GEDIR contemporary routing approaches. Thus, EGR performs better than the compared contemporary routing approaches.

(c) Performance Analysis Under Vehicle Density

Herein, we examine the effect of vehicle density on the effectiveness of the proposed approach. In this simulation, geometrical localization for GSP outage is used with EGR.

According to the result shown in Figure 14, the EGR has the minimum end-to-end delay when compared to other contemporary routing approaches. Due to its distinctive forwarding vehicle selection procedure, the end-to-end delay of the both EGR and SLD-GEDIR is not affected much by increasing vehicle density. However, EGR takes less time for the transmission of the packets from a source vehicle to the destination vehicle because it uses residual link lifetime, which reduces the number of unnecessary attempts of sending a packet between two vehicles that ultimately reduces the transmission delay. Contemporary routing approaches J-GEDIR and P-GEDIR pick many forwarding vehicles, which results in longer end-to-end delays. It is also significant that contemporary routing approaches reduce the area in which possible forwarding vehicles belong, and they choose these forwarding vehicles, which results in approximately equal end-to-end latency. The proposed EGR gives an average 35% lower end-to-end delay compared to SLD-GEDIR, wherein the end-to-end delay for EGR is about 70% to 75% lower than that of J-GEDIR and P-GEDIR, respectively. Thus, it is evident that EGR performs contemporary routing approaches.

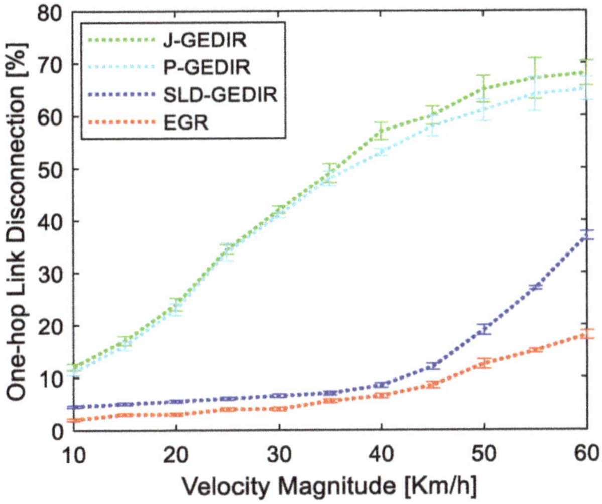

Figure 13. Effect of Velocity on one-hop link disconnection.

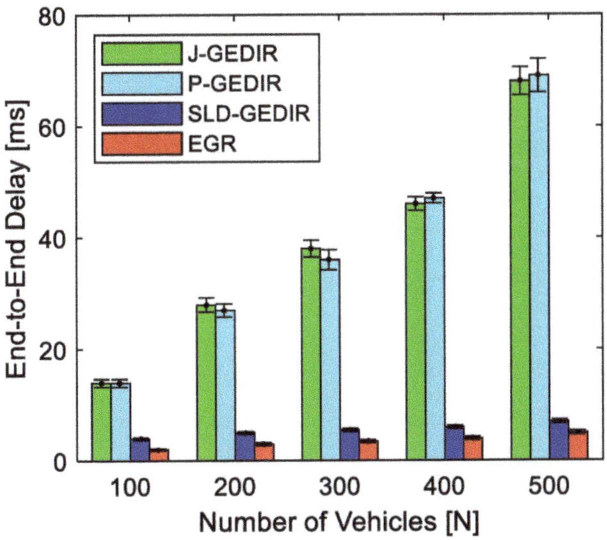

Figure 14. Effect of vehicle density on end-to-end delay.

The results shown in Figure 15 reveal that, with 100 vehicles in the forward transmission region, the throughput of EGR is almost the same as in the contemporary routing approaches. However, as the number of vehicle increases, the throughput of the contemporary routing approaches J-GEDIR and P-GEDIR declines more quickly than in both EGR and SLD-GEDIR. The rate of decrement in SLD-GEDIR is more compared to EGR. This is due the fact that EGR picks forwarding vehicle with the highest reliable link between itsself and the sender, reducing the amount of packet loss during transmission. The proposed EGR gives an average 10% higher throughput compared to SLD-GEDIR, wherein, for EGR, it is about 25% to 30% higher than that of J-GEDIR and P-GEDIR, respectively. Hence, EGR outperforms contemporary routing approaches in terms of throughput.

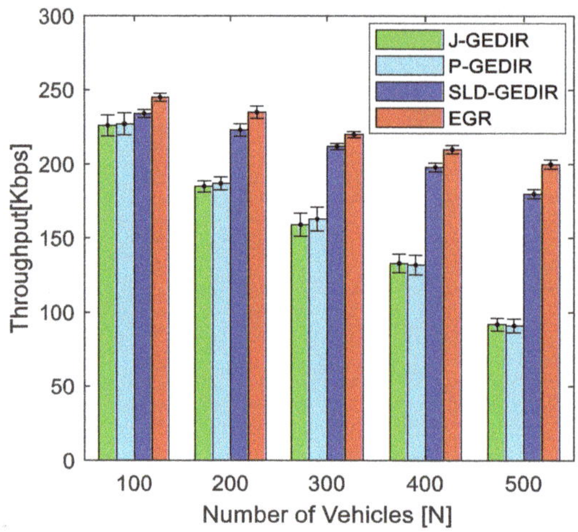

Figure 15. Effect of vehicle density on throughput.

The simulation results for one-hop link disconnection with increasing vehicles density are shown in Figure 16. It is evident that EGR has fewer one-hop link disconnections than the routing approaches that are considered to be contemporary routing approaches. It is also interesting to note that, in contrast to contemporary routing approaches, the link disconnections of both EGR and SLD-GEDIR do not change much with increasing vehicle density but rather slightly decrease. This is because both routing approaches only ever chooses one forwarding vehicle, but in other two contemporary routing approaches, J-GEDIR and P-GEDIR, the number of forwarding vehicles rises as vehicle density rises. For the proposed EGR, the average rate of one-hop link disconnection is about 3%, whereas it is about 5% for SLD-GEDIR, and it ranges from 20% to 30% for the rest of the J-GEDIR and P-GEDIR contemporary routing approaches. As a result, it can be seen that the proposed EGR exhibits less link disconnection than most contemporary routing approaches.

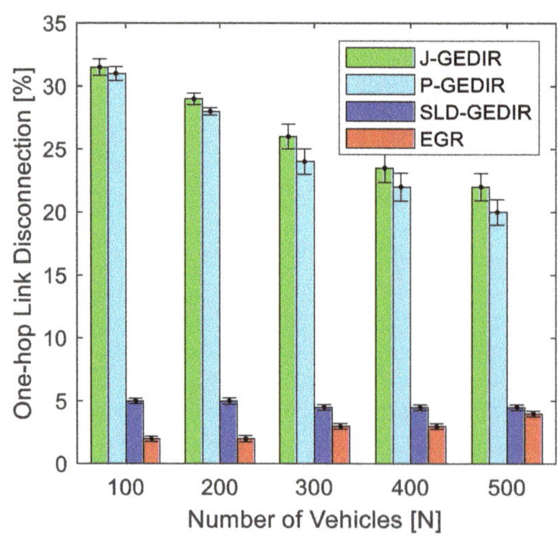

Figure 16. Effect of vehicle density on one-hop link disconnection.

5. Conclusions and Future Work

In this paper, an efficient geocast routing (EGR) approach using the temporal dependence of link quality for a highway traffic environment is presented. A geometrical localization for GPS outage problem and a temporal link quality estimation model are presented to support the EGR. We develop a framework for analysis that quantifies the likelihood of choosing the longest-lasting link out of a list of predetermined links between a transmitting vehicle and its surrounding vehicles. The EGR performs better under the shadowing fading channel situation while choosing a next forwarding vehicle using the probability measure. The performance of EGR in highway traffic environments is measured for both increasing vehicle density and increasing vehicle velocity. A comparative performance study based on simulations has been done to assess the efficacy and scalability of EGR in terms of resolving the issue of wireless communication link breakage and assuring message delivery in a highway traffic environment. It is evident from the examination of simulation results that EGR is superior than contemporary routing approaches for different vehicle traffic settings. In the future, we measure the performance of the proposed EGR in the presence of interfering vehicles for a highway traffic environment.

Author Contributions: Conceptualization R.Y.; formal analysis S.K.; investigation, R.Y.; methodology, R.Y.; supervision, S.K. and O.K.; validation, A.K. and D.; writing, R.Y.; review and editing, S.K., O.K., M.A. and J.L. All authors have read and agreed to the published version of the manuscript.

Funding: The research is funded by the School of Computer and Systems Science, Jawaharlal Nehru University, India.

Data Availability Statement: Research data will be available on individual requests to the corresponding author considering collaboration possibilities with the researcher or research team and with restrictions that the data will be used only for further research in the related literature progress.

Acknowledgments: The research is supported by the Computing and Informatics Research Center, Department of Computer Science, Nottingham Trent University, UK.

Conflicts of Interest: The authors declare that they have no conflict of interest.

References

1. Kaiwartya, O.; Abdullah, A.H.; Cao, Y.; Altameem, A.; Prasad, M.; Lin, C.-T.; Liu, X. Internet of Vehicles: Motivation, Layered Architecture, Network Model, Challenges, and Future Aspects. *IEEE Access* **2016**, *4*, 5356–5373. [CrossRef]
2. Li, Y.; Barthelemy, J.; Sun, S.; Perez, P.; Moran, B. Urban Vehicle Localization in Public LoRaWan Network. *IEEE Internet Things J.* **2021**, *9*, 10283–10294. [CrossRef]
3. Kaiwartya, O.; Kumar, S. Guaranteed Geocast Routing Protocol for Vehicular Adhoc Networks in Highway Traffic Environment. *Wirel. Pers. Commun.* **2015**, *83*, 2657–2682. [CrossRef]
4. Kumar, S.; Dohare, U.; Kumar, K.; Dora, D.P.; Qureshi, K.N.; Kharel, R. Cybersecurity Measures for Geocasting in Vehicular Cyber Physical System Environments. *IEEE Internet Things J.* **2018**, *6*, 5916–5926. [CrossRef]
5. Kaiwartya, O.; Cao, Y.; Lloret, J.; Kumar, S.; Aslam, N.; Kharel, R.; Abdullah, A.H.; Shah, R.R. Geometry-Based Localization for GPS Outage in Vehicular Cyber Physical Systems. *IEEE Trans. Veh. Technol.* **2018**, *67*, 3800–3812. [CrossRef]
6. Blum, J.; Eskandarian, A.; Hoffman, L. Challenges of Intervehicle Ad Hoc Networks. *IEEE Trans. Intell. Transp. Syst.* **2004**, *5*, 347–351. [CrossRef]
7. Sofra, N.; Gkelias, A.; Leung, K.K. Route Construction for Long Lifetime in VANETs. *IEEE Trans. Veh. Technol.* **2011**, *60*, 3450–3461. [CrossRef]
8. U.S. Department of Transportation. ITS Strategic Plan 2015–2019. Available online: http://www.its.dot.gov/research_areas/strategicplan2015.htm (accessed on 10 October 2022).
9. Fawcett, J.; Robinson, P. Adaptive routing for road traffic. *IEEE Comput. Graph. Appl.* **2000**, *20*, 46–53. [CrossRef]
10. Ammoun, S.; Nashashibi, F.; Laurgeau, C. Crossroads risk assessment using GPS and inter-vehicle communications. *IET Intell. Transp. Syst.* **2007**, *1*, 95–101. [CrossRef]
11. Wu, D.; Zhang, Y.; Bao, L.; Regan, A.C. Location-Based Crowdsourcing for Vehicular Communication in Hybrid Networks. *IEEE Trans. Intell. Transp. Syst.* **2013**, *14*, 837–846. [CrossRef]
12. Cui, Y.; Ge, S.S. Autonomous vehicle positioning with GPS in urban canyon environments. *IEEE Trans. Robot. Autom.* **2003**, *19*, 15–25.

13. Kaiwartya, O.; Kumar, S. Geocasting in vehicular adhoc networks using particle swarm optimization. In Proceedings of the International Conference on Information Systems and Design of Communication Lisbon, Lisboa, Portugal, 16–17 May 2014; pp. 62–66.
14. Tsiachris, S.; Koltsidas, G.; Pavlidou, F.-N. Junction-Based Geographic Routing Algorithm for Vehicular Ad hoc Networks. *Wirel. Pers. Commun.* **2012**, *71*, 955–973. [CrossRef]
15. Raw, R.S.; Das, S. Performance analysis of P-GEDIR protocol for vehicular adhoc network in urban traffic environments. *Wirel. Pers. Commun.* **2013**, *68*, 65–78. [CrossRef]
16. Stojmenovic, I.; Ruhil, A.P.; Lobiyal, D.K. Voronoi diagram and convex hull based geocasting and routing in wireless networks. *Wirel. Commun. Mob. Comput.* **2006**, *6*, 247–258. [CrossRef]
17. Kaiwartya, O.; Kumar, S.; Lobiyal, D.K.; Abdullah, A.H.; Hassan, A.N. Performance Improvement in Geographic Routing for Vehicular Ad Hoc Networks. *Sensors* **2014**, *14*, 22342–22371. [CrossRef] [PubMed]
18. Sangho, O.; Jaewon, K.; Gruteser, M. Location-based flooding techniques for Vehicular emergency messaging (D-Flooding). In Proceedings of the 2006 Third Annual International Conference on Mobile and Ubiquitous Systems: Networking & Services, San Jose, CA, USA, 17–21 July 2006; Springer: Berlin/Heidelberg, Germany, 2006; pp. 1–9.
19. Maihofer, C.; Eberhardt, R. Geocast in vehicular environments: Caching and transmission range control for improved efficiency (CTRC). In Proceedings of the 2004 Intelligent Vehicles Symposium, Parma, Italy, 14–17 June 2004; IEEE: Piscataway, NJ, USA, 2004; pp. 951–956.
20. Raw, R.S.; Lobiyal, D.K. B-MFR routing protocol for vehicular ad hoc networks. In Proceedings of the 2010 International Conference on Networking and Information Technology (ICNIT), Manila, Philippines, 11–12 June 2010; pp. 420–423.
21. Sommer, C.; Tonguz, O.K.; Dressler, F. Traffic information systems: Efficient message dissemination via adaptive beaconing. *IEEE Commun. Mag.* **2011**, *49*, 173–179. [CrossRef]
22. Qing, Y.; Lim, A.; Shuang, L.; Jian, F.; Agrawal, P. ACAR: Adaptive Connectivity Aware Routing Protocol for Vehicular Ad Hoc Networks. In Proceedings of the 17th International Conference on Computer Communications and Networks, ICCCN'08, Saint Thomas, VI, USA, 3–7 August 2008; pp. 1–6.
23. Kaiwartya, O.; Kumar, S. Cache agent-based geocasting in VANETs. *Int. J. Inf. Commun. Technol.* **2015**, *7*, 562. [CrossRef]
24. Kasana, R.; Kumar, S. Multimetric Next Hop Vehicle Selection for Geocasting in Vehicular Ad-hoc Networks. In Proceedings of the 2015 IEEE International Conference on Computational Intelligence & Communication Technology, Ghaziabad, India, 13–14 February 2015; pp. 400–405.
25. Li, Z.; Wu, P. Intermittent geocast routing in urban vehicular delay tolerant networks. *Tsinghua Sci. Technol.* **2016**, *21*, 630–642. [CrossRef]
26. Karimi, R.; Shokrollahi, S. PGRP: Predictive geographic routing protocol for VANETs. *Comput. Netw.* **2018**, *141*, 67–81. [CrossRef]
27. Kasana, R.; Kumar, S.; Kaiwartya, O.; Yan, W.; Cao, Y.; Abdullah, A.H. Location error resilient geographical routing for vehicular ad-hoc networks. *IET Intell. Transp. Syst.* **2017**, *11*, 450–458. [CrossRef]
28. Oliveira, H.A.; Nakamura, E.F.; Loureiro, A.A.; Boukerche, A. Error analysis of localization systems for sensor networks. In Proceedings of the 13th Annual ACM International Workshop on Geographic Information Systems, New York, NY, USA, 4–5 November 2005; ACM: New York, NY, USA, 2005; pp. 71–78.
29. Qureshi, K.N.; Islam, F.U.; Kaiwartya, O.; Kumar, A.; Lloret, J. Improved Road Segment-Based Geographical Routing Protocol for Vehicular Ad-hoc Networks. *Electronics* **2020**, *9*, 1248. [CrossRef]
30. Bai, X.; Chen, S.; Shi, Y.; Liang, C.; Lv, X. Blockchain-based Authentication and Proof-of-Reputation Mechanism for Trust Data Sharing in Internet of Vehicles. *Adhoc Sens. Wirel. Netw.* **2022**, *53*, 85–113.
31. Pandey, P.K.; Kansal, V.; Swaroop, A. ALMR: Alternate Link Based Multipath Reactive Routing Protocol for Vehicular Ad Hoc Networks (VANETs). *Adhoc Sens. Wirel. Netw.* **2021**, *50*, 27–53.
32. Rathore, R.; Sangwan, S.; Kaiwartya, O. Towards Trusted Green Computing for Wireless Sensor Networks: Multi Metric Optimization Approach. *Ad Hoc Sens. Wirel. Netw.* **2021**, *49*, 131–171.
33. Hizal, S.; Zengin, A. QEAODV: A New Routing Protocol based on Quality of Service in MANETs. *Adhoc Sens. Wirel. Netw.* **2021**, *49*, 81–109.
34. Prasad, M.; Liu, Y.-T.; Li, D.-L.; Lin, C.-T.; Shah, R.R.; Kaiwartya, O.P. A New Mechanism for Data Visualization with Tsk-Type Preprocessed Collaborative Fuzzy Rule Based System. *J. Artif. Intell. Soft Comput. Res.* **2017**, *7*, 33–46. [CrossRef]

Disclaimer/Publisher's Note: The statements, opinions and data contained in all publications are solely those of the individual author(s) and contributor(s) and not of MDPI and/or the editor(s). MDPI and/or the editor(s) disclaim responsibility for any injury to people or property resulting from any ideas, methods, instructions or products referred to in the content.

electronics

MDPI

Article

Adaptive Truck Platooning with Drones: A Decentralized Approach for Highway Monitoring

J. de Curtò [1,2,3,*], I. de Zarzà [1,2,3], Juan Carlos Cano [2], Pietro Manzoni [2] and Carlos T. Calafate [2]

1 Informatik und Mathematik, GOETHE-University Frankfurt am Main, 60323 Frankfurt am Main, Germany; dezarza@em.uni-frankfurt.de
2 Departamento de Informática de Sistemas y Computadores, Universitat Politècnica de València, 46022 València, Spain; jucano@disca.upv.es (J.C.C.); pmanzoni@disca.upv.es (P.M.); calafate@disca.upv.es (C.T.C.)
3 Estudis d'Informàtica, Multimèdia i Telecomunicació, Universitat Oberta de Catalunya, 08018 Barcelona, Spain
* Correspondence: decurto@em.uni-frankfurt.de

Abstract: The increasing demand for efficient and safe transportation systems has led to the development of autonomous vehicles and vehicle platooning. Truck platooning, in particular, offers numerous benefits, such as reduced fuel consumption, enhanced traffic flow, and increased safety. In this paper, we present a drone-based decentralized framework for truck platooning in highway monitoring scenarios. Our approach employs multiple drones, which communicate with the trucks and make real-time decisions on whether to form a platoon or not, leveraging Model Predictive Control (MPC) and Unscented Kalman Filter (UKF) for drone formation control. The proposed framework integrates a simple truck model in the existing drone-based simulation, addressing the truck dynamics and constraints for practical applicability. Simulation results demonstrate the effectiveness of our approach in maintaining the desired platoon formations while ensuring collision avoidance and adhering to the vehicle constraints. This innovative drone-based truck platooning system has the potential to significantly improve highway monitoring efficiency, traffic management, and safety. Our drone-based truck platooning system is primarily designed for implementation in highway monitoring and management scenarios, where its enhanced communication and real-time decision-making capabilities can significantly contribute to traffic efficiency and safety. Future work may focus on field trials to validate the system in real-world conditions and further refine the algorithms based on practical feedback and evolving vehicular technologies.

Keywords: truck platooning; MPC; UKF; drones; V2V communication; connected vehicles

1. Introduction

The rapid growth of the transportation industry has led to increased demand for efficient and safe vehicular systems. In particular, truck platooning has emerged as a promising solution to enhance fuel efficiency, reduce traffic congestion, and improve road safety [1,2]. Truck platooning involves a group of trucks traveling in close proximity to each other, taking advantage of the reduced air resistance and improved traffic flow. However, the current centralized approaches to truck platooning have limitations, such as reliance on infrastructure-based communication systems and the need for extensive pre-planning [3,4]. On the other hand, drones have become increasingly popular due to their versatile applications in various domains, including monitoring, surveillance, and communication [5,6]. Drones' ability to operate in complex environments, combined with their real-time data acquisition capabilities, make them ideal candidates for assisting truck platooning systems.

In this paper, we propose a decentralized drone-based truck platooning framework that leverages the capabilities of drones to improve the efficiency and safety of highway monitoring. Our approach integrates Model Predictive Control (MPC) and Unscented Kalman Filter (UKF) techniques for drone formation control while considering the dynamics and constraints of both trucks and drones. The A* [7] search algorithm is a well-known pathfinding and graph traversal method that is primarily used for waypoint navigation in our context.

The main objective of this study is to develop a decentralized framework for drone-based truck platooning that addresses the challenges posed by the dynamic highway environment. The proposed framework should be able to:

- Maintain a desired formation for the drones while avoiding collisions.
- Adapt to the changes in truck dynamics and constraints.
- Make decentralized decisions for platooning based on real-time data acquired by the drones.

The primary contributions of this paper are as follows:

- We propose a novel decentralized drone-based truck platooning framework that integrates MPC and UKF techniques for drone formation control.
- We develop a simulation model that considers the dynamics and constraints of both trucks and drones, providing a realistic evaluation of our proposed framework.
- We present a comprehensive performance analysis of the proposed framework, demonstrating its effectiveness in improving the efficiency and safety of highway monitoring.

Recent advancements have broadened the Internet of Things (IoT) applications beyond conventional domains, unveiling a multitude of innovative use cases. These applications, by offering new data collection and processing methodologies, enhance connectivity and efficiency across several sectors. This paper delves into the realm of highway monitoring and transportation, examining drone-based truck platooning as a novel IoT application. Our proposed model harnesses the robust connectivity and adaptability of IoT devices, facilitating decentralized, real-time collaboration between drones and trucks. Moreover, employing sophisticated control and estimation methods like MPC and the UKF underlines the capacity of IoT-driven solutions to address intricate issues, propelling the evolution of more intelligent and sustainable transportation systems. In this work, we introduce a unique combination of MPC, the UKF, and drone-IoT technology to optimize highway truck platooning.

This paper is organized as follows: Section 2 presents a comprehensive literature review of the relevant background and prior works in the areas of truck platooning, drone-based monitoring, and decentralized control approaches. Section 3 introduces the system model and provides a detailed mathematical representation of the truck and drone dynamics. In Section 4, the UKF and MPC techniques are discussed along with the potential integration of A* for waypoint navigation. Section 5 describes the proposed decentralized drone-based truck platooning framework, including its formulation and algorithmic implementation. Section 6 presents the simulation setup and results, demonstrating the performance and effectiveness of the proposed framework. Finally, Section 7 provides a comparison with other V2V approaches, and Section 8 concludes the paper and gives possible directions for future work in this area.

2. Related Work

Truck platooning [3,8] has been widely studied, aiming to improve efficiency and safety. Initial research focused on Adaptive Cruise Control (ACC) systems for maintaining safe distances between vehicles [9], while newer studies have shifted toward cooperative adaptive cruise control (CACC), exploiting vehicle-to-vehicle (V2V) communication for enabling closer truck spacing, and thereby enhancing performance [10–12]. Benefits such as reduced fuel consumption [13] and improved traffic flow [14] have also been highlighted.

However, most of the existing methods depend on centralized control systems and extensive pre-planning, which may limit adaptability to dynamic environments. This is where the decentralized control, studied in various domains like robotics [15] and multi-agent systems [16], comes into the picture. The decentralized control, relying on local interactions and decision-making processes, ensures adaptability and resilience. Within this context, the subfield of formation control coordinates multiple agents to achieve a desired formation or pattern, using techniques like MPC [17] and UKF [18,19], which are known for their robustness and efficiency.

To further enhance the dynamic adaptability [20,21] of this approach, we incorporate drones, which are recognized for their versatility and real-time data acquisition capabilities. Drones have already been utilized in transportation applications like traffic monitoring [5], infrastructure inspection [6], and emergency response [22], indicating their potential to improve the efficiency and safety of transportation systems. Thus, their application in truck platooning provides a promising solution to overcome the limitations of existing methodologies, contributing to a more effective and sustainable transportation system.

The integration of drones in truck platooning systems has been relatively unexplored in the literature. A few studies have proposed the use of drones for monitoring and communication purposes in platooning systems [23]. However, these works mainly focus on centralized control approaches and do not address the challenges associated with decentralized decision making and formation control. In summary, while there has been extensive research on truck platooning and drone applications in transportation, the potential of combining these technologies in a decentralized framework has not been fully realized. Our proposed drone-based truck platooning framework addresses this gap by integrating MPC and UKF techniques for decentralized formation control while considering the dynamics and constraints of both trucks and drones.

The prevalent limitations in current truck platooning systems stem from their reliance on centralized control mechanisms [24] and lack of robustness in dynamic environments. Centralized systems often struggle with scalability and are vulnerable to unique points of failure, hindering their adaptability to unpredictable road conditions and traffic patterns. Our work addresses these challenges by proposing a decentralized, drone-assisted framework for truck platooning. By leveraging the agility and real-time data acquisition capabilities of drones, combined with the robustness of MPC and UKF techniques, we enable more dynamic and resilient platoon formation and maintenance. This approach not only enhances the adaptability of truck platoons to changing traffic scenarios but also improves the overall efficiency and safety of the transportation system. Moreover, the integration of drones in the platooning process fills a significant gap in the existing literature, providing a novel perspective on leveraging aerial support for ground vehicle coordination.

Furthermore, an emerging area of interest in truck platooning research is the stability of platoons under variable communication conditions. The impact of communication quality on the safety and efficiency of truck platoons is crucial, particularly in dynamic highway environments. Studies such as those in [25,26] have focused on this aspect, examining how fluctuations in V2V communication can affect platoon stability and proposing mechanisms to mitigate these effects. These works highlight the importance of robust communication frameworks and adaptive control strategies in maintaining platoon integrity, especially when faced with communication delays or disruptions. The integration of such considerations into a decentralized, drone-assisted framework could further enhance its resilience and adaptability, aligning with the growing emphasis on reliable and flexible platooning systems in contemporary transportation research.

3. System Model

In this section, we introduce the system model for our proposed decentralized drone-based truck platooning system. The bicycle model for trucks is a well-adopted approach in automotive engineering, offering a balance between simplicity and accuracy in capturing essential vehicular movements [27,28]. For the dynamics of drones, the double integrator model is widely recognized for its effectiveness in representing the fundamental aspects of quadcopter flight [29].

3.1. Dynamics of the Truck

We model the dynamics of each truck using a simple bicycle model [27,28], which captures the longitudinal and lateral motion of the vehicle. The state of the zth truck at time t is represented by its position $p_z(t) \in \mathbb{R}^2$, velocity $v_z(t) \in \mathbb{R}$, and heading angle $\theta_z(t) \in \mathbb{R}$. The control inputs for each truck are the acceleration $a_z(t) \in \mathbb{R}$ and the steering angle $\delta_z(t) \in \mathbb{R}$. The dynamics of the truck can be described by the following equations:

$$\dot{p}_z(t) = v_z(t) \begin{bmatrix} \cos(\theta_z(t)) \\ \sin(\theta_z(t)) \end{bmatrix}, \tag{1}$$

$$\dot{v}_z(t) = a_z(t), \tag{2}$$

$$\dot{\theta}_z(t) = \frac{v_z(t)}{L} \tan(\delta_z(t)), \tag{3}$$

where L is the effective wheelbase of the truck. The bicycle model is a common approach in vehicular dynamics that strikes a balance between simplicity and realism. Despite being a simplified representation, it captures the essential dynamics of the truck, including its longitudinal (forward/backward) and lateral (left/right) motion. It characterizes a vehicle as a rigid body moving in a 2D plane, representing the two axles of a truck as one axle, hence the term 'bicycle'.

3.2. Dynamics of the Drone

We assume that the drones have a simplified double integrator model [29], which is commonly used to represent the dynamics of quadcopters. The state of the oth drone at time t is represented by its position $q_o(t) \in \mathbb{R}^2$ and velocity $w_o(t) \in \mathbb{R}^2$. The control input for each drone is the acceleration $b_o(t) \in \mathbb{R}^2$. The dynamics of the drone can be described by the following equations: $\dot{q}_o(t) = w_o(t)$, and $\dot{w}_o(t) = b_o(t)$.

4. Advanced Control and Navigation Techniques

In this section, we briefly describe the UKF and MPC techniques used in our framework as well as the possibility of integrating A* for waypoint navigation.

4.1. Unscented Kalman Filter

The UKF [18] is a state estimation technique that extends the traditional Kalman Filter to nonlinear systems. The UKF uses a deterministic sampling approach called Unscented Transformation, which propagates a set of sigma points through the nonlinear system to obtain an approximation of the posterior distribution. The UKF equations are given as follows:

1. Sigma point generation: Generate $2n+1$ sigma points $\mathcal{X}_k^{(z)}$ and weights $W_k^{(z)}$ from the current state estimate $\mathbf{x}_k$ and covariance matrix $\mathbf{P}_k$.
2. Prediction: Propagate the sigma points through the nonlinear process model, $f(\cdot)$, to obtain the predicted sigma points $\mathcal{X}_{k+1|k}^{(z)} = f(\mathcal{X}_k^{(z)}, \mathbf{u}_k)$.
3. State prediction: Compute the predicted state estimate, $\mathbf{x}_{k+1|k} = \sum_{z=0}^{2n} W_k^{(z)} \mathcal{X}_{k+1|k}^{(z)}$, and covariance, $\mathbf{P}_{k+1|k} = \sum_{z=0}^{2n} W_k^{(z)} (\mathcal{X}_{k+1|k}^{(z)} - \mathbf{x}_{k+1|k})(\mathcal{X}_{k+1|k}^{(z)} - \mathbf{x}_{k+1|k})^\top$.
4. Update: Incorporate the measurements $\mathbf{z}_{k+1}$ using the nonlinear measurement model $h(\cdot)$, the predicted sigma points $\mathcal{X}_{k+1|k}^{(z)}$, and the Kalman gain $\mathbf{K}_{k+1}$ to update the state estimate $\mathbf{x}_{k+1}$ and covariance $\mathbf{P}_{k+1}$.

In the UKF framework, the primary input variables include the current state estimate $\mathbf{x}_k$, covariance matrix $\mathbf{P}_k$, control input $\mathbf{u}_k$, and the latest measurements $\mathbf{z}_{k+1}$. The state estimate $\mathbf{x}_k$ represents the best guess of the system's current state, whereas $\mathbf{P}_k$ quantifies the uncertainty in this estimate. The control input $\mathbf{u}_k$ pertains to external actions applied to the system, and $\mathbf{z}_{k+1}$ are the new measurements obtained from sensors or other sources. The output variables of the UKF process are the updated state estimate $\mathbf{x}_{k+1}$ and covariance matrix $\mathbf{P}_{k+1}$, which reflect the new, refined estimates of the system's state and associated uncertainty after considering the latest measurements and the dynamics of the system. This iterative process allows the UKF to adaptively refine its understanding of the system state over time, even in the presence of nonlinearities and uncertainties.

4.2. Model Predictive Control

MPC [17] is an advanced control technique that computes the optimal control inputs by solving an optimization problem over a finite prediction horizon. The objective is to minimize a cost function subject to the constraints of the system's dynamics and limitations. The MPC equations are formulated as follows:

$$\min_{\mathbf{u}_{k:k+N-1}} \sum_{z=0}^{N-1} \begin{bmatrix} (\mathbf{x}_{k+z|k} - \mathbf{x}_{ref})^{\top} \mathbf{Q} (\mathbf{x}_{k+z|k} - \mathbf{x}_{ref}) \\ +(\mathbf{u}_{k+z} - \mathbf{u}_{ref})^{\top} \mathbf{R} (\mathbf{u}_{k+z} - \mathbf{u}_{ref}) \end{bmatrix} \tag{4}$$

$$\text{s.t.} \quad \begin{aligned} \mathbf{x}_{k+z+1|k} &= f(\mathbf{x}_{k+z|k}, \mathbf{u}_{k+z}), \\ z &= 0, \ldots, N-1 \end{aligned} \tag{5}$$

$$\mathbf{x}_{k|k} = \mathbf{x}_k \tag{6}$$

$$\mathbf{x}_{min} \leq \mathbf{x}_{k+z|k} \leq \mathbf{x}_{max}, \quad z = 1, \ldots, N \tag{7}$$

$$\mathbf{u}_{min} \leq \mathbf{u}_{k+z} \leq \mathbf{u}_{max}, \quad z = 0, \ldots, N-1 \tag{8}$$

After solving the optimization problem, the first control input $\mathbf{u}_k^*$ in the optimal control sequence is applied to the system. The process is repeated at each time step, making MPC a receding horizon control strategy.

The Sequential Least Squares Quadratic Programming (SLSQP) method is an optimization algorithm used to solve nonlinear optimization problems with constraints. It is particularly well-suited for MPC problems, where the goal is to find a sequence of control inputs that minimize a cost function while satisfying system constraints. In the context of MPC, the SLSQP algorithm works as follows: (1) linearize the nonlinear constraints at the current solution estimate, (2) solve the Quadratic Programming (QP) problem formed by the linearized constraints and the quadratic objective function, (3) update the solution estimate using the solution of the QP problem, and (4) check the convergence criteria. If the criteria are met, stop the algorithm and return the solution. If not, go back to step 1 and continue the iterations. In our drone-based truck platooning problem, we used the SLSQP algorithm to solve the MPC problem for each drone. The algorithm helps determine the optimal accelerations for the drones to maintain their assigned formation positions relative to the trucks while minimizing the control effort and ensuring constraints on drone dynamics are satisfied.

The SLSQP algorithm is integral to our MPC formulation, enabling the efficient computation of optimal control actions for drones in the platooning scenario. Table 1 summarizes the steps of the SLSQP algorithm along with their respective inputs and outputs.

Table 1. SLSQP algorithm steps with input and output variables in the MPC framework for drone-based truck platooning.

Step	Description	Input Variables	Output Variables
(1)	Linearize constraints	Current solution, Nonlinear constraints	Linearized constraints
(2)	Solve QP problem	Linearized constraints, Quadratic objective	QP solution
(3)	Update solution	QP solution	Updated solution estimate
(4)	Check convergence	Updated solution, Convergence criteria	Convergence status

Each step of the SLSQP algorithm plays a crucial role in ensuring that the drones in our system effectively maintain formation while navigating around trucks. By iteratively refining the solution based on the linearized constraints and quadratic objective function, the algorithm adeptly balances the need for precision in control with the constraints imposed by drone dynamics and platooning requirements. This results in an efficient and effective control strategy for our complex, multi-agent system.

4.3. Integration with A* for Waypoint Navigation

In addition to UKF and MPC, the framework can be extended to incorporate the A* algorithm [30] for waypoint navigation. A* is an informed search algorithm that computes the shortest path between an initial position and a goal position while considering the constraints of the environment. A* uses a cost function $f(n) = g(n) + h(n)$ to evaluate the nodes in the search space, where $g(n)$ is the cost from the start node to the current node n, and $h(n)$ is the heuristic cost from node n to the goal node. The heuristic function should be admissible and consistent to guarantee optimality.

In our framework, the A* algorithm can be used to generate a sequence of waypoints that guide the drones through the environment while avoiding obstacles. The generated waypoints can serve as intermediate reference positions for the MPC controller, allowing the drones to navigate complex environments while maintaining the desired formation and tracking the trucks. In summary, our proposed framework integrates UKF, MPC, and A* to enable decentralized drone-based truck platooning in various scenarios. The combination of these techniques allows the drones to estimate their states accurately, compute optimal control inputs considering system constraints, and navigate the environment efficiently while avoiding obstacles.

4.4. Formation Control

We define the desired formation of drones relative to the trucks using a set of formation positions $\{f_k \in \mathbb{R}^2\}_{k=1}^{N}$, where N is the number of drones. Each formation position f_k is associated with a specific truck and a relative position in the truck's local coordinate frame. The objective of the formation control is to minimize the distance between the actual drone positions and the target positions derived from the formation positions and the current truck positions:

$$\min_{b_o(t)} \sum_{o=1}^{N} \|q_o(t) - (p_{\text{truck}(o)}(t) + f_{\text{rel}(o)}(t))\|^2, \tag{9}$$

subject to the drone dynamics and the constraints on the maximum drone acceleration. $b_o(t)$ represents the control input that should be chosen such that it adjusts the drone's position to follow the target positions while respecting the drone's dynamics and constraints on maximum acceleration.

4.5. Model Predictive Control and Unscented Kalman Filter

To achieve decentralized formation control, we employ MPC and UKF techniques for each drone. MPC is used to optimize the drone's acceleration based on its current state and the target position, considering a finite time horizon and constraints on the maximum acceleration. UKF is used to estimate the state of the trucks and drones based on noisy measurements, providing a robust and efficient solution for the state estimation problem. In our framework, each drone computes its optimal acceleration using MPC and updates its state estimate using UKF based on the available measurements. This decentralized approach allows the drones to adapt their behavior in real time based on the changing positions and velocities of the trucks as well as the other drones in the formation.

In our decentralized approach, each drone communicates with the trucks and other drones using a wireless communication system. This system allows the drones to exchange information about their states and the trucks' states while maintaining the decentralized structure of the control algorithm. The communication model is based on a periodic broadcast of state information, including position and velocity measurements, from the trucks and drones. Each drone receives the measurements from the trucks and other drones in its communication range and then uses this information to update its state estimate with the UKF and determine its optimal acceleration using MPC.

To elucidate the interactions and workflows of the UKF, MPC, and A* algorithm within our framework, we present a sequence diagram in Figure 1. This diagram visually represents the flow of data and control among distinct components of the system, illustrating a clear understanding of how these advanced techniques collaboratively contribute to the effective functioning of the decentralized drone-based truck platooning system.

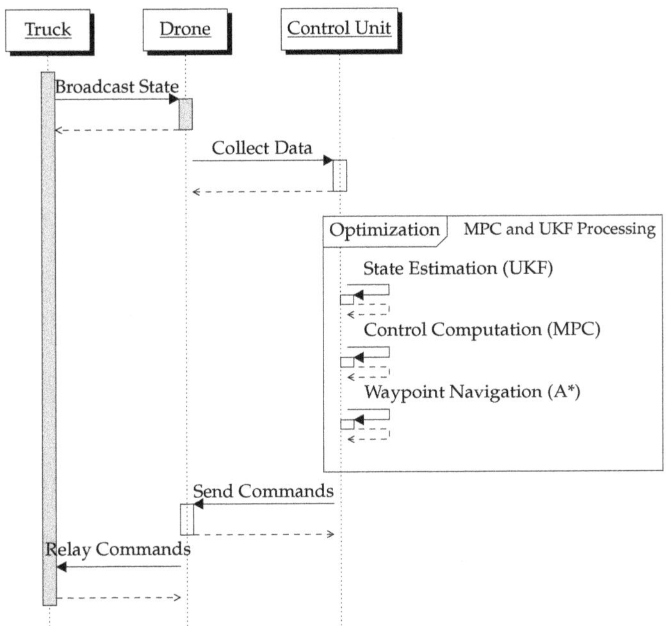

Figure 1. Sequence diagram of the advanced control and navigation techniques in the decentralized truck platooning system.

5. Decentralized Truck Platooning with Drones

In this section, we present the Decentralized Drone-Based Truck Platooning (DDTP) algorithm, which combines the decentralized formation control of drones with the truck platooning problem. The DDTP takes into account the truck dynamics and constraints, the drone dynamics, and the communication model between the trucks and drones.

The truck platooning problem aims to optimize the trucks' behavior, such as their velocities and distances, to improve fuel efficiency, traffic flow, and safety. To achieve this, we define a cost function O_t for the trucks, which takes into account their velocities, accelerations, and the distances between them:

$$O_t = \sum_{z=1}^{N_t} \left[\alpha_1 (v_z - v_{ref})^2 + \alpha_2 (a_z)^2 + \alpha_3 (d_z - d_{ref})^2 \right], \tag{10}$$

where N_t is the number of trucks, v_z and a_z are the velocity and acceleration of truck z, d_z is the distance between truck z and the preceding truck, v_{ref} and d_{ref} are the reference velocity and distance, and α_1, α_2, and α_3 are weighting factors.

The trucks use a distributed control algorithm to minimize O_t, taking into account the constraints on their velocities, accelerations, and distances:

$$v_{min} \leq v_z \leq v_{max}, \; a_{min} \leq a_z \leq a_{max}, \; d_{safe} \leq d_z, \tag{11}$$

where v_{min}, v_{max}, a_{min}, a_{max}, and d_{safe} are the minimum and maximum velocity, acceleration, and safe distance constraints.

The drones' formation control problem aims to maintain a desired formation relative to the trucks while taking into account the truck dynamics, the communication model, and the drone dynamics. To achieve this, we define a cost function O_d for the drones, which takes into account their velocities, accelerations, and the distances to their target positions:

$$O_d = \sum_{o=1}^{N_d} \left[\beta_1 (\mathbf{v}_o - \mathbf{v}_{ref_o})^2 + \beta_2 (\mathbf{a}_o)^2 + \beta_3 (\mathbf{p}_o - \mathbf{p}_{t_o})^2 \right], \tag{12}$$

where N_d is the number of drones, $\mathbf{p}_o$, $\mathbf{v}_o$, and $\mathbf{a}_o$ are the position, velocity, and acceleration of drone o, $\mathbf{p}_{t_o}$ and $\mathbf{v}_{ref_o}$ are the target position and reference velocity for drone o, and β_1, β_2, and β_3 are weighting factors.

The drones use a decentralized MPC algorithm to minimize O_d, taking into account the constraints on their velocities, accelerations, and the communication model:

$$\mathbf{v}_{min} \leq \mathbf{v}_o \leq \mathbf{v}_{max}, \tag{13}$$

$$\mathbf{a}_{min} \leq \mathbf{a}_o \leq \mathbf{a}_{max}, \tag{14}$$

where $\mathbf{v}_{min}$, $\mathbf{v}_{max}$, $\mathbf{a}_{min}$, and $\mathbf{a}_{max}$ are the minimum and maximum velocity and acceleration constraints for the drones.

The communication between drones and trucks is essential for the decentralized framework. Each drone receives information about the position, velocity, and acceleration of its assigned truck. The drones also exchange information with their neighboring drones to maintain the formation. We model the communication as a time-varying graph $G = (V, E)$, where V is the set of vertices representing the drones and trucks, and E is the set of edges representing the communication links.

We assume a communication range R_c for each drone and truck. An edge $(z, o) \in E$ exists if the distance between vertices z and o is within the communication range:

$$|\mathbf{p}_z - \mathbf{p}_o| \leq R_c. \tag{15}$$

Based on the communication graph, the drones and trucks can exchange information and update their states using the distributed algorithms described in the previous subsections.

5.1. DDTP Algorithm

The DDTP algorithm combines the truck platooning, drone formation control, and communication model to provide a decentralized solution for the problem of truck platooning with drones. The algorithm is summarized as follows:

1. Initialize the states of the trucks and drones, including their positions, velocities, and accelerations.
2. Update the communication graph G based on the current positions of the trucks and drones.
3. Each truck computes its optimal acceleration to minimize O_t using the distributed control algorithm, taking into account the constraints and the information received from the drones.
4. Each drone computes its optimal acceleration to minimize O_d using the decentralized MPC algorithm, taking into account the constraints and the information received from the trucks and neighboring drones.
5. Update the states of the trucks and drones based on their optimal accelerations and the system dynamics.
6. Repeat steps 2–5 until the desired time horizon is reached.

By following the DDTP algorithm, the drones can adapt their behavior in real time based on the trucks' dynamics and constraints, maintaining the desired formation, and assisting the trucks in platooning efficiently and safely.

The DDTP algorithm delineates the dynamics of both drones and trucks in the platooning system, incorporating decentralized decision making for optimal coordination. Table 2 identifies the input and output variables for each step of the DDTP algorithm.

Table 2. Input and output variables for each step of the DDTP algorithm.

Step	Description	Input Variables	Output Variables
1	Initialize states	Initial conditions	Positions, velocities, accelerations
2	Update communication graph	Current positions	Updated graph G
3	Truck optimal acceleration	G, Truck constraints, Drone data	Truck accelerations
4	Drone optimal acceleration	G, Drone constraints, Truck data	Drone accelerations
5	Update states	Optimal accelerations, System dynamics	Updated positions, velocities
6	Iterative process	Current states, G	Final platooning state

This detailed analysis of the DDTP algorithm shows the intricate interplay between trucks and drones in achieving efficient and safe platooning. By continuously updating the communication graph and adapting to the dynamic environment, the algorithm ensures that both trucks and drones respond optimally to the evolving platooning scenario.

Figure 2 represents the DDTP scenario, where trucks (T_o) are forming a platoon on the road and are being monitored and assisted by drones (D_o). The dashed lines represent communication links between trucks and drones. The dashed circles around the drones illustrate their communication range. The drones can communicate with each other and with the trucks within this range.

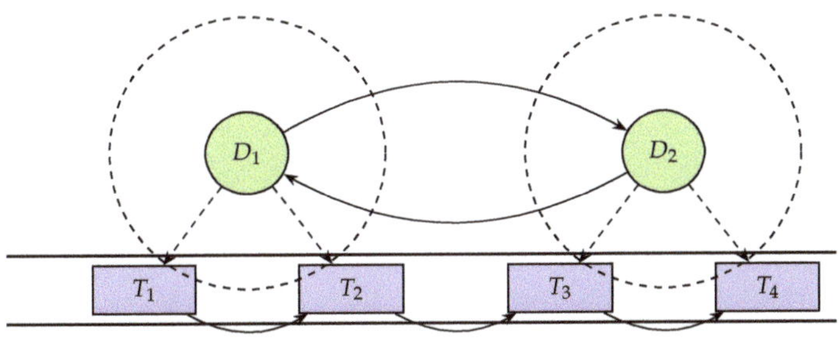

Figure 2. Decentralized Drone-Based Truck Platooning (DDTP). Trucks (T_o) form a platoon on the road while being monitored and assisted by drones (D_o). Dashed lines represent communication links between trucks and drones. Drones communicate with each other and with the trucks within their communication range, which is illustrated by the dashed circles.

The DDTP algorithm is designed to manage the coordination between autonomous trucks and drones for efficient platooning. A formal description is proposed in Algorithm 1.

The algorithm starts by initializing the states of the trucks and drones. In each iteration, the communication graph G is updated based on the current positions of the trucks and drones. Each truck then computes its optimal acceleration to minimize the truck cost function O_t, taking into account the constraints and the information received from the drones. Each drone computes its optimal acceleration to minimize the drone cost function O_d, considering the constraints and the information received from the trucks and neighboring drones. The states of the trucks and drones are then updated based on their computed optimal accelerations. This process is repeated until the desired time horizon is reached, resulting in an optimal platoon configuration.

Algorithm 1 Decentralized Drone-Based Truck Platooning (DDTP)

Input: Number of trucks N_t, number of drones N_d, initial states of trucks and drones, time horizon T
Output: Optimal platoon configuration
Initialize: Truck and drone states (positions, velocities, accelerations)
for $t = 1$ to T **do**
  Update the communication graph G
  for all Trucks $o = 1$ to N_t **do**
    Compute optimal acceleration a_o^* to minimize O_t using distributed control
    Update truck state based on a_o^*
  end for
  for all Drones $z = 1$ to N_d **do**
    Compute optimal acceleration $\mathbf{a}_z^*$ to minimize O_d using decentralized MPC
    Update drone state based on $\mathbf{a}_z^*$
  end for
end for
return Final platoon configuration

5.2. Addressing Realism and Justification for Drones in DDTP

Recent advancements in UAVs and vehicular communications have opened new avenues for enhancing vehicle platooning strategies. While traditional V2V communication systems offer robust solutions for platooning, the integration of drones in a DDTP system presents unique advantages. Drones, due to their aerial perspective and maneuverability, can provide comprehensive real-time data about the platoon and its surrounding environment, which may not be fully captured by ground-based sensors. This aerial surveillance facilitates more informed decision making and enhanced safety measures, particularly in complex traffic situations or hazardous road conditions [31].

Moreover, the inclusion of drones allows for the mitigation of challenges posed by line-of-sight issues and signal obstruction common in V2V communications. Drones can serve as dynamic relay points, ensuring consistent and reliable communication links among trucks, especially in areas with poor network coverage or in scenarios where direct communication between vehicles is hindered by obstacles or terrain. Furthermore, drones equipped with advanced sensors can provide additional data on environmental conditions, traffic patterns, and potential hazards, contributing to more efficient route planning and energy management within the platoon [6,32].

However, it is crucial to acknowledge that while drones can significantly enhance data accuracy, they are not immune to measurement errors and uncertainties inherent in any real-world system. To address these challenges, robust sensor fusion techniques and sophisticated estimation algorithms, such as Extended Kalman Filters or Particle Filters, can be employed to integrate data from multiple sources, thereby improving the reliability and accuracy of the system's state estimation [33,34]. Additionally, the development of advanced algorithms for real-time data processing and decision making is essential to handle the dynamic and uncertain nature of real-world driving environments.

5.3. Energy Considerations in DDTP

The practicality and realism of implementing a drone-assisted truck platooning system, such as DDTP, raise several significant concerns, particularly regarding energy management and operational complexities. Drones, as integral parts of this system, are inherently constrained by their battery life and energy consumption patterns, which directly impact their operational duration and efficiency. To enhance the realism of our model, incorporating energy considerations into the mathematical framework is essential for realistic decision making regarding drone trajectories.

One approach to integrate energy constraints is by modifying the cost function O_d for drones, including a term that accounts for energy consumption. This could be related to the distance traveled, time in operation, or specific maneuvers performed by the drone. Such a term would require the drones to not only maintain the desired formation and assist in platooning but also optimize their flight paths for energy efficiency. This consideration ensures that the drones can operate effectively throughout the required duration of the platooning mission without necessitating frequent recharging or replacements.

Additionally, the potential complexities associated with drone operations, such as regulatory constraints, varying weather conditions, and the need for sophisticated control algorithms, must be acknowledged. These factors can significantly influence the feasibility and reliability of deploying drones for truck platooning support. For instance, adverse weather conditions can impede drone operations, requiring the system to have contingency plans or alternative communication strategies.

Furthermore, while drones offer distinct advantages in terms of providing an aerial perspective and potentially overcoming line-of-sight communication issues common in V2V systems, it is worth considering whether a connected vehicle solution with advanced V2V communication could provide comparable performance. Such a solution might involve trucks communicating directly with a centralized system, which could coordinate the platoon without the need for aerial support. This approach could potentially offer similar benefits in terms of platoon coordination and safety while avoiding the complexities and energy constraints associated with drone usage.

Enhanced Energy-Aware Cost Function for Drones: To incorporate energy considerations into the drone cost function, we can introduce an energy term E_o for each drone. The energy consumption E_o can be modeled as a function of its velocity $\mathbf{v}_o$, acceleration $\mathbf{a}_o$, and time in operation t:

$$E_o = \gamma_1 f_v(\mathbf{v}_o) + \gamma_2 f_a(\mathbf{a}_o) + \gamma_3 t, \tag{16}$$

where γ_1, γ_2, and γ_3 are weighting factors. The revised drone cost function O'_d becomes:

$$O'_d = \sum_{o=1}^{N_d} \left[\beta_1(\mathbf{v}_o - \mathbf{v}_{ref_o})^2 + \beta_2(\mathbf{a}_o)^2 + \beta_3(\mathbf{p}_o - \mathbf{p}_{t_o})^2 + \delta E_o\right], \tag{17}$$

Operational Complexity Considerations: To account for operational complexities, we introduce a stochastic component $\xi_o(t)$ representing environmental factors affecting drone o at time t. The acceleration constraint can be revised as shown below:

$$\mathbf{a}_o = \mathbf{a}_{cmd_o} + \xi_o(t), \tag{18}$$

Comparative Effectiveness Analysis with V2V Systems: To compare the DDTP system with a V2V communication-based system, we can define a performance metric Π for both systems:

$$\Pi_{DDTP} = \eta_{eff} \sum_{o=1}^{N_d} O'_d(o) + \eta_{safety} S_{DDTP}, \tag{19}$$

$$\Pi_{V2V} = \eta_{eff} \sum_{o=1}^{N_t} O_t(o) + \eta_{safety} S_{V2V}, \tag{20}$$

where η_{eff} and η_{safety} are factors prioritizing efficiency and safety. S_{DDTP} and S_{V2V} are safety performance indicators for each system.

6. Simulation and Results

To evaluate the performance of the DDTP algorithm and analyze its effectiveness under various scenarios, we conducted a series of simulations using the Mesa multi-agent simulation framework. Mesa is a powerful Python library that allows for the creation, analysis, and visualization of agent-based models. In our case, the agents are the trucks and drones, each with their own set of characteristics and behaviors. The simulation environment facilitates the interaction between these agents, allowing us to observe the emergent behavior of the system as a whole.

The implementation of the agents and their interactions is based on object-oriented programming principles, making it easy to define the various components of our model and their interactions. The trucks and drones are represented as agents with their own state and behavior, and the environment is represented as a grid where these agents interact. The agents update their state in each time step of the simulation based on their own behavior and the state of the environment and other agents.

In the following use cases, we will delve into the specifics of the simulation scenarios, the behavior of the agents, and the results obtained. The goal is to provide a comprehensive understanding of how the DDTP algorithm performs in different conditions and to highlight its strengths and potential areas for improvement.

We first test the UKF in the following scenario: we track the position and velocity of a drone in 2D space with some process and measurement noise, as depicted in Figure 3. Steps:

- Initialize the UKF with a certain state and covariance.
- Simulate the drone's movement with a constant velocity and some noise.
- Use the UKF to estimate the drone's state over time.
- Compare the estimated state with the actual state.

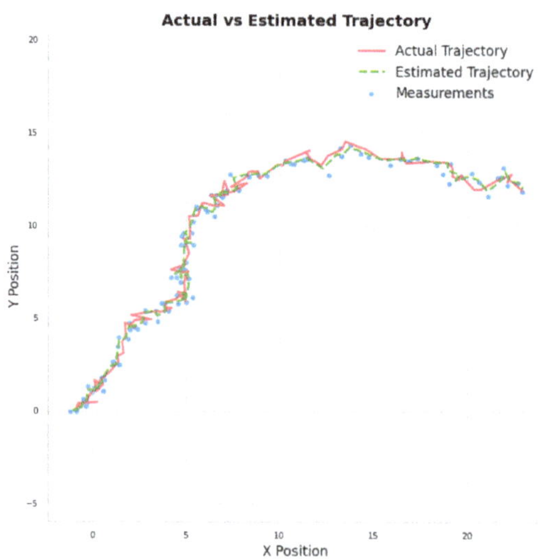

Figure 3. Actual vs. estimated trajectory of the drone. The red line represents the actual trajectory, which is plotted in 2D space with units in meters (m). The green dashed line shows the trajectory estimated by the UKF, which is also in meters. The blue dots symbolize the noisy measurements with their positions reflecting the drone's coordinates in meters (m).

The UKF seems to be doing a good job of estimating the drone's state despite the presence of process and measurement noise. The estimated trajectory closely follows the actual trajectory, and the measurements are scattered around the actual positions, as expected due to the added noise.

In Figure 3, the actual trajectory is depicted by a red line, representing the true path that the drone has taken. This trajectory shows a smooth and continuous movement, reflecting the simulated physics of the drone's motion. The estimated trajectory, shown as a dashed green line, represents the UKF's best estimate of the drone's path given the noisy measurements. The close alignment of the estimated trajectory with the actual trajectory demonstrates the effectiveness of the UKF in accurately tracking the drone's state even in the presence of noise. The scattered points in blue represent the noisy measurements of the drone's position. These points are visibly scattered around the actual trajectory, showcasing the impact of noise on the measurements. Despite this noise, the UKF is able to filter out the noise and provide a smooth and accurate estimate of the drone's trajectory, showcasing its robustness and reliability.

To evaluate the performance of DDTP, we conducted a series of simulations using various scenarios. In this section, we describe the simulation setup, which is followed by the results and analysis.

The simulation environment consists of highway segments with multiple lanes, where a group of trucks aims to form a platoon assisted by a swarm of drones. We consider the following parameters: number of trucks, number of drones, truck length, maximum truck velocity, maximum truck acceleration, maximum truck steering angle, maximum drone velocity, maximum drone acceleration, and communication range. The trucks and drones are initialized with random positions and velocities within the specified constraints. The desired platoon formation is set, and the drones are assigned to follow specific trucks. The simulations run for a predefined time horizon. Figure 4 illustrates the UKF-Based Multi-Drone Collision Avoidance.

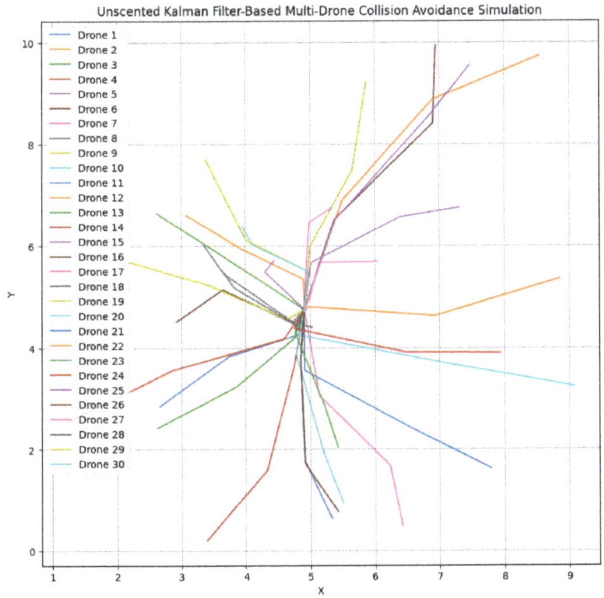

Figure 4. UKF-based multi-drone collision avoidance simulation: This plot visualizes the performance of a multi-drone system using UKFs for state estimation and a collision avoidance algorithm. The trajectories of each drone (indicated by different colored lines; a total of 30 drones) demonstrate effective collision avoidance in the presence of simulated wind effects and noisy measurements of drone positions. The drones' positions and velocities are measured in meters and meters per second (m/s), respectively. The decentralized approach enables drones to adapt their behavior in real time, ensuring safe operation in a dynamic environment.

Figure 4 is a comprehensive visualization of the paths taken by 30 drones in a two-dimensional space over 100 time units. The trajectories are influenced by both the consensus algorithm for target position calculation and the UKF for state estimation and collision avoidance. Each drone is represented by a unique trajectory, which is distinguished by color and labeled for identification. The trajectories showcase the effectiveness of the collision avoidance algorithm, as they do not intersect, indicating that drones successfully avoid each other throughout the simulation. The grid provides a clear reference for assessing the positions of the drones at any given time, and the square shape of the plot ensures an undistorted view of their movements. The enhanced plot size improves visibility and readability, ensuring that even with 30 drones, their paths are distinguishable and can be analyzed individually.

Next, in Figure 5, the simulation incorporates a dynamic wind field to evaluate the drones' performance under varying environmental conditions. The wind field is generated using a combination of sine and cosine functions, ensuring a smooth and continuous variation in wind speed and direction over time and space. Specifically, the wind components in the X and Y directions (W_x and W_y) are calculated as follows:

$$W_x(t, X) = W_{\max} \cdot \sin\left(0.1 \cdot t + \frac{0.5 \cdot \pi \cdot X}{L_x}\right), \tag{21}$$

$$W_y(t, Y) = W_{\max} \cdot \cos\left(0.1 \cdot t + \frac{0.5 \cdot \pi \cdot Y}{L_y}\right), \tag{22}$$

where $W_{\max}$ represents the maximum wind speed, t is the time, X and Y are the spatial coordinates, and L_x and L_y are the dimensions of the simulation grid. The factors of 0.1 and 0.5π in the sine and cosine functions introduce temporal and spatial periodicity, creating a varying wind pattern that influences the drones' movement.

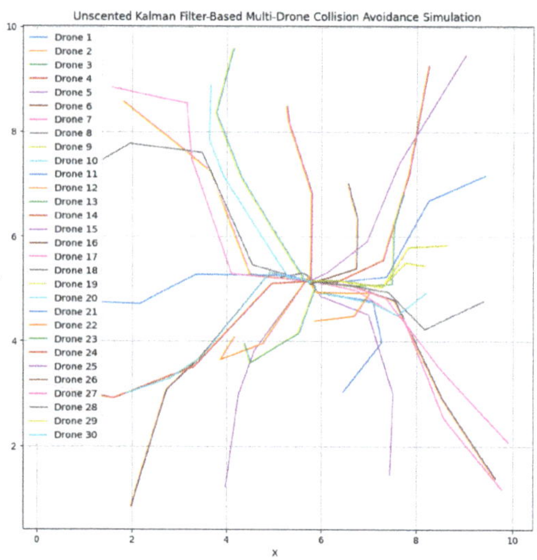

Figure 5. UKF-based multi-drone collision avoidance simulation with dynamic wind field: This plot illustrates the trajectories of 30 drones navigating in a 2D simulation space (dimensions in meters, m) over a 100-second time horizon. Each drone's trajectory, represented by a unique colored line, demonstrates its ability to adapt to a sinusoidal wind field (wind speed measured in meters per second, m/s) and avoid collisions with other drones.

The wind velocities at the specific positions of the drones are then interpolated from the generated wind field, providing a continuous and smooth variation in wind influence across the simulation grid. This dynamic and spatially varying wind field introduces an additional layer of complexity to the simulation, challenging the drones' navigation algorithms and highlighting their ability to maintain stable and collision-free trajectories under perturbed conditions.

Figure 5 provides a detailed visualization of 30 drones navigating through a dynamic environment influenced by wind and avoiding collisions using UKF for state estimation. Each drone's trajectory is distinctly presented, demonstrating their ability to adapt to environmental conditions while ensuring safe navigation. The sinusoidal wind field introduces varying challenges across the space, testing the robustness of the drones' control and estimation algorithms. The figure's size and square shape provide a clear and undistorted view of the drones' paths, highlighting the effectiveness of the collision avoidance mechanism. The grid enhances the visualization, making it easier to follow the drones' movements and assess their responses to the wind field and each other. The legend ensures clarity in distinguishing between the different drones.

Figure 6 depicts the advanced coordination strategy involving truck platooning. Figure 7 shows the performance metrics: distance traveled by each drone, time to reach goal positions, and average inter-drone distance over time.

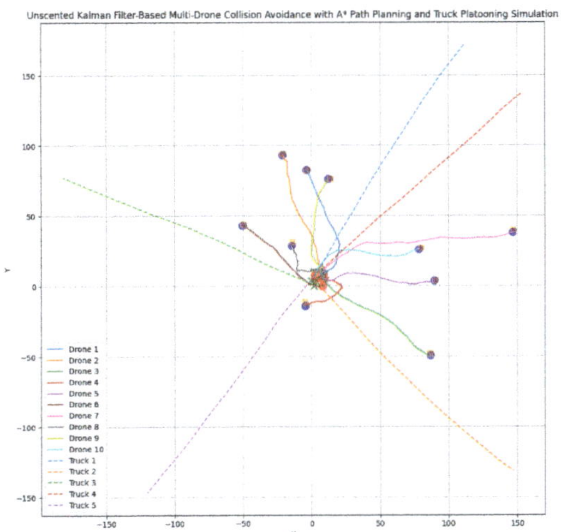

Figure 6. Advanced coordination using UKF and MPC: multi-drone collision avoidance with A* path planning and truck platooning simulation with 10 drones and 5 trucks.

The collision avoidance strategy for the multi-drone system is implemented using MPC in conjunction with the UKF for state estimation. The drones are assumed to operate in a two-dimensional space, and their states are represented by their positions and velocities. The UKF is employed to estimate these states in the presence of process and measurement noise, enhancing the robustness of the system.

The MPC is formulated to minimize the distance between the drones' positions and their respective target positions while ensuring collision avoidance. The optimization problem is subject to constraints that prevent the drones from coming closer than a predefined safety distance. The optimization problem is solved using the SLSQP method.

The simulation is set up with a variable number of drones operating in a defined space with initial random positions and velocities. The drones are tasked to reach target positions while avoiding collisions, which is influenced by a dynamic wind field.

The A* path-planning algorithm is integrated to guide the drones toward their target positions while avoiding obstacles. The space is discretized into a grid, and the algorithm finds the shortest path from the drone's current position to its target.

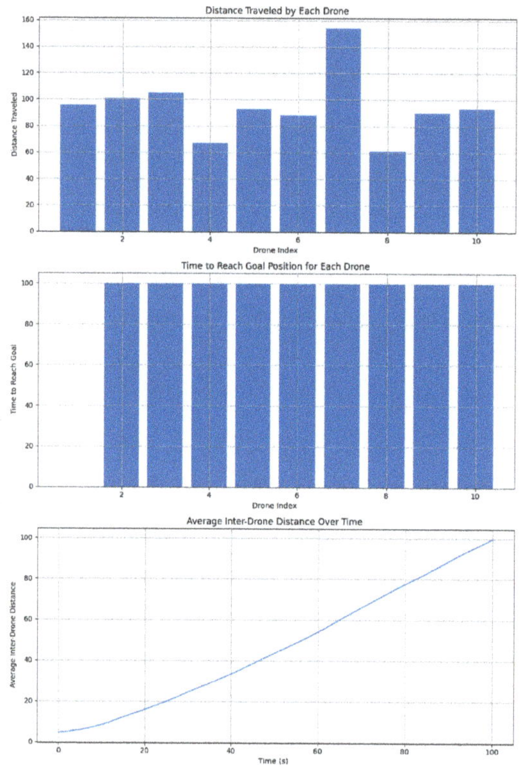

Figure 7. Performance analysis of the multi-drone collision avoidance system with A* path planning and truck platooning. The figure depicts three key metrics with distances measured in meters (m): (1) the total distance traveled by each drone, represented as individual bars for each drone indexed from 1 to 10; (2) the time taken, measured in seconds, by each drone to reach its assigned goal position, which is also displayed as bars corresponding to each drone; and (3) the average distance maintained between drones over time, which is plotted as a continuous line. The x-axis in the first two subplots represents the drone index, while in the third subplot, it indicates elapsed time in seconds.

The drones are configured to fly in a formation, and performance metrics such as distance traveled, time to reach the goal, and average inter-drone distance are calculated and visualized.

This implementation and simulation demonstrate the effectiveness of the integrated MPC and UKF approach in ensuring the collision-free navigation of multiple drones in a dynamic environment.

Figures 6 and 7 provide a comprehensive view of the multi-drone navigation scenario, showcasing the path planning, collision avoidance, and drone dynamics handling capabilities of the implemented algorithms. They also offer insights into the performance of the system, helping to evaluate its efficiency and safety.

In particular, Figure 6 showcasing the drone trajectories and positions provides a vivid representation of the complex dynamics involved in multi-drone navigation and collision avoidance. The different colored lines illustrate the unique path taken by each drone, revealing how they navigate through space to reach their respective goal positions while adhering to the constraints imposed by the UKF and A* path-planning algorithms. The red circles, representing the initial positions, indicate a random starting point for each drone. The trajectories emanate from these points, showcasing the movement of the drones as they progress through the simulation. The blue circles at the end of each trajectory signify the final positions of the drones, providing a clear visual indication of where each drone ended up at the conclusion of the simulation. The green crosses mark the goal positions, serving as the target destinations for the drones. The trajectories show how each drone maneuvers toward its goal, taking into account the need to avoid collisions with other drones. The orange squares represent the formation positions, which are intermediary targets for the drones as they make their way toward their final destinations. These positions help maintain an organized structure among the drones, ensuring a coordinated movement that minimizes the risk of collisions. The grid overlay adds a layer of precision to the visualization, allowing for a more accurate interpretation of the drones' positions in the 2D space. The legend helps distinguish the paths of different drones, providing clarity in the midst of the complex web of trajectories.

The performance metrics in Figure 7, divided into three subplots, offers a comprehensive analysis of the efficiency and safety of the drone navigation system. Each subplot focuses on a specific aspect of performance, providing insights that are crucial for evaluating the success of the algorithms and identifying areas for improvement. In the first subplot, the distance traveled by each drone is displayed, allowing for a comparison across all drones. The variation in the distances traveled highlights the individual challenges faced by each drone in navigating to its goal position. Drones that had to travel longer distances or navigate around obstacles and other drones would have longer trajectories, which are reflected in the higher bars. The second subplot provides a look into the time efficiency of the system, displaying the time taken by each drone to reach its goal position. Shorter times indicate a more efficient path, while longer times may suggest the need for detours to avoid collisions or navigate around obstacles. This metric is crucial for applications where time efficiency is of the essence, such as in delivery drones or emergency response scenarios. The third subplot shows the average inter-drone distance over time, serving as a key indicator of the safety of the system. Maintaining a safe distance between all pairs of drones reduces the risk of collisions and ensures a safer operational environment. A consistent high average distance throughout the simulation is indicative of the effectiveness of the collision avoidance algorithm.

To sum up, our simulation environment implemented in Python allowed us to evaluate the DDTP under various conditions. We analyzed the performance of the DDTP algorithm in terms of drone formation maintenance, which is evaluated by measuring the deviation of the drones from their assigned formation relative to the trucks. The results indicate that the drones successfully maintained their formation with minor deviations from their target positions. The average deviation stayed within acceptable limits, suggesting that the drones effectively adapted their behavior based on the trucks' dynamics and constraints.

While our actual simulations provide a detailed analysis of the DDTP algorithm, focusing primarily on drone behavior and their interaction with trucks, we acknowledge the need for a broader evaluation that includes additional critical aspects. The current framework has not explicitly addressed the nuances of truck platooning dynamics, particularly in terms of platoon stability under varying conditions and the impact of uncertainties like communication delays and packet loss. These elements are crucial for a comprehensive assessment of the DDTP system in real-world scenarios.

Future enhancements to our simulation could include the implementation of more sophisticated models to simulate communication challenges, such as delays and packet loss, which are inherent in real-world networks. Incorporating these aspects would provide a more realistic representation of the communication constraints and their impact on the system's performance. Moreover, a deeper focus on platoon stability, considering external disturbances and the effectiveness of control algorithms, would significantly enhance our understanding of the system's robustness and reliability.

Additionally, exploring the integration of energy consumption and battery life constraints in drone operations would add another layer of realism, reflecting the practical limitations and operational challenges in drone-swarm coordination. The dynamic nature of the environment and the system's uncertainties also suggest the potential benefit of implementing adaptive control algorithms. These would allow the system to adjust to changing conditions and maintain optimal performance. Lastly, the dynamic changes in network topology, due to drone movements or communication range limits, present an interesting avenue for future research to evaluate the resilience and efficiency of the communication network among the drones.

Consideration on System Dynamics and Drone Assistance

To provide a more comprehensive understanding of the DDTP algorithm's performance in realistic scenarios, we have expanded our simulation framework qualitatively to include factors such as traffic density, environmental conditions, and the efficacy of drone-assisted communication. This enhanced approach allows us to evaluate the system's robustness under varying conditions and to assess the potential improvements offered by drone technology in communication systems.

The simulation considers three levels of traffic density (Low, Medium, High) and three types of environmental conditions (Clear, Moderate, High). We investigate the impact of these factors on two critical communication metrics: communication delay and packet loss rate. These metrics are crucial in assessing the efficiency and reliability of the DDTP system in real-world conditions.

The communication delay, measured in milliseconds, is assumed to increase with higher traffic density and worse environmental conditions. The delay is formulated to be dependent on both the number of trucks and the conditions of the environment. The base delay is set for clear conditions and low traffic with additional increments for each level of traffic density and environmental condition. This approach provides a realistic representation of the challenges faced in high-density traffic scenarios and adverse environmental conditions.

Similarly, the packet loss rate, represented in percentage, is influenced by traffic density. The base packet loss rate is defined for clear conditions, with additional increments for higher traffic densities. This metric is essential to understand how reliable the communication is among the trucks and drones under different traffic conditions.

Recognizing the potential of drones in enhancing communication systems, we introduce reduction factors to simulate the improvements brought about by drone-assisted communication. These factors represent the anticipated enhancements in communication delay and packet loss rate due to the efficiency of drone communication systems. The simulation compares the original and drone-assisted communication metrics to highlight the impact of drones in improving the system's performance.

The simulation results are visualized through a series of plots, comparing the communication delay and packet loss rate across different traffic densities, environmental conditions, and numbers of trucks. The plots distinctly illustrate the improvement in communication metrics when drone assistance is factored in, showcasing the potential of integrating drone technology by means of the DDTP system.

In Figure 8, the first set of plots presents the original communication delay and packet loss rate without drone assistance. Following this, another set of plots demonstrates the expected reduced communication delay and packet loss rate when drone assistance is incorporated. This comparative analysis shows by means of visualization the benefits of using drones for communication purposes in complex traffic and environmental scenarios.

Figure 9 incorporates confidence intervals while focusing on a side-by-side comparison of the communication metrics with and without the intervention of UAV technology.

In developing our communication model for the DDTP system, several key assumptions were made to ensure a realistic representation of real-world scenarios. Firstly, we assumed that the communication range between drones and trucks is limited, reflecting the practical constraints of wireless communication systems. This range limitation plays a crucial role in the formation and maintenance of the communication network, particularly in dense traffic conditions or complex environments.

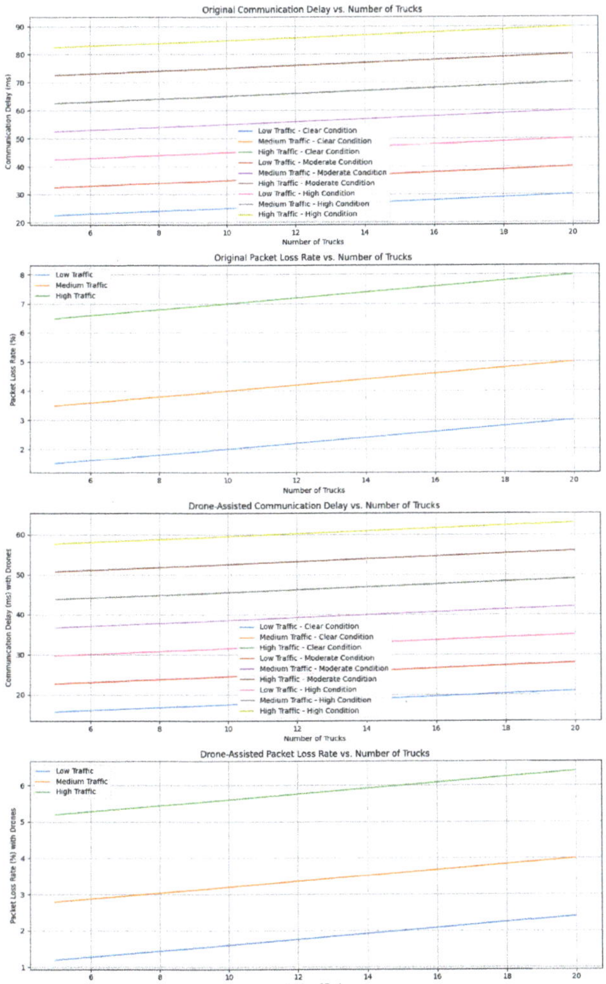

Figure 8. Visual comparative analysis of communication metrics in a simplified DDTP system with and without drone assistance: The figure comprises four plots illustrating the impact of traffic density and environmental conditions on communication delay and packet loss rate for different numbers of trucks. The top two plots display the original communication delay and packet loss rate without drone assistance, revealing how these metrics escalate with increasing traffic density and deteriorating environmental conditions. The bottom two plots demonstrate the expected improvements achieved through drone-assisted communication with noticeable reductions in both communication delay and packet loss rate.

Secondly, we considered the impact of environmental factors, such as weather conditions and physical obstructions, on communication reliability. These factors can cause fluctuations in signal strength and quality, leading to increased communication delays and packet loss rates. By incorporating these variables into our model, we aimed to mirror the challenges encountered in real-world vehicular communication networks.

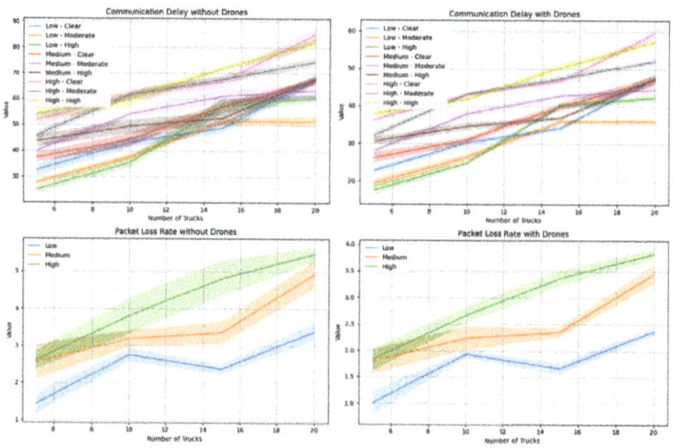

Figure 9. Impact of drone assistance on communication delay and packet loss rate: This figure presents a comparative analysis of communication delay and packet loss rate in a DDTP system under varying traffic densities and environmental conditions both with and without drone assistance. The top row illustrates the communication delay, while the bottom row focuses on the packet loss rate. Each column represents the metrics without (left) and with (right) drone assistance, respectively. The shaded areas around each line indicate confidence intervals.

Furthermore, our model assumed a time-variant nature of the communication links, acknowledging that the quality of communication can change dynamically over time. This aspect is particularly relevant in mobile networks, where the relative positions of trucks and drones continuously evolve, affecting the connectivity and stability of the communication links.

In terms of accuracy, our communication model strives to realistically capture the complexities and uncertainties inherent in vehicular networks. The introduction of drone-assisted communication is based on the premise that drones can provide more stable and reliable communication links, especially in scenarios where direct line-of-sight communication between trucks is not feasible. However, it is important to note that while our model incorporates key factors affecting communication, it remains a simplified representation. Real-world communication systems may face additional challenges, such as spectrum congestion, interference from other electronic devices, and regulatory constraints on communication frequencies.

The results from our simulation, as depicted in Figure 9, offer valuable insights into the potential improvements in communication efficiency through drone assistance. However, these results should be interpreted with an understanding of the model's assumptions and inherent limitations. Future research could focus on refining the communication model by integrating more detailed environmental data, advanced signal propagation models, and adaptive communication protocols to enhance its realism and predictive accuracy. This approach would not only strengthen the validity of the simulation results but also provide a more comprehensive understanding of the potential impact of drone-assisted communication in real-world truck platooning scenarios.

7. Discussion

In this section, we present a comparative analysis of various V2V communication and coordination approaches in the context of platooning and traffic management. The focus is on contrasting the DDTP method with existing strategies, highlighting key differences in control algorithms, communication strategies, scalability, performance metrics, limitations, and applicable scenarios. Table 3 summarizes these comparisons.

Table 3. Comparative analysis of V2V communication and coordination approaches.

Method/Approach	Control Algorithm	Communication Strategy	Scalability	Performance Metrics	Limitations	Application Scenarios
DDTP	MPC and UKF	Decentralized	High	Collision avoidance, Formation accuracy	Cost, Real-world validation	Highway truck platooning
CACC-Based Platooning [35,36]	Cooperative Adaptive Cruise Control (CACC)	V2V	Moderate	Traffic flow efficiency, Safety	Communication reliability	Highway and urban traffic management
V2V Communication [37,38]	Linear Quadratic Regulator (LQR)	Decentralized	Moderate	Fuel efficiency, Safety	Communication latency, Range	Smart city transportation, Platooning
Autonomous Drone-Assisted Traffic Monitoring [39]	Simple Heuristics	Drone-to-Vehicle	Moderate	Traffic congestion analysis, Response time	Drone battery life, Data processing	Urban traffic surveillance

The DDTP approach, which combines MPC and UKF, is characterized for its decentralized communication strategy and high scalability, making it well-suited for highway truck platooning. The performance metrics primarily focus on collision avoidance and formation accuracy with ongoing challenges related to cost and the need for real-world validation.

Comparatively, CACC-based platooning, as explored in [35,36], utilizes V2V communication to improve traffic flow efficiency and safety. While this approach has moderate scalability, its reliance on communication reliability can be a limiting factor, especially in dense urban environments.

The use of Linear Quadratic Regulator (LQR) in V2V communication systems [37,38] offers improvements in fuel efficiency and safety for smart city transportation and platooning. However, challenges related to communication latency and range can impact its effectiveness in dynamic traffic conditions.

Lastly, the autonomous drone-assisted traffic monitoring approach described in [39] uses simple heuristic algorithms and drone-to-vehicle communications. This method is particularly useful for urban traffic surveillance, analyzing traffic congestion, and improving response times. The limitations of this approach include drone battery life and the need for efficient data processing.

In conclusion, the study of diverse V2V communication and coordination strategies reveals the multifaceted nature of implementing technology in traffic management and platooning. While each approach offers unique benefits, the DDTP method shines in its adaptability and potential for addressing the specific challenges of highway truck platooning. Our analysis highlights the importance of considering a broad range of factors, including scalability, communication strategy, and control algorithms, in the development and application of these technologies. The future of V2V communication and coordination in vehicular networks holds significant promise, and approaches like DDTP represent a step forward in harnessing this potential. Emphasizing the continuous evolution and adaptation of these systems, further research and real-world implementations will be essential in optimizing their effectiveness and addressing existing limitations.

8. Conclusions and Future Work

In this paper, we presented a Decentralized Drone-Based Truck Platooning (DDTP) that combines Model Predictive Control (MPC) and Unscented Kalman Filter (UKF) techniques to achieve efficient and safe platooning on highways. The proposed framework is capable of adapting in real time to the trucks' dynamics and constraints, maintaining the desired formation, and assisting the trucks in platooning efficiently and safely. The simulation results demonstrated the effectiveness of the DDTP algorithm in terms of drone formation maintenance. As future work, we plan to improve the proposed framework by addressing (1) handling complex scenarios, such as dynamic traffic and varying environmental conditions, requiring advanced control algorithms and additional sensor data; (2) the implementation of multi-objective optimization for additional objectives, like fuel consumption minimization and collision risk reduction; (3) improvements in scalability and robustness for larger fleets while tackling communication failures and uncertainties through advanced consensus algorithms and fault-tolerant strategies; and (4) a validation of the framework via real-world experiments to assess performance, safety, and reliability under realistic conditions.

In practical terms, the results of our research have significant implications for the future of automated transportation systems particularly in the realm of long-haul trucking and logistics. The DDTP framework, leveraging advanced MPC and UKF techniques, is poised for potential implementation in scenarios where highways are equipped with smart infrastructure, enabling real-time communication and coordination between drones and trucks. However, it is important to acknowledge the limitations of our current work. Our simulation-based study primarily focuses on ideal conditions and controlled environments. In reality, the performance of DDTP may be influenced by factors such as unpredictable traffic patterns, diverse weather conditions, and the varying reliability of communication networks. Additionally, the complexity and cost of the required technology and infrastructure pose challenges for widespread adoption. As we move forward, it is crucial to address these limitations through comprehensive real-world testing, refinement of the control algorithms to handle dynamic and unpredictable environments, and the exploration of cost-effective and scalable solutions for technology deployment.

Author Contributions: Author contributions: conceptualization, J.d.C. and I.d.Z.; funding acquisition, C.T.C., P.M. and J.C.C.; investigation, I.d.Z. and J.d.C.; methodology, I.d.Z. and J.d.C.; software, J.d.C. and I.d.Z.; supervision, C.T.C., J.C.C. and P.M., writing—original draft, J.d.C. and I.d.Z.; writing—review and editing, C.T.C., J.C.C., P.M., J.d.C. and I.d.Z. All authors have read and agreed to the published version of the manuscript.

Funding: We thank the following funding sources from GOETHE-University Frankfurt am Main; 'DePP—Dezentrale Plannung von Platoons im Straßengüterverkehr mit Hilfe einer KI auf Basis einzelner LKW', 'Center for Data Science & AI' and 'HessianAI–AI Biology'. We acknowledge the support of R&D project PID2021-122580NB-I00 funded by MCIN/AEI/10.13039/501100011033 and ERDF.

Data Availability Statement: Data is contained within the article.

Conflicts of Interest: The authors declare that they have no conflicts of interest. The funders had no role in the design of the study, in the collection, analyses, or interpretation of data, in the writing of the manuscript, or in the decision to publish the results.

Abbreviations

The following abbreviations are used in this manuscript:

MPC	Model Predictive Control
UKF	Unscented Kalman Filter
IoT	Internet of Things
ACC	Adaptive Cruise Control
CACC	Cooperative Adaptive Cruise Control
V2V	Vehicle-to-Vehicle
SLSQP	Sequential Least Squares Quadratic Programming
QP	Quadratic Programming
DDTP	Decentralized Drone-Based Truck Platooning

References

1. van de Hoef, S.; Johansson, K.H.; Dimarogonas, D.V. Computing feasible vehicle platooning opportunities for transport assignments. *IFAC-PapersOnLine* **2016**, *49*, 43–48. [CrossRef]
2. Boysen, N.; Briskorn, D.; Schwerdfeger, S. The identical-path truck platooning problem. *Transp. Res. Part Methodol.* **2018**, *109*, 26–39. [CrossRef]
3. Zeng, Y.; Wang, M.; Rajan, R.T. Decentralized coordination for truck platooning. *Comput.-Aided Civ. Infrastruct. Eng.* **2022**, *37*, 1997–2015. [CrossRef]

4. Larsson, E.; Sennton, G.; Larson, J. The vehicle platooning problem: Computational complexity and heuristics. *Transp. Res. Part Emerg. Technol.* **2015**, *60*, 258–277. [CrossRef]
5. Bisio, I.; Garibotto, C.; Haleem, H.; Lavagetto, F.; Sciarrone, A. systematic review of drone based road traffic monitoring system. *IEEE Access* **2022**, *10*, 101537–101555. [CrossRef]
6. Shi, W.; Zhou, H.; Li, J.; Xu, W.; Zhang, N.; Shen, X. Drone assisted vehicular networks: Architecture, challenges and opportunities. *IEEE Netw.* **2018**, *32*, 130–137. [CrossRef]
7. Russell, S.J.; Norvig, P. *Artificial Intelligence: A Modern Approach*, 4th ed.; Pearson: London, UK, 2021.
8. Elbert, R.; Roeper, F.; Rammensee, M. Decentralized coordination of platoons–A conceptual approach using deep reinforcement learning. In Proceedings of the World Conference on Transport Research—WCTR, Montreal, QC, Canada, 17–21 July 2023.
9. Vahidi, A.; Eskandarian, A. Research advances in intelligent collision avoidance and adaptive cruise control. *IEEE Trans. Intell. Transp. Syst.* **2003**, *4*, 143–153. [CrossRef]
10. Dey, K.C.; Yan, L.; Wang, X.; Wang, Y.; Shen, H.; Chowdhury, M.; Yu, L.; Qiu, C.; Soundararaj, V. A review of communication, driver characteristics, and controls aspects of cooperative adaptive cruise control (CACC). *IEEE Trans. Intell. Transp. Syst.* **2016**, *17*, 491–509. [CrossRef]
11. Park, S.; Nam, S.; Sankar, G.S.; Han, K. Evaluating the Efficiency of Connected and Automated Buses Platooning in Mixed Traffic Environment. *Electronics* **2022**, *11*, 3231. [CrossRef]
12. Aslam, A.; Santos, P.M.; Santos, F.; Almeida, L. Empirical performance models of MAC protocols for cooperative platooning applications. *Electronics* **2019**, *8*, 1334. [CrossRef]
13. Liang, K.Y.; Mårtensson, J.; Johansson, K.H. Heavy-duty vehicle platoon formation for fuel efficiency. *IEEE Trans. Intell. Transp. Syst.* **2016**, *17*, 1051–1061. [CrossRef]
14. Chang, X.; Li, H.; Rong, J.; Zhao, X.; Li, A. Analysis on traffic stability and capacity for mixed traffic flow with platoons of intelligent connected vehicles. *Phys. Stat. Mech. Its Appl.* **2020**, *557*, 124829. [CrossRef]
15. Amato, C.; Konidaris, G.; Cruz, G.; Maynor, C.A.; How, J.P.; Kaelbling, L.P. Planning for decentralized control of multiple robots under uncertainty. In Proceedings of the 2015 IEEE International Conference on Robotics and Automation (ICRA), Seattle, WA, USA, 26–30 May 2015; pp. 1241–1248.
16. Weyns, D.; Holvoet, T.; Schelfthout, K.; Wielemans, J. Decentralized control of automatic guided vehicles: Applying multi-agent systems in practice. In Proceedings of the Companion to the 23rd ACM SIGPLAN Conference on Object-Oriented Programming Systems Languages and Applications, Nashville, TN, USA, 19–23 October 2008; pp. 663–674.
17. Camacho, E.F.; Alba, C.B. *Model predictive Control*; Springer Science & Business Media: Berlin/Heidelberg, Germany, 2013.
18. Wan, E.A.; Van Der Merwe, R. The unscented kalman filter for nonlinear estimation. In Proceedings of the IEEE 2000 Adaptive Systems for Signal Processing, Communications, and Control Symposium (Cat. No. 00EX373), Lake Louise, AB, Canada, 4 October 2000; pp. 153–158.
19. Menegaz, H.M.; Ishihara, J.Y.; Borges, G.A.; Vargas, A.N. systematization of the unscented kalman filter theory. *IEEE Trans. Autom. Control.* **2015**, *60*, 2583–2598. [CrossRef]
20. Mushtaq, A.; Haq, I.U.; Nabi, W.U.; Khan, A.; Shafiq, O. Traffic flow management of autonomous vehicles using platooning and collision avoidance strategies. *Electronics* **2021**, *10*, 1221. [CrossRef]
21. Ahmad, M.; Khan, Z.; Koubaa, A.; Boulila, W. A microscopic platoon stability model using vehicle-to-vehicle communication. *Electronics* **2022**, *11*, 1994. [CrossRef]
22. López, L.B.; van Mane, N.; van der Zee, E.; Bos, S. *DroneAlert: Autonomous Drones for Emergency Response*; Springer International Publishing: Cham, Switzerland, 2017; pp. 303–321.
23. Chung, S.H. Applications of smart technologies in logistics and transport: A review. *Transp. Res. Part Logist. Transp. Rev.* **2021**, *153*, 102455. [CrossRef]
24. Guo, G.; Wen, S. Communication scheduling and control of a platoon of vehicles in VANETs. *IEEE Trans. Intell. Transp. Syst.* **2015**, *17*, 1551–1563. [CrossRef]
25. Zhang, Y.; Xu, Z.; Wang, Z.; Yao, X.; Xu, Z. Impacts of communication delay on vehicle platoon string stability and its compensation strategy: A review. *J. Traffic Transp. Eng.* **2023**, *10*, 508–529.
26. Godinho, D.A.; Neto, A.A.; Mozelli, L.A.; de Oliveira Souza, F. A Strategy for Traffic Safety of Vehicular Platoons Under Connection Loss and Time-Delay. *IEEE Trans. Intell. Transp. Syst.* **2023**, *24*, 6627–6638. [CrossRef]
27. Rajamani, R. *Vehicle Dynamics and Control*; Springer Science & Business Media: Berlin/Heidelberg, Germany, 2011.
28. Barth, M.; Boriboonsomsin, K. Real-world carbon dioxide impacts of traffic congestion. *Transp. Res. Rec.* **2008**, *2058*, 163–171. [CrossRef]
29. Hoffmann, G.; Huang, H.; Waslander, S.; Tomlin, C. Quadrotor helicopter flight dynamics and control: Theory and experiment. In Proceedings of the AIAA Guidance, Navigation and Control Conference and Exhibit, Hilton Head, SC, USA, 20–23 August 2007.
30. Hart, P.E.; Nilsson, N.J.; Raphael, B. A formal basis for the heuristic determination of minimum cost paths. *IEEE Trans. Syst. Sci. Cybern.* **1968**, *4*, 100–107. [CrossRef]
31. Uhlemann, E. Platooning: Connected vehicles for safety and efficiency [Connected Vehicles]. *IEEE Veh. Technol. Mag.* **2016**, *11*, 13–18. [CrossRef]
32. Niu, Z.; Shen, X.S.; Zhang, Q.; Tang, Y. Space-air-ground integrated vehicular network for connected and automated vehicles: Challenges and solutions. *Intell. Converg. Netw.* **2020**, *1*, 142–169. [CrossRef]

33. Gupta, A.; Fernando, X. Simultaneous localization and mapping (slam) and data fusion in unmanned aerial vehicles: Recent advances and challenges. *Drones* **2022**, *6*, 85. [CrossRef]
34. Hu, J.; Chen, C.; Cai, L.; Khosravi, M.R.; Pei, Q.; Wan, S. UAV-assisted vehicular edge computing for the 6G internet of vehicles: Architecture, intelligence, and challenges. *IEEE Commun. Stand. Mag.* **2021**, *5*, 12–18. [CrossRef]
35. Shladover, S.; Lu, X.Y.; Yang, S.; Ramezani, H.; Spring, J.; Nowakowski, C.; Nelson, D. Cooperative Adaptive Cruise Control (CACC) for Partially Automated Truck Platooning. Technical Report, Department of Transportation, State of California, 2018. Available online: https://escholarship.org/content/qt260060w4/qt260060w4.pdf (accessed on 3 December 2023)
36. Partani, S.; Weinand, A.; Schotten, H.D. A Controller for Network-Assisted CACC based Platooning. In Proceedings of the Mobile Communication-Technologies and Applications; 24. ITG-Symposium, Osnabrueck, Germany, 15–16 May 2019; pp. 1–6.
37. Alam, A.; Besselink, B.; Turri, V.; Mårtensson, J.; Johansson, K.H. Heavy-duty vehicle platooning for sustainable freight transportation: A cooperative method to enhance safety and efficiency. *IEEE Control. Syst. Mag.* **2015**, *35*, 34–56.
38. Kaku, A.; Mukai, M.; Kawabe, T. A centralized control system for ecological vehicle platooning using linear quadratic regulator theory. *Artif. Life Robot.* **2012**, *17*, 70–74. [CrossRef]
39. Motlagh, N.H.; Kortoçi, P.; Su, X.; Lovén, L.; Hoel, H.K.; Haugsvær, S.B.; Srivastava, V.; Gulbrandsen, C.F.; Nurmi, P.; Tarkoma, S. Unmanned aerial vehicles for air pollution monitoring: A survey. *IEEE Internet Things J.* **2023**. [CrossRef]

Disclaimer/Publisher's Note: The statements, opinions and data contained in all publications are solely those of the individual author(s) and contributor(s) and not of MDPI and/or the editor(s). MDPI and/or the editor(s) disclaim responsibility for any injury to people or property resulting from any ideas, methods, instructions or products referred to in the content.

electronics

MDPI

Article

Vehicular Localization Framework with UWB and DAG-Based Distributed Ledger for Ensuring Positioning Accuracy and Security

Kiseok Kim †, Sangmin Lee †, Taehoon Yoo † and Hwangnam Kim *

School of Electrical Engineering, Korea University, Seoul 02841, Republic of Korea; kisuk528@korea.ac.kr (K.K.); lsm5505@korea.ac.kr (S.L.); yth921@korea.ac.kr (T.Y.)

* Correspondence: hnkim@korea.ac.kr; Tel.: +82-2-3290-4821

† These authors contributed equally to this work.

Abstract: In performing various missions using various types of vehicles or other moving objects, the positioning of each agent within the swarm is essential. In particular, for missions that require precise location estimation, the case of malicious attacks through data forgery cannot be excluded. In this paper, we propose a highly secure and accurate localization framework utilizing a Directed-Acyclic-Graph (DAG)-based distributed ledger as an intelligent vehicle network for an Ultra-Wideband (UWB) positioning system. When performing UWB positioning, the data generated from each node are used to calculate the Time of Flight (ToF), and if any of them are tampered with, the overall positioning performance is greatly reduced. We prevented positioning performance degradation by ensuring the safety and integrity of the data by applying a chain-based logical network between each node utilized for UWB positioning. The experimental results indicated that the proposed framework was effective at providing system stability and security without affecting the UWB positioning performance. In addition, the performance of the framework was verified by presenting defense indicators for various attack scenarios.

Keywords: vehicle network; UWB; accurate positioning; DAG; consensus algorithm

1. Introduction

Vehicular positioning systems are actively used in various fields such as autonomous vehicle driving, production, the distribution industry, and the military [1]. Generally, the Global Positioning System (GPS) [2], which utilizes satellites, is used for the positioning of ground vehicles or aerial vehicles. However, GPS has the disadvantage that the reception signal is weak in complex urban terrain and cannot be used in indoor environments. This is because GPS estimates the location through direct communication between a mobile node and a subset of satellites, and in environments with many obstacles or indoors, where radio reception is not possible, it may not be possible to accurately measure the distance between them due to the straightness of radio waves [3].

Furthermore, the accuracy of GPS is generally less accurate than that of indoor positioning systems. In automotive services that require a high precision for the indoor positioning system, metrics other than GPS are often adopted to meet the strict requirements for accuracy and reliability [4]. In these environments, various vehicular positioning systems that build their own infrastructure to estimate the location or the target are actively used [5]. In particular, the Vehicular Ad hoc Network (VANET) positioning system [6] addressed in this paper is mainly used when performing missions with a specific purpose in a robot or vehicle network or performing support tasks for other vehicles in the swarm. Application examples include positioning for autonomous parking in an underground parking lot, drone shows using a swarm of micro drones, real-time traffic information sharing, various missions using unmanned vehicles in spaces where human have difficulty entering, and smart factories and roads [7].

Vehicular positioning systems can utilize various sensors and wireless signals, such as Inertial-Measurement-Unit (IMU)-based Dead Reckoning (DR) [8], Bluetooth [9], ultrasonic waves [10], WiFi [11], and Ultra-Wideband (UWB) [12]. Among them, UWB is the metric that has recently received the most attention in VANET positioning and is actively used and researched in various fields. Compared to other systems, the UWB positioning system has strengths in various aspects, such as high location estimation accuracy, reasonable configuration price, and high system robustness [13].

Distance-based UWB positioning systems generally use the Time Difference of Arrival (TDOA) [14–16] and Two-Way Ranging (TWR) [17,18] in the ranging process, and both methods utilize device timeline data. Distance-based positioning has advantages such as no major constraints on the infrastructure configuration, high accuracy, and no severe interference from multiple paths, but it has a fatal disadvantage in that the location estimation error is greatly affected by distance. Generally, in order to perform n-dimensional positioning, n+1 pieces of distance information are used, and if even one distance has an unexpected value, the overall positioning performance will be seriously degraded.

The core strength of the UWB system is high-accuracy positioning based on distance measurement with high time resolution by using pulse waves. In other words, the precise measurement of timestamp, which is the device timeline data, is possible. In missions where positioning accuracy is key, if the transmitting device ID or receiving timestamp data for the UWB signal are unclear or have an error in their values, the probability of the task failing can greatly increase. Even if an attacker with a malicious intent manipulates the data of only one device among the nodes that make up the system, serious degradation of the positioning performance may occur, leading to task failure. As a result, distance-based ad hoc device indoor positioning systems such as UWB systems have the disadvantage of being quite vulnerable to external attacks.

When positioning data (in our case, the transmitting device ID for the received signal and the device timestamp data at the time of signal reception) are forged or altered due to a malicious attack on mission performance, the overall performance of the positioning system is greatly reduced, and the mission success rate is also significantly reduced. So, we propose a localization framework applying a Directed-Acyclic-Graph (DAG)-based distributed ledger to build a positioning system that not only securely processes and manages the sensor data generated from the positioning nodes, but also guarantees deterministic performance without degradation, depending on the network scale.

Most distributed ledger systems applied to IoT networks, such as positioning systems, are based on blockchain and, in many cases, adopt the Proof-of-Work (PoW) as the consensus algorithm [19]. This method basically performs a hashing operation with a nonce value as the input to create a block, so it requires high computing power and causes significant delay due to the complex verification method. Moreover, the data handled in the positioning system comprise a micro-transaction with a simple structure. However, the unique nature of blockchain, which requires block generation, makes it difficult to create micro-transactions in the positioning system [20]. Our proposal adopts a DAG-based distributed ledger technology. Compared to blockchain, it has a relatively simple verification method and does not require a separate hashing operation to process the transactions, resulting in high processing speed. Additionally, it is more suitable for handling micro-transactions because it processes and manages each transaction as an individual task. Moreover, due to the nature of DAG technology, the processing speed increases as the number of participating nodes increases. This fact can give our proposal many advantages in solving scalability problems in large-scale network environments such as smart cities, smart grids, and vehicular networks [21].

Since positioning systems are mainly used for tracking or monitoring purposes, high real-time performance is required. These systems are inevitably sensitive to latency. Even with DAG-based technology, the delay time and workload that occur in the process of creating, verifying, and confirming transactions and synchronizing transaction information cannot be ignored [22]. For these reasons, we propose a new consensus algorithm suitable for the positioning environment to reduce the delay time that occurs in the ledger layer during the positioning process and improve the transaction processing speed. It is a global state consensus based on a threshold and proposes a method to quickly confirm micro-transactions for positioning.

Based on the descriptions so far about using distributed ledger technology in a localization environment, the contributions of this article are summarized as follows:

- Distance-based positioning methods that use multiple distance values and the fixed coordinates of anchors have default errors and serious distortion problems in noisy environments. To solve this, we propose an enhanced localization technique using a nonlinear least-squares optimization technique and verified its performance.
- We propose a new consensus algorithm applicable to DAG-based distributed ledgers and suitable for positioning systems. Due to the nature of existing DAG-based distributed ledgers, individual nodes manage their own local state, which can cause excessive computing load and delay the time to reach consensus. We applied a new consensus algorithm to improve the efficiency during the consensus-reaching process and describe in detail how to apply it in the target environment.
- We designed a complete localization framework with a DAG-based distributed ledger equipped with the proposed localization technology and consensus algorithm. Through this, we systematized and concretized our suggestions. In addition, we provide an environment for the libraries and tools to utilize related technologies along with fully implemented functions. This sets the stage for development expansion and reduces future development costs.

The remaining part of this paper is organized as follows. We introduce the related work in Section 2. In Section 3, we give a detailed explanation of our positioning system and consensus algorithm. Section 4 evaluates the performance of the proposed localization framework. Finally, Section 5 concludes this paper.

2. Related Work

2.1. UWB Localization

UWB localization is a technology that performs precise location estimation using ultra-wideband. UWB localization generally uses the Time of Arrival (ToA) [23–25] and Angle of Arrival (AoA) [26–28] to perform distance and angle-based positioning, so the positioning principle is the same as other wireless communication positioning systems. However, UWB technology has the characteristic of enabling more-accurate positioning by utilizing a highly wide bandwidth and transmitting pulse signals with a fairly short width [16,29].

UWB positioning has the following three major strengths compared to other systems. First, unlike packet-type messages used in other RF-based positioning, the UWB system uses pulse signals with a very short time width, so it performs accurate positioning based on a high time resolution [30,31]. Second, the UWB positioning system has a relatively weak degree of multipath interference [32,33]. Therefore, it has the strengths of robustness in environments with many obstacles and high noise levels. Since the UWB system communicates with very short pulse signals of tens of picoseconds to several nanoseconds, it is possible to distinguish between signals even if there is only a slight difference in the path distance between direct and reflected waves. In theory, in a UWB system using pulses with a width of 500 ps, if the signal path difference is more than 15 cm, the two pulse signals can be distinguished and do not cause mutual interference. Third, the UWB system utilizes ultra-wideband and has a low energy level, so it has good signal penetration. Therefore, it has the advantage of being strong against signal decline even in an environment with many obstacles, smooth connection between devices, and high-speed data transmission.

However, the UWB positioning system does not have only advantages. Designing and implementing a positioning system can be more expensive and complex compared to RFID,

Bluetooth, and other sensor-based networks [34,35]. Additionally, technical regulations and standardization are different in many countries, which can cause problems with frequency spectrum management [28,36]. In terms of security, it shows weaknesses like other wireless-communication-based positioning systems [37,38]. In particular, malicious users can spoof UWB signals to forge or alter data, which can seriously impair positioning performance. Therefore, in order to apply UWB positioning to large-scale applications, the cost, regulatory, and security aspects must be considered.

2.2. DAG-Based Distributed Ledger

DAG generally refers to a directed acyclic graph data structure, but in the blockchain industry, it is commonly used as a term for distributed ledger technology utilizing the DAG structure [39,40]. DAG-based distributed ledger technology is used to implement a distributed system to process and store data transactions [41]. In a DAG-based distributed ledger system, all transactions are connected simultaneously and in parallel in the form of a directed graph. Since the DAG structure itself, where transactions are connected, acts as a kind of chain, a distributed ledger can be implemented by eliminating the blocks of the existing blockchain system and leaving only the chain format [42].

Blockchain-based distributed ledgers have scalability problems when applied to IoT-based networking systems that handle many data because they require complex verification and block creation for transaction processing [43,44]. Additionally, considering the limited computing power on vehicle networks using low-power devices such as UAVs, high latency may occur, which may not be helpful in applying distributed ledger technology in real scenarios. Therefore, we intended to solve the existing problems by facilitating data processing on vehicle networks using DAG-based distributed ledger technology.

The following are ongoing studies involving DAG-based distributed ledger systems. A study was conducted comparing the differences between blockchain and DAG-based distributed ledger structures. Research has been progressed to analyze various DAG-based commercial blockchain technologies [45]. Additionally, extensive research has been conducted to apply DAG-based distributed ledger technology to IoT networks in various fields such as smart grids [46], industrial IoT [47], the 6G network [48], energy trading [49], and smart communities [50]. Furthermore, studies related to advanced DAG-based distributed ledger systems are emerging, such as considering the efficient consensus process of DAG-based distributed ledger technology [51], flexibly adapting to transaction demands [52], and developing a private distributed ledger system [53]. Lastly, research has also progressed on the security of DAG-based distributed ledgers, such as resilience in IoT applications [54] and the stability of unverified transactions [55].

In summary, many studies have been conducted to apply DAG-based distributed ledger technology in various fields. In addition, using DAG-based distributed ledger technology has advantages in terms of transaction processing time and computing power compared to existing blockchain-based systems. Therefore, we introduced DAG-based distributed ledger technology into a localization system composed of multiple IoT devices. This approach promotes the parallel processing of data in the localization network, ensuring fast transaction processing and solving existing scalability problems. Moreover, the proposed system can prevent data forgery and falsification through transparent processing of localization data transactions. Consequently, our design enables secure data transmission and storage in a localization system where each node collects and processes the positioning information of its neighbors.

3. System Design

3.1. Overview

Figure 1 shows the schematic of our proposal applied in a real environment. It has a general design that performs positioning using UWB in a local network environment consisting of a server, anchors, and tags, and this reflects the intention to minimize the cost of infrastructure modification due to the application of a new technology. To list the differences, firstly, our proposal utilizes an improved localization technique. When measuring location based on distance, multiple distance values and fixed coordinates of the anchor are used as the inputs. These inputs have default errors and can cause significant distortion in noisy environments. These changes in input values can be a factor that significantly reduces the quality of positioning. We solved this problem by improving the conventional positioning method. Second, our proposal designs a new positioning system using a DAG-based distributed ledger. In vehicle networks comprised of manned/unmanned aerial vehicles, excellent positioning techniques are essential for accurate position measurement, and it is very important to protect the system from malicious attacks such as forgery and falsification, which can significantly reduce the accuracy of position measurement. Typically, high levels of security incur significant overhead in networking and computation. On the other hand, positioning systems require high real-time performance. In short, positioning performance and system safety are a trade-off. In summary, the security technology applied to the positioning system must satisfy both low system overhead and excellent safety. For this reason, this paper applied a DAG-based distributed ledger with improved consensus to the positioning system.

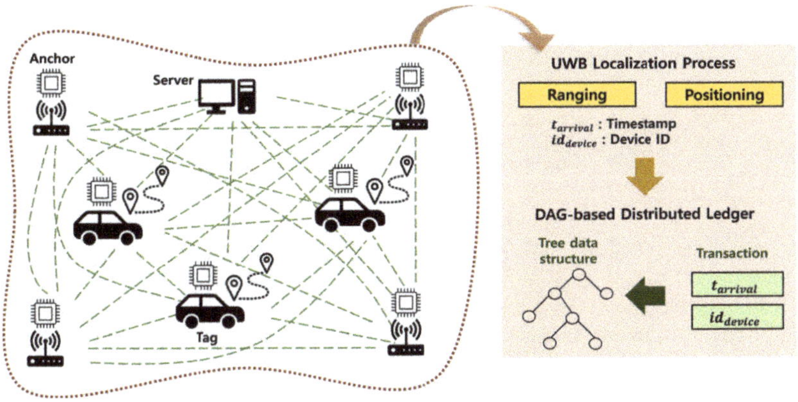

Figure 1. This is a system schematic of the proposed localization framework with a distributed ledger.

Based on the two new proposals mentioned above, we ultimately propose a complete localization framework. Our proposed localization framework provides a positioning system based on trust. All the positioning nodes operate as participants on the DAG-based distributed ledger. This means that transactions containing data necessary for positioning are created, verified, and confirmed according to a consensus, and transaction information is distributed throughout the network. Figure 2 shows the system architecture of the proposed localization framework. The description of the components can be listed as follows:

- The **device layer** is a device abstraction layer that generalizes resources for the purpose of managing and controlling the resources available in the framework. This method reduces the development costs for heterogeneous devices and provides an intuitive interface to the users.
- The **P2P network layer** manages P2P connections between nodes and defines security-related tasks.
- The **service layer** verifies the transaction information collected through the network and records it in the ledger. The recorded transaction includes not only the UWB sensor data, but also the details of the devices participating in the network, such as the device ID and status information, and based on this, the main services such as generating the location information of the target devices and generating device metadata are operated.
- The **application layer** includes the implementation of various functions through the API, such as the framework's positioning service, device control, and distributed ledger

query. Based on this, it provides various positioning services to the users and useful functions necessary for positioning developers to develop their own technologies.

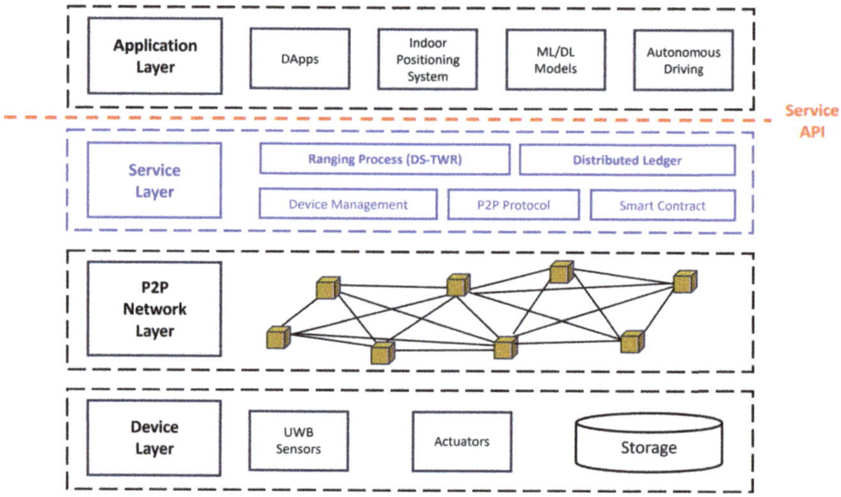

Figure 2. This is a system architecture of the proposed localization framework with a distributed ledger.

At the device layer, the data for positioning are generated in real-time by positioning devices such as UWB sensors. Not only that, the data necessary to operate various services along with the location information, such as the device status information and actuator information, are generated. The proposed framework manages all these data as transactions and transmits them to the P2P network. The service layer verifies the transactions received from the device layer and updates the verified transactions to the DAG of the distributed ledger. In this way, the transactions recorded in the DAG, more precisely the transaction information, are used to operate various service processes at the service layer. For example, the main services include a positioning service process, which calculates the location information of the target devices using positioning techniques such as Double-Sided TWR (DS-TWR), and a device manager, which dynamically monitors and controls the devices participating in the network. We designed the framework so that these services can be added or expanded as needed, and the implementation of various functions for these services can be exposed to users or developers through the API.

Over the next two subsections, we describe in detail the implementation of the improved localization technique and the distributed ledger with the new consensus mechanism in the proposed framework.

3.2. *UWB Localization*

3.2.1. Ranging Module for Framework

In our framework, the TWR method was used among the UWB distance measurement techniques. TWR is a ranging method that measures the Time of Flight (ToF), which is the transmission and reception time between devices, through two-way message exchange. TWR is divided into Single-Sided TWR (SS-TWR) and DS-TWR depending on the number of message exchanges. SS-TWR is a method of sending and receiving a total of two messages once from each device, and the ToF is calculated by dividing the difference between the round-trip time and the reply time in half. This method has the advantage of having a simple structure with a small total number of messages exchanged between devices. However, SS-TWR has a clear disadvantage in that, as the clock offset and response time to the first received message increase, the ToF estimation error can increase to a level where the distance estimation becomes significantly inaccurate. DS-TWR, an extension of basic SS-TWR, uses two round-trip time measurements and combines them to produce the ToF results with reduced error even in the case of quite long response delays. The proposed system uses DS-TWR, which uses a total of 3 messages to estimate the ToF for ranging between nodes, and this method does not need to perform clock synchronization between the nodes [56].

In the proposed framework, the estimated ToF in the timestamp can be calculated using the following equation:

$$t_{ToF} = \frac{T_{round1} \times T_{round2} - T_{reply1} \times T_{reply2}}{T_{round1} + T_{round2} + T_{reply1} + T_{reply2}}, \tag{1}$$

where all the parameters are shown in Figure 3, which shows the process of communication between the nodes.

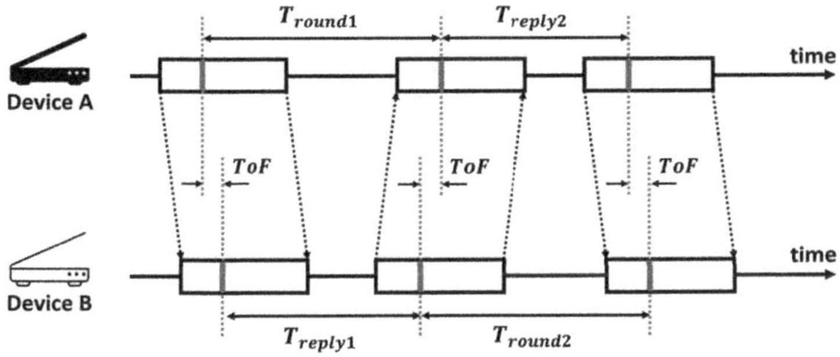

Figure 3. Diagram of asymmetric DS-TWR.

Then, the measured timestamp can be straightforwardly converted into the distance using the following equation:

$$d = \alpha \times t_{ToF} + \beta. \tag{2}$$

In this equation, d represents the estimated distance between the ranging nodes, t_{ToF} is the estimated ToF from the timestamp including the antenna delay correction factor, and α is a proportional factor. β is a constant value utilized in the calculation of the estimated distance from the ToF.

The DW1000, which is the UWB sensor used in our system implementation, is equipped with a sampling clock operating at 63.8976 GHz [57]. In other words, 63.8976×10^9 clock cycles occur per second, which means that the clock cycle is $(63.8976 \times 10^9)^{-1}$ s. Therefore, the clock period is approximately 15.65 ps. This high sampling frequency allows timestamps to be measured at very short time intervals, enabling highly precise distance measurement and location tracking. Considering the propagation speed of UWB signals of 300,000 km/s, the UWB signal propagates 0.0047 m per clock cycle, which means that it travels 0.0047 m per timestamp of the ToF calculated with DS-TWR. Therefore, we established the proportional factor α at 0.0047, and through subsequent experiments, we empirically derived the optimal β value of 0.2321.

3.2.2. Nonlinear Least Squares for 3D Position Estimation

The proposed framework utilizes trilateration, a distance-based technique, for the positioning operations. Trilateration is a technique to determine an unknown point using distance information from three or more known points. In general, n + 1 anchor information is required to perform n-dimensional positioning. By solving the resulting system of nonlinear equations, the target's position can be effectively estimated.

One of the core strengths of trilateration lies in its simplicity and efficiency. When the distances from the anchors are precisely known, and the anchors are non-collinear, trilateration yields highly accurate positional estimates. Additionally, it is scalable and adaptable, making it a preferred choice for applications ranging from indoor positioning systems to Global Navigation Satellite Systems (GNSSs). In the context of 3D positioning, the process becomes somewhat intricate due to the introduction of a third dimension.

We employed a nonlinear least-squares approach to enhance the accuracy and robustness of positional estimates, even in the presence of measurement noise and errors. In the proposed framework, the conventional trilateration technique is augmented through an algorithm that harnesses the power of nonlinear least-squares optimization, particularly the Levenberg–Marquardt algorithm [58], for enhanced accuracy and reliability in 3D position estimation. This enhancement is pivotal in scenarios where conventional trilateration might be susceptible to errors induced by various factors, including distance measurement inaccuracies and environmental noise.

Algorithm 1 shows the detailed process of the trilateration technique we applied. The algorithm is initiated by defining a cost function $f(\mathbf{p})$, which is the summation of the squared differences between the actual measured distances and the computed Euclidean distances from the estimated tag position to the anchors:

$$f(\mathbf{p}) = \sum_{i=1}^{4}\left(\sqrt{(\mathbf{a}_i - \mathbf{p})^\top(\mathbf{a}_i - \mathbf{p})} - d_i\right)^2. \tag{3}$$

This cost function serves as the objective to be minimized, facilitating an accurate estimation of the tag's 3D coordinates $\mathbf{p}$.

Algorithm 1 The 3D trilateration using nonlinear least squares.

Input: Anchor positions $\mathbf{A} = \{\mathbf{a}_1, \mathbf{a}_2, \mathbf{a}_3, \mathbf{a}_4\}$, distances $d = \{d_1, d_2, d_3, d_4\}$
Output: Tag position $\mathbf{p}$ and residual $\mathbf{r}$
Function `Main`:

- **Data:** Define the cost function $f(\mathbf{p})$ as the sum of squared differences between the measured distances and the distances computed from the estimated tag position:
  $f(\mathbf{p}) = \sum_{i=1}^{4}\left(\sqrt{(\mathbf{a}_i - \mathbf{p})^\top(\mathbf{a}_i - \mathbf{p})} - d_i\right)^2$
- **Data:** Define the Jacobian matrix $\mathbf{J}(\mathbf{p})$ of partial derivatives of f with respect to $\mathbf{p}$:
  $\mathbf{J}(\mathbf{p}) = \left[\frac{\partial f}{\partial p_x}, \frac{\partial f}{\partial p_y}, \frac{\partial f}{\partial p_z}\right]$
- Choose an initial guess $\mathbf{p}_0$ for the tag position
- **Data:** Set optimization settings: maximum iterations, tolerance, etc.
- Solve the nonlinear least-squares problem using Levenberg–Marquardt optimization algorithm to find $\mathbf{p} = \arg\min_{\mathbf{p}} f(\mathbf{p})$
- **if** *optimization converges* **then**
  - **Data:** Compute the residual $r(\mathbf{p})$ to evaluate the quality of the solution:
    $r(\mathbf{p}) = \sum_{i=1}^{4}\left(\sqrt{(\mathbf{a}_i - \mathbf{p})^\top(\mathbf{a}_i - \mathbf{p})} - d_i\right)^2$
  - **Result:** Return $\mathbf{p}$ and $\mathbf{r}$
- **end**
- **else**
  - **Result:** Return an error indication
- **end**

The optimization process employs the Jacobian matrix $\mathbf{J}(\mathbf{p})$, consisting of the partial derivatives of the cost function with respect to the coordinates of the estimated tag position. The Levenberg–Marquardt algorithm optimizes the tag's position by iteratively refining an initial estimate $\mathbf{p}_0$. This iterative refinement ensures convergence towards a solution that minimizes the residual error, adhering to pre-established tolerance levels and maximum iteration constraints.

Then, the residual $r(\mathbf{p})$ to evaluate the quality of the position estimate uses the Root Mean Square (RMS) value of the prediction error and is calculated as follows.

$$r(\mathbf{p}) = \sqrt{\frac{1}{4}\sum_{i=1}^{4}\left(\sqrt{(\mathbf{a}_i - \mathbf{p})^\top(\mathbf{a}_i - \mathbf{p})} - d_i\right)^2}. \tag{4}$$

A lower residual value indicates a higher accuracy of the estimated position, ensuring alignment with the measured distances from the anchors.

This enhanced trilateration algorithm is particularly advantageous in environments where precision and reliability are paramount. The integration of the Levenberg–Marquardt optimization algorithm compensates for potential errors and noise in the distance measurements, delivering more-accurate and -consistent results. Applications of this refined trilateration technique span across various domains including indoor positioning, robot navigation, and satellite-based positioning systems. Its adaptability and scalability cater to the diverse and dynamic requirements of these applications, offering a robust solution for real-time, accurate 3D position estimation in complex environments.

3.2.3. Dataset for DAG-Based Distributed Ledger

As previously mentioned, in tasks that require precise positioning accuracy, the positioning system can collapse even if only a part of the data of the nodes that make up the cluster is slightly tampered with by a malicious attack. Furthermore, in the case of 3D trilateration, the accuracy of the distance estimate is directly related to the positioning performance, and the task may fail even if the error in only one of the four distances increases. Therefore, to ensure the security and reliability of the positioning system, a specific policy must be established to prevent distance precision degradation from attackers.

In this regard, the proposed framework establishes a security policy to ensure the integrity of the data. We created a transaction with a dataset consisting of the reception timestamp and transmitting device ID for the received UWB signal, which comprises the raw data of the ranging module that outputs the distance, and recorded it in the distributed ledger according to the verification and consensus procedure.

3.3. Distributed Ledger for Positioning

The localization framework we propose can provide strong reliability as distributed ledger technology is applied. Since all decisions made in the distributed ledger are made through consensus among network participants, actions that violate consensus are not possible. In addition, if a malicious node arbitrarily modifies the data or attempts to hack the system, this is almost impossible because the hashes of all blocks connected to a chain in the ledger must be modified. This allows us to keep the system safe from external threats. In addition, in a positioning system that operates in a conventional centralized manner, the main authority is usually concentrated on the main server, so if this main server goes down, there is a risk that the entire system will be paralyzed. However, our proposed decentralized localization framework solves the side effects of conventional centralized methods by decentralizing between nodes that smartly distribute trust. Moreover, this system of components with equal rights provides strong sustainability. If a single node is incapacitated by an external attack or a system error, the centralized system risks performance degradation or, in serious cases, the paralysis of the entire system. However, our proposed system has an environment that can solve problems arising from the temporary absence of a single node through dynamic allocation. For example, when an arbitrary anchor becomes incapacitated, a tag with the lowest priority from a mission perspective takes over the role of that anchor to prevent performance degradation. On the other hand, these features may also have an advantage in terms of adaptively applying to heterogeneous environments. In an environment where various topologies are formed and roles are dynamically assigned depending on the mission or environment, it is possible to respond to system development or modification without large costs.

It makes sure that the applied distributed ledger does not have a negative impact on previous positioning performance, so we designed a new consensus algorithm to reduce the task load in terms of latency. In the nature of the DAG structure, where each node maintains a global state rather than maintaining a global state across the network, the number of pending transactions may increase. To improve this, we devised a new method that defines verification nodes and reaches consensus based on scores awarded through mutual approval between these nodes (in the existing method, the created transaction verifies the pending transaction). First, we assumed some facts as follows:

1. All nodes participating in the network are connected to each other.
2. The tag is connected to at least one of the anchors or servers.
3. Anchors and servers are functionally separate expressions according to their positioning roles, but are the same logical node in a distributed network.

The proposed consensus algorithm operates by two components called the *miner* and *prover*. A *miner* verifies transactions created in real-time and updates the verified transactions in the DAG. Afterwards, the updated information is broadcast to all *miners* existing in the network. *Provers* approve and confirm transactions created by *miners* according to defined global state consensus. Global state consensus refers to the process of increasing the cumulative score, which is a newly defined parameter at a vertex of the DAG that represents a single transaction, as a *prover* approves a single transaction. If the score exceeds the specified threshold, the global state of the transaction is updated to confirmation. The cumulative score can be expressed as:

$$CS = 2(2n - n_{total})/n_{total} \tag{5}$$

where n is the number of approved *provers* and n_{total} is the total number of provers. *provers* are all or some of the *miners* selected, approve pending transactions in the DAG, and broadcast the fact that the transaction has been approved to all *miners* participating in the network. When a node that is both a *miner* and a *prover* receives this fact, it indexes its DAG with approved transaction information (e.g., transaction ID), accesses a vertex storing the transaction, and updates the cumulative score. This process is repeated, and when the cumulative score of the transaction exceeds the threshold, the transaction is confirmed.

4. Evaluation

After the construction of the proposed framework, we conducted a series of experiments to verify its operations and evaluate its performance. First, we verified the performance of the positioning system that makes up the framework and, then, verified how data tampering by malicious attacks affects the positioning performance. Then, in the distributed ledger technology applying a new consensus algorithm, we verified the impact of the confirmation threshold on transaction processing performance and how system performance varied according to changes in the network scale. Lastly, we evaluated the security performance of the proposed framework by modeling a race attack scenario, one of the common blockchain attack methods.

4.1. Verification of the Proposed Localization System Accuracy

We implemented a positioning device using a Raspberry Pi 4 and DWM1000 UWB module in a real experimental environment. Detailed specifications for the positioning module implementation and experiment are shown in Table 1. Each hardware component has the following characteristics.

The Raspberry Pi 4 is a powerful computing device with a 1.5 GHz quad-core CPU and up to 4 GB RAM, making it ideal for rapid data processing and extensive sensor processing via GPIO pins. However, the Linux-based operating systems we utilized have common security vulnerabilities and require regular updates and strong security measures.

Table 1. Specifications for positioning module implementation and experiment.

Component	Description
Computing Board	Raspberry Pi 4
Operating System	Ubuntu MATE 22.04
UWB Module	DWM 1000
Based Standard (UWB)	IEEE 802.15.4a/z
Utilized Channel	UWB Channel 5 (6240.0–6739.2 MHz)
Sampling Clock Frequency	63.8976 GHz
Modulation Scheme	BPM/BPSK

The DWM1000 module, which uses the DW1000 UWB chipset, has a distance measurement accuracy of several tens of centimeters and shows precise location tracking performance, making it suitable for use in experiments sensitive to spatial data. While the module itself is not vulnerable to traditional malware, integration with other systems requires secure data transmission protocols to prevent wireless signal interference or tampering, highlighting the need for encryption and security practices.

To evaluate the performance of the proposed localization system, we first verified the positioning performance in an environment without malicious attacks. Using the implemented positioning device, we set the anchors and tags and performed the 3D localization of three tag nodes. The ranging and positioning modules mentioned earlier were employed for indoor localization.

Figure 4 and Table 2 show the results of the localization performance experiment. The enhanced positioning module demonstrated the excellent performance of the UWB system in real-world environments where noise can occur. Each of the three tag nodes showed localization errors of 10.50 cm, 10.82 cm, and 13.10 cm and standard deviations of 5.10 cm, 4.41 cm, and 6.07 cm, respectively. Thus, the proposed localization system exhibited outstanding performance when there were no external attacks.

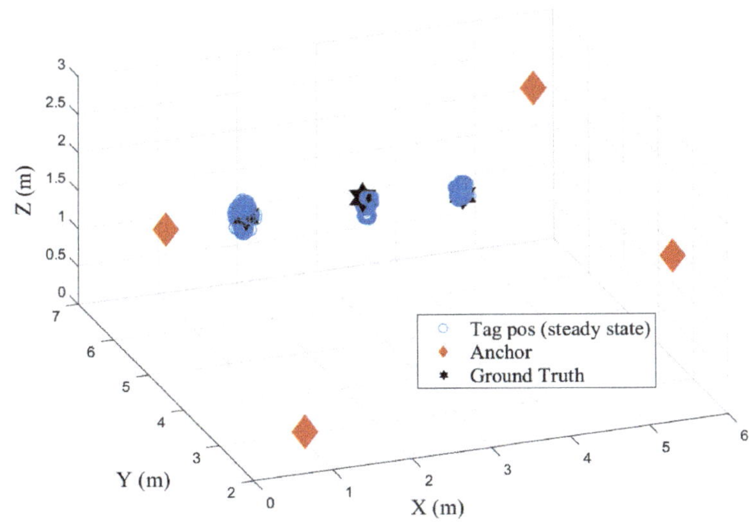

Figure 4. The 3D positioning experiment in a multi-tag environment of the proposed positioning system.

Table 2. Positioning performance of the proposed localization system.

Node	Tag 1	Tag 2	Tag 3
Average Positioning Error (cm)	10.50	10.82	13.10
Standard Deviation (cm)	5.10	4.41	6.07

4.2. *Effect of Malicious Attacks on Positioning Accuracy*

In Section 4.1, we verified the performance of the positioning system within the service layer of the proposed framework using the enhanced positioning module. In this section, we demonstrate the impact of malicious attacks on the performance of a positioning system through experiments in a real environment. In this section, we show how the performance of an entire positioning system can be seriously damaged when a malicious attack is inflicted on one of its constituent nodes. In other words, it shows that measures to strengthen security in the positioning system are necessary.

The experimental configuration can then be summarized in terms of localization as follows. In Section 4.3, we evaluated the data processing performance of the consensus algorithm of the proposed framework. There is a risk of delay in the positioning system due to additional system load caused by the blockchain, which was applied to defend against malicious attacks. We showed that this risk is not a problem in practice through the verification of the new consensus algorithm in terms of the time to reach consensus. Section 4.4 evaluated the consensus algorithm using the race attack scenario, which is a common malicious attack scenario considered in DAG-based distributed ledger systems. Our simulations demonstrated that our proposed solution can disable malicious transactions faster than IOTA [59], the most-widely adopted DAG-based distributed ledger system. Afterwards, we verified that the proposed framework is a security-stable system by showing that it maintained excellent data processing performance even when the intensity of the attack increased.

We conducted two experiments to examine the impact of tampering with the data of one node in the cluster, formed in the experimental environment of Section 4.1, on the positioning performance.

4.2.1. Timestamp Data Tampering

Initially, we tampered with the reception timestamp, which is recorded on the device when an anchor node receives a UWB message. As mentioned in Section 3.2.1, the implemented module records the timestamp data that the DW1000 chipset generates and records internally. Therefore, the minimum interval of recordable timestamps was 15.65 ps, and we refer to this value as a time slot. We generated tampered timestamp data by adding Gaussian noise of various scales to one anchor node. We conducted experiments for a total of eight cases, including the steady state, and collected data for three minutes for each case. Table 3 and Figure 5 present the experimental results.

Table 3. Positioning performance under normal and timestamp data tampering conditions.

State	Steady	500 Time Slots	1000 Time Slots	2000 Time Slots	5000 Time Slots	10,000 Time Slots	50,000 Time Slots	100,000 Time Slots
Average Error (m)	0.115	5.839	17.925	42.895	170.638	552.038	1.287×10^4	4.345×10^4

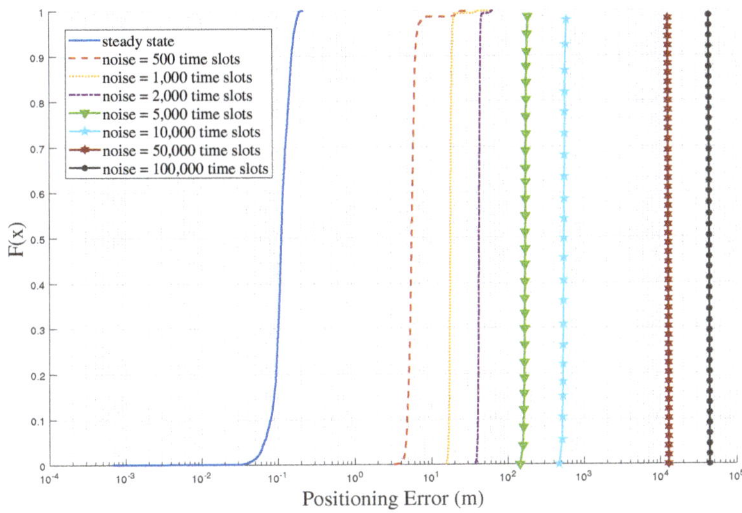

Figure 5. CDF of positioning errors in normal and timestamp data tampered situations.

Table 3 first shows the average error, representing the mean error of the localization data collected from the three tag nodes. The steady state case performed normal localization without data tampering. The seven noise-generating cases added Gaussian noise with the respective mean and 10% of the numerical standard deviation to each of the message reception timestamp data. The experimental results showed that even a slight time data manipulation of an average of 500 time slots can increase the error to nearly 6 m, an unacceptable error range for most tasks requiring precise localization. Additionally, as a result of performing timestamp modulation over 1000 time slots, it can be seen that the position estimation error was too large to say that positioning was being performed.

To the average error values shown in Table 3, we graphed the cumulative distribution function for the positioning error data collected from a total of eight cases, including the steady state. Figure 5 shows the CDF of the positioning error data for each case from the same experiment, with the x-axis representing the positioning error in a log scale. As seen in the table and figure, when a time data attack occurred, the positioning performance deteriorated severely, not only in cases of significant numerical manipulation, but also in cases of minor alterations. As can be seen in the graph, the steady state showed stable results with very low positioning errors for most data. For the remaining seven cases that included noise, most of the data collected over 3 min showed highly large positioning errors at a similar level.

4.2.2. Device ID Data Tampering

Each node that makes up the positioning system receives UWB signals from other nodes. At this time, the UWB message includes the ID of the transmitting device and transmission timestamp information. Nodes in the positioning system are given device IDs starting from Number 1 according to the registration order after system operation. Therefore, all nodes in the system have different device IDs, and each device pair performs DS-TWR-based ranging using the node ID and transmission and reception timestamp data.

In this subsection, we randomly tampered with the transmitting device's ID. When one specific node out of the total of seven constituting nodes received a UWB message, we randomly reset the transmitting node's ID to one of the remaining six nodes (excluding itself), performed ranging with that distance information, and executed positioning. We implemented this environment in reality and checked the performance.

As a result, the positioning errors of the three tags were 8.85 e + 12 m, 185.95 m, and 7.86 e + 12 m, respectively. The results of the experiment showed how seriously positioning performance can be damaged if the identity of the transmitting device is unknown or the data are damaged due to a malicious attack.

The basis on which this result was derived is as follows. The device ID data tampering experiment evaluated the positioning performance by randomly altering the transmitting device ID for a specific target among the system nodes each time a signal was received. Therefore, the positioning error value for each tag may significantly vary with each repetition of the experiment. The key point, however, is that the potential positioning error can be quite large. All time data for calculating the ToF based on DS-TWR were obtained within the timeline of each device. Therefore, using this method to determine the ToF did not require a synchronization process for aligning timelines between devices. However, if only the device ID were manipulated, the receiving node may interpret the signal as coming from the same device, while the actual timestamp is from a device with a completely different timeline. This would lead to a distortion of all time data used for the ToF calculation, thereby disrupting the overall ranging performance.

4.3. Transaction Measurement

This subsection evaluates in detail the performance of the distributed ledger applied to the proposed localization framework through two experiments. For these experiments, we implemented a simple Proof-of-Concept (PoC) system and evaluated our proposal through a simulation in an environment where transactions were scheduled to be generated similar to real situations. The experiment was largely divided into two parts.

First, we measured the change in transaction throughput according to the change in the confirmation threshold and analyzed the results. According to the proposed consensus algorithm, each transaction recorded in the DAG had a score accumulated each time it was approved by the *prover*, and when this score was greater than the threshold, the transaction was judged to be in the confirmation state. For this reason, how the threshold is set can greatly affect the transaction processing performance of the proposed localization framework. We conducted a simulation by changing the threshold from 10 to 50 in 10 increments in an environment with 30 tags and 20 fixed nodes composed of a single server and multiple anchors, and the result appears as Figure 6. The x-axis of the graph represents the size of the threshold, and the y-axis represents the confirmed transaction rate. According to the results, it can be seen that the smaller the threshold, the faster transaction confirmations reached saturation. Analyzing the results, as the threshold became smaller, the proposed localization framework could process transactions faster, but the security level was lowered, making it vulnerable in protecting the system from malicious transactions. Conversely, as the threshold increased, our system provided stronger security performance, but the delays occurring during transaction processing increased, which, in turn, lowered the quality of the positioning service. When the threshold was high, a single transaction must be approved by more *provers* to increase the cumulative score above the threshold, and in this process, factors that cause performance degradation, such as processing delay, which occurs while the *prover* approves the transaction, and network delay, which occurs in the process of broadcasting a newly created transaction or a transaction indicating that the *prover* has approved the transaction, their influence increased. In other words, safety and positioning performance are a trade-off, and it is important to set the optimal value that can properly maintain both.

Second, we measured the change in system performance as we varied the network scale of the localization framework. In an environment such as a vehicular network where the topology of participating nodes changes in real-time and the network scale increases or decreases irregularly, it is a very important consideration whether the system maintains constant performance and stably provides services. We measured the average transaction processing time of the proposed localization framework, which changes, while simultaneously increasing the number of *miners* and *provers* at a constant rate, and the results appear as Figure 7. According to the results, the average processing time was between 0.09 and 0.11 s, and the difference between the maximum and minimum average times was only 0.02 s. For reference, as the network scale increased, the processing speed tended to become slightly faster. This is because the speed of approving transactions increased as the number of *provers* increased, so it had nothing to do with what we wanted to suggest through this experiment.

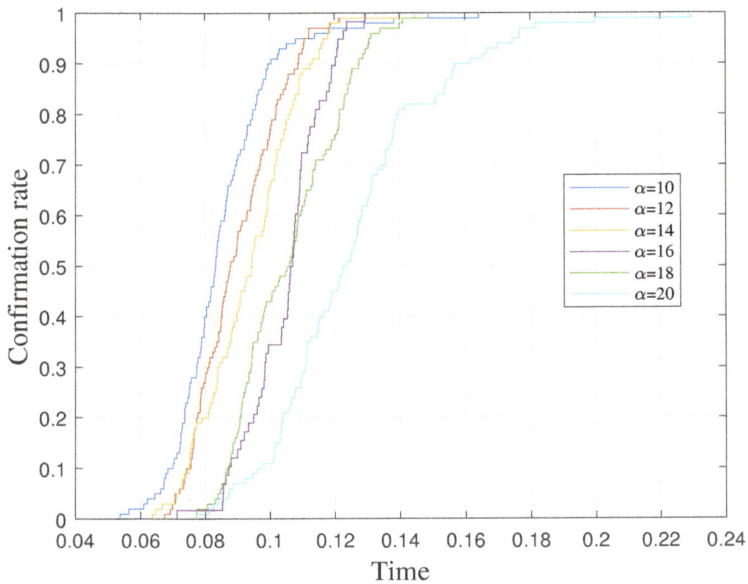

Figure 6. The confirmation saturation rate at different confirmation thresholds (α).

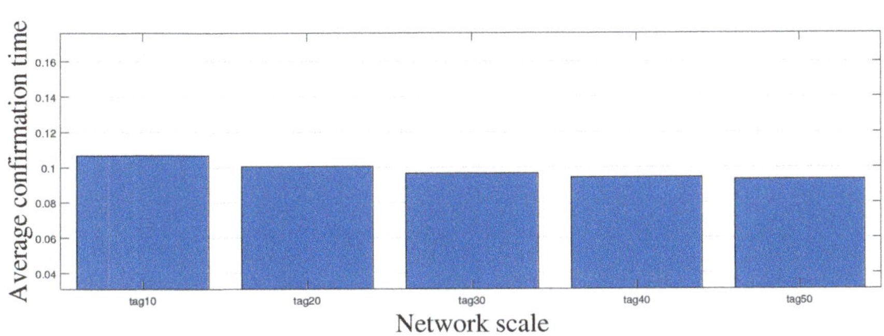

Figure 7. The average confirmation time at different network scales.

4.4. Security Performance

This subsection evaluates the security-related performance of the distributed ledger applied in the proposed framework. To evaluate the security performance, we implemented a simulation environment for race attacks, a representative attack that can occur in general distributed ledger systems. A race attack is a type of double-spending attack, a common distributed ledger attack scenario, and refers to an attack with the intention of disrupting the verification system of the distributed ledger using multiple transactions. In a simulation environment, some of the total transactions that occur over a period of time are defined as transactions with malicious intent to attack the entire system. When malicious transactions are verified by the *prover* and recorded in the DAG, the attack is ultimately successful.

Therefore, users with malicious intent strive to add to the distributed ledger by continuously generating attack transactions.

First, we measured the average deactivation time for the malicious transactions that occurred and validated the performance of the distributed ledger framework. In order to compare the performance of the proposed framework according to the race attack probability change in the simulation, we used IOTA [59], the most-representative DAG-based distributed ledger system. IOTA is a representative example of a cryptocurrency based on DAG-based distributed ledger technology, which, unlike blockchain, does not use blocks. IOTA uses a structure called a Tangle to build a graph by connecting each transaction with previous transactions. Additionally, this graph-creation method has an advantageous structure for supporting micro-transactions between small devices, so it has high potential for use in the vehicular network field. Therefore, in this experiment, we would like to evaluate the security performance of the proposed system by comparing the proposed system with IOTA, the most-commonly used DAG-based distributed ledger system.

Both the proposed system and IOTA have a DAG-based data storage structure, but there is a significant difference in the consensus algorithm that verifies the generated transactions. The proposed system updates transactions through two components: *miner* and *prover*. During the update process, the cumulative score for a single transaction is increased, and the transaction is verified when the transaction exceeds a certain threshold. The verification process based on this consensus algorithm is advantageous for processing multiple transactions created simultaneously, and we intended to verify this through experiments. In the simulation environment, general network conditions such as the transaction creation speed, network communication delay, and bandwidth settings were reflected using empirical values [46].

Figure 8 shows the results of a simulation performed by changing the race attack probability from 10% to 50% in an environment with a total of 20 nodes using 10 tags, a single server, and 10 anchors. In a simulation environment consisting of a total of 20 nodes, each tag communicates with a fixed anchor and continuously generates transactions related to the location data. We assumed an attack by a malicious node and added some incorrect data transaction information to these periodically generated normal transactions. Attacks caused by these malicious transactions can be identified through the verification process of the distributed ledger system. Deactivation time refers to the time it takes to deactivate a transaction that occurs in a distributed ledger system by determining it to be a malicious transaction. In this experiment, we evaluated the security-related performance of each distributed ledger system by averaging the time it takes to identify and disable malicious transactions.

In the case of the verification process based on IOTA's Tangle structure, when 10% of all transactions that occurred were attacked, it took about 3.8 s to determine and deactivate them. These results confirmed that even if the race attack probability was increased from 10% to 50%, there was a similar average deactivation time of about 3.2 s. In comparison, the average deactivation time of the proposed framework at a race attack probability of 10% was about 1.3 s, which can be seen to be much faster than the existing system. Additionally, the proposed framework found that the average deactivation time of about 1.3 s was maintained even by increasing the probability of competing attacks from 10% to 50%. IOTA distributed ledger system uses a special DAG architecture called "tags" to group and distinguish transactions. A "tag" contains a reference to the two immediately preceding transactions and is used to categorize transactions together with a specific topic or purpose. "Tags" are a special feature of IOTA, but the process of forming these "tags" increases the time it takes to verify and propagate transactions. Additionally, this accumulation of transaction propagation and verification time delays ultimately increases the average deactivation time. This means that the verification process using the *miner* and *prover* of the proposed framework can verify malicious transactions much faster than existing systems. As a result, the proposed framework is more suitable for transaction processing in vehicular networks, where real-time updates are important.

Second, Figure 9 measures the transaction throughput of the proposed framework according to changes in the race attack probability. Through previous experiments, it was confirmed that the existing distributed ledger system IOTA has a transaction deactivation time that is approximately twice as long as the proposed system. The reason why the transaction deactivation time is longer than the proposed system is because the IOTA system has slower transaction propagation, verification, and confirmation times. Therefore, in the second experiment, the performance of the proposed system was confirmed through the detailed verification of transaction throughput according to the attack scenario of the proposed system, rather than a comparative experiment with the existing system, IOTA. The results showed that the distribution of the final transaction confirmation time was between approximately 4.5 and 6 s. As can be seen in the experiment above, since it took about 1.3 s to verify a malicious transaction, it can be seen that more time was needed to confirm one transaction. Additionally, we can see that the confirmation rate distribution had almost no difference even as the attack transaction rate increased from 10% to 50%. This means that the proposed distributed ledger framework can support a certain level of transaction processing regardless of the proportion of malicious transactions among newly created transactions. Since the proposed framework guaranteed stable transaction processing against transaction attacks by malicious attack nodes, it can be confirmed that it is a secure system.

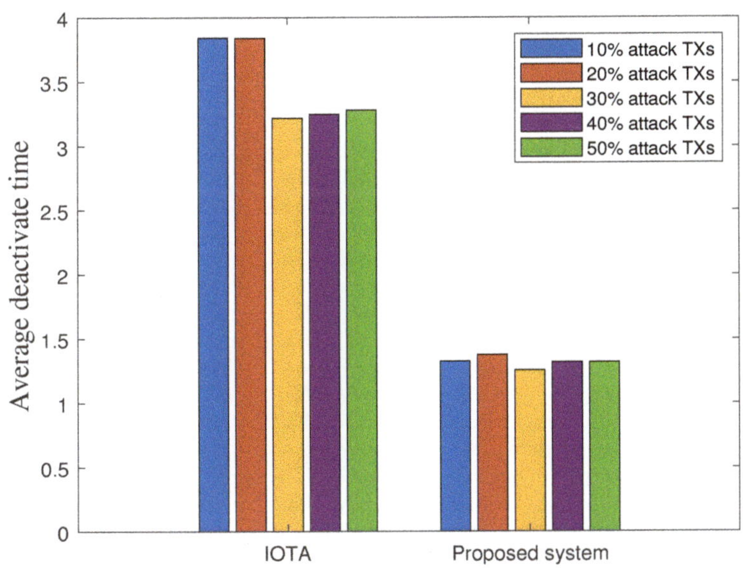

Figure 8. The security performance evaluation result of the average deactivation time according to the probability of malicious transactions.

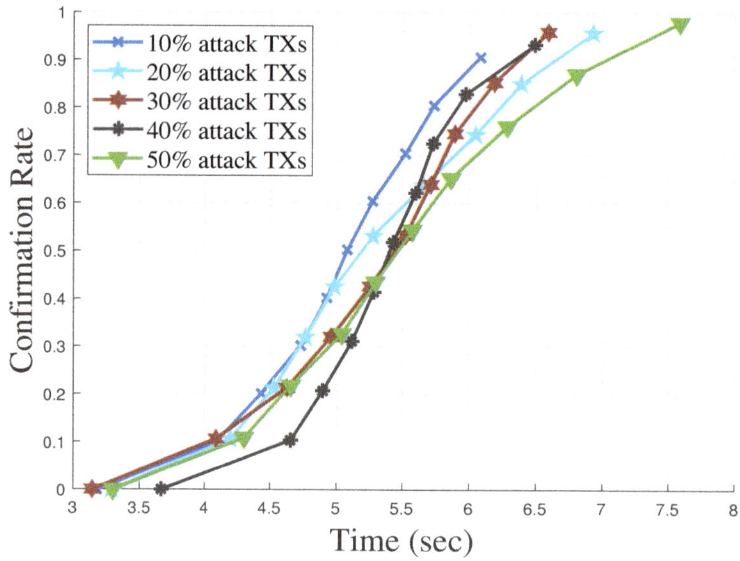

Figure 9. The result of measuring the confirmation rate according to the probability of a malicious transaction.

5. Conclusions

In the process of performing various missions using a swarm of vehicles, the possibility of malicious attacks such as data forgery and falsification must be considered, especially in situations where accurate location estimation is required. In the UWB positioning system, data generated from each node are used to calculate the ToF. Data safety is very important because, if even a single piece of data is forged or altered, the accuracy of location measurement can be greatly reduced. Therefore, by introducing a DAG-based logical network between each node, data integrity and safety are guaranteed, and location measurement performance is prevented from deteriorating. In this paper, we proposed a safe and accurate location measurement framework using improved UWB positioning techniques and a DAG-based distributed ledger. We proposed a DS-TWR-based ranging module optimized through empirical calibration and a positioning module. which improved the existing trilateration by applying the Levenberg–Marquardt algorithm, a nonlinear least-squares optimization technique. Moreover, by applying a new consensus algorithm to the distributed ledger of the proposed localization framework, the transaction processing performance per unit time was improved, and the delay time was reduced, thereby satisfying the real-time requirements of location technology. Ultimately, we proposed a complete positioning framework that applies improved positioning techniques and distributed ledger technology and conducted various performance evaluations for each detailed technology. We demonstrated the performance of the localization system constituting our framework through experiments in a real environment and validated how the localization performance changes when the actual received data are tampered with by malicious attacks. In addition, the distributed ledger technology using a new consensus algorithm was verified in terms of threshold dependence and network scale dependence. As a result, the proposed system reduced the speed of confirmation saturation as the threshold increased, and consistent transaction processing performance can be achieved regardless of network scale. Finally, we verified the security performance of the proposed framework by measuring the average deactivation time and confirmation rate through a race attack scenario on the distributed ledger system. The proposed distributed ledger technology had a reduced average deactivation time compared to existing DAG systems and showed constant transaction processing performance regardless of malicious transaction creation.

We have several research plans for future work. First, we plan to improve the positioning performance through machine learning or a more-advanced optimization model. Second, we will apply a smart contract and off-chain solution for the efficiency and scalability of our proposal. This will allow us to improve the positioning accuracy and strengthen the decentralization of the proposed positioning framework.

Author Contributions: Conceptualization, K.K., S.L. and T.Y.; methodology, K.K., S.L., T.Y. and H.K.; software, K.K. and S.L.; validation, S.L., T.Y. and H.K.; formal analysis, S.L., T.Y. and H.K.; investigation, K.K., S.L. and T.Y.; resources, K.K., S.L. and T.Y.; data curation, K.K., S.L. and T.Y.; writing—draft preparation, K.K., S.L., T.Y. and H.K.; writing—review and editing, K.K., S.L., T.Y. and H.K.; visualization, K.K., S.L. and T.Y.; supervision, H.K.; project administration, H.K.; funding acquisition, H.K. All authors have read and agreed to the published version of the manuscript.

Funding: This work was supported through the National Research Foundation of Korea (NRF) using the financial resources granted by the Ministry of Science and ICT (MSIT) (No. RS-2022-00143994) and this research was supported by the MSIT (Ministry of Science and ICT), Korea, under the ITRC (Information Technology Research Center) support program (IITP-2021-0-01835) supervised by the IITP (Institute of Information & Communications Technology Planning & Evaluation).

Data Availability Statement: Data is contained within the article.

Conflicts of Interest: The authors declare no conflict of interest.

Abbreviations

The following abbreviations are used in this manuscript:

VANET	Vehicular Ad hoc Network
UWB	Ultra-Wideband
ToF	Time of Flight
DAG	Directed Acyclic Graph

References

1. Lee, H.; Tak, J.; Choi, J. Wearable antenna integrated into military berets for indoor/outdoor positioning system. *IEEE Antennas Wirel. Propag. Lett.* **2017**, *16*, 1919–1922. [CrossRef]
2. El-Rabbany, A. *Introduction to GPS: The Global Positioning System*; Artech House: Norwood, MA, USA, 2002.
3. Kim Geok, T.; Zar Aung, K.; Sandar Aung, M.; Thu Soe, M.; Abdaziz, A.; Pao Liew, C.; Hossain, F.; Tso, C.P.; Yong, W.H. Review of indoor positioning: Radio wave technology. *Appl. Sci.* **2020**, *11*, 279. [CrossRef]
4. Wing, M.G.; Eklund, A.; Kellogg, L.D. Consumer-grade global positioning system (GPS) accuracy and reliability. *J. For.* **2005**, *103*, 169–173. [CrossRef]
5. Gao, J. *Development of a Precise GPS/INS/on-Board Vehicle Sensors Integrated Vehicular Positioning System*; University of Calgary: Calgary, AB, Canada, 2007.
6. Singh, S.; Agrawal, S. VANET routing protocols: Issues and challenges. In Proceedings of the 2014 Recent Advances in Engineering and Computational Sciences (RAECS), Chandigarh, India, 6–8 March 2014; pp. 1–5.
7. Biagioni, J.; Eriksson, J. Inferring road maps from global positioning system traces: Survey and comparative evaluation. *Transp. Res. Rec.* **2012**, *2291*, 61–71. [CrossRef]
8. Tsai, C.C. A localization system of a mobile robot by fusing dead-reckoning and ultrasonic measurements. *IEEE Trans. Instrum. Meas.* **1998**, *47*, 1399–1404. [CrossRef]
9. Hallberg, J.; Nilsson, M.; Synnes, K. Positioning with bluetooth. In Proceedings of the 10th International Conference on Telecommunications, ICT 2003, Tahiti, Papeete, 23 Febraury–1 March 2003; IEEE: Piscataway, NJ, USA, 2003; Volume 2, pp. 954–958.
10. Suzuki, A.; Iyota, T.; Choi, Y.; Kubota, Y.; Watanabe, K.; Yamane, A. Measurement accuracy on indoor positioning system using spread spectrum ultrasonic waves. In Proceedings of the 2009 4th International Conference on Autonomous Robots and Agents, ICARA 2009, Wellington, New Zealand, 10–12 February 2009; IEEE: Piscataway, NJ, USA, 2009; pp. 294–297.
11. Shin, B.J.; Lee, K.W.; Choi, S.H.; Kim, J.Y.; Lee, W.J.; Kim, H.S. Indoor WiFi positioning system for Android-based smartphone. In Proceedings of the 2010 International Conference on Information and Communication Technology Convergence (ICTC), Jeju, Republic of Korea, 17–19 November 2010; IEEE: Piscataway, NJ, USA, 2010; pp. 319–320.
12. Sahinoglu, Z.; Gezici, S.; Guvenc, I. *Ultra-Wideband Positioning Systems*; Cambridge University Press: Cambridge, NY, USA, 2008.
13. Alarifi, A.; Al-Salman, A.; Alsaleh, M.; Alnafessah, A.; Al-Hadhrami, S.; Al-Ammar, M.A.; Al-Khalifa, H.S. Ultra wideband indoor positioning technologies: Analysis and recent advances. *Sensors* **2016**, *16*, 707. [CrossRef]

14. Gustafsson, F.; Gunnarsson, F. Positioning using time-difference of arrival measurements. In Proceedings of the 2003 IEEE International Conference on Acoustics, Speech, and Signal Processing, Proceedings (ICASSP'03), Hong Kong, China, 6–10 April 2003; IEEE: Piscataway, NJ, USA, 2003; Volume 6, pp. 6–553.
15. Díez-González, J.; Álvarez, R.; Verde, P.; Ferrero-Guillén, R.; Perez, H. Analysis of reliable deployment of TDOA local positioning architectures. *Neurocomputing* **2022**, *484*, 149–160. [CrossRef]
16. Zhao, W.; Goudar, A.; Schoellig, A.P. Finding the right place: Sensor placement for UWB time difference of arrival localization in cluttered indoor environments. *IEEE Robot. Autom. Lett.* **2022**, *7*, 6075–6082. [CrossRef]
17. Jiang, Y.; Leung, V.C. An asymmetric double sided two-way ranging for crystal offset. In Proceedings of the 2007 International Symposium on Signals, Systems and Electronics, Montreal, QC, Canada, 30 July–2 August 2007; IEEE: Piscataway, NJ, USA, 2007; pp. 525–528.
18. Li, D.; Wang, X.; Chen, D.; Zhang, Q.; Yang, Y. A precise ultra-wideband ranging method using pre-corrected strategy and particle swarm optimization algorithm. *Measurement* **2022**, *194*, 110966. [CrossRef]
19. Lasla, N.; Al-Sahan, L.; Abdallah, M.; Younis, M. Green-PoW: An energy-efficient blockchain Proof-of-Work consensus algorithm. *Comput. Net.* **2022**, *214*, 109118. [CrossRef]
20. Pradhan, N.R.; Singh, A.P.; Verma, S.; Kavita; Wozniak, M.; Shafi, J.; Ijaz, M.F. A blockchain based lightweight peer-to-peer energy trading framework for secured high throughput micro-transactions. *Sci. Rep.* **2022**, *12*, 14523. [CrossRef] [PubMed]
21. Wang, S.; Li, H.; Chen, J.; Wang, J.; Deng, Y. DAG blockchain-based lightweight authentication and authorization scheme for IoT devices. *J. Inf. Secur. Appl.* **2022**, *66*, 103134. [CrossRef]
22. Nguyen, D.C.; Hosseinalipour, S.; Love, D.J.; Pathirana, P.N.; Brinton, C.G. Latency optimization for blockchain-empowered federated learning in multi-server edge computing. *IEEE J. Sel. Areas Commun.* **2022**, *40*, 3373–3390. [CrossRef]
23. Falsi, C.; Dardari, D.; Mucchi, L.; Win, M.Z. Time of arrival estimation for UWB localizers in realistic environments. *Eurasip J. Adv. Signal Process.* **2006**, *2006*, 032082. [CrossRef]
24. Panwar, K.; Katwe, M.; Babu, P.; Ghare, P.; Singh, K. A majorization-minimization algorithm for hybrid TOA-RSS based localization in NLOS environment. *IEEE Commun. Lett.* **2022**, *26*, 1017–1021. [CrossRef]
25. Pan, M.; Liu, P.; Liu, S.; Qi, W.; Huang, Y.; You, X.; Jia, X.; Li, X. Efficient joint DOA and TOA estimation for indoor positioning with 5G picocell base stations. *IEEE Trans. Instrum. Meas.* **2022**, *71*, 1–19. [CrossRef]
26. Peng, R.; Sichitiu, M.L. Angle of arrival localization for wireless sensor networks. In Proceedings of the 2006 3rd Annual IEEE Communications Society on Sensor and Ad Hoc Communications and Networks, Reston, VA, USA, 25–28 September 2006; IEEE: Piscataway, NJ, USA, 2006; Volume 1, pp. 374–382.
27. Florio, A.; Avitabile, G.; Coviello, G. Multiple source angle of arrival estimation through phase interferometry. *IEEE Trans. Circuits Syst. II Express Briefs* **2022**, *69*, 674–678. [CrossRef]
28. Shih, S.L.; Wen, C.K.; Jin, S.; Wong, K.K. Angle-of-Arrival Estimation with Practical Phone Antenna Configurations. *IEEE Internet Things J.* **2023**, *10*, 20006–20020. [CrossRef]
29. Dabove, P.; Di Pietra, V.; Piras, M.; Jabbar, A.A.; Kazim, S.A. Indoor positioning using Ultra-wide band (UWB) technologies: Positioning accuracies and sensors' performances. In Proceedings of the 2018 IEEE/ION Position, Location and Navigation Symposium (PLANS), Monterey, CA, USA, 23–26 April 2018; IEEE: Piscataway, NJ, USA, 2018; pp. 175–184.
30. Mahfouz, M.R.; Fathy, A.E.; Kuhn, M.J.; Wang, Y. Recent trends and advances in UWB positioning. In Proceedings of the 2009 IEEE MTT-S International Microwave Workshop on Wireless Sensing, Local Positioning, and RFID, Cavtat, Croatia, 24–25 September 2009; IEEE: Piscataway, NJ, USA, 2009; pp. 1–4.
31. Lee, S.; Yoo, S.; Lee, J.Y.; Park, S.; Kim, H. Drone positioning system using UWB sensing and out-of-band control. *IEEE Sens. J.* **2021**, *22*, 5329–5343. [CrossRef]
32. Lee, J.Y.; Yoo, S. Large error performance of UWB ranging in multipath and multiuser environments. *IEEE Trans. Microw. Theory Tech.* **2006**, *54*, 1887–1895.
33. Mazhar, F.; Khan, M.G.; Sällberg, B. Precise indoor positioning using UWB: A review of methods, algorithms and implementations. *Wirel. Pers. Commun.* **2017**, *97*, 4467–4491. [CrossRef]
34. Song, Z.; Jiang, G.; Huang, C. A survey on indoor positioning technologies. In Proceedings of the Theoretical and Mathematical Foundations of Computer Science: Second International Conference, ICTMF 2011, Singapore, 5–6 May 2011; Selected Papers; Springer: Berlin/Heidelberg, Germany, 2011; pp. 198–206.
35. Soganci, H.; Gezici, S.; Poor, H.V. Accurate positioning in ultra-wideband systems. *IEEE Wirel. Commun.* **2011**, *18*, 19–27. [CrossRef]
36. Coppens, D.; Shahid, A.; Lemey, S.; Van Herbruggen, B.; Marshall, C.; De Poorter, E. An overview of UWB standards and organizations (IEEE 802.15. 4, FiRa, Apple): Interoperability aspects and future research directions. *IEEE Access* **2022**, *10*, 70219–70241. [CrossRef]
37. Yang, Y.; Wu, L.; Yin, G.; Li, L.; Zhao, H. A survey on security and privacy issues in Internet-of-Things. *IEEE Internet Things J.* **2017**, *4*, 1250–1258. [CrossRef]
38. Gu, Y.; Lo, A.; Niemegeers, I. A survey of indoor positioning systems for wireless personal networks. *IEEE Commun. Surv. Tutor.* **2009**, *11*, 13–32. [CrossRef]

39. Benčić, F.M.; Žarko, I.P. Distributed ledger technology: Blockchain compared to directed acyclic graph. In Proceedings of the 2018 IEEE 38th International Conference on Distributed Computing Systems (ICDCS), Vienna, Austria, 2–5 July 2018; IEEE: Piscataway, NJ, USA, 2018; pp. 1569–1570.
40. Park, S.; Oh, S.; Kim, H. Performance analysis of DAG-based cryptocurrency. In Proceedings of the 2019 IEEE International Conference on Communications Workshops (ICC Workshops), Shanghai, China, 20–24 May 2019; IEEE: Piscataway, NJ, USA, 2019; pp. 1–6.
41. Zhu, Q.; Loke, S.W.; Trujillo-Rasua, R.; Jiang, F.; Xiang, Y. Applications of distributed ledger technologies to the internet of things: A survey. *Acm Comput. Surv. (Csur)* **2019**, *52*, 1–34. [CrossRef]
42. Park, S.; Lee, J.; Kim, H. Efficient computation offloading for ethereum DApps. *J. Ind. Inf. Integr.* **2023**, *31*, 100411. [CrossRef]
43. Wang, T.; Wang, Q.; Shen, Z.; Jia, Z.; Shao, Z. Understanding characteristics and system implications of dag-based blockchain in iot environments. *IEEE Internet Things J.* **2021**, *9*, 14478–14489. [CrossRef]
44. Lee, J.; Lee, B.; Jung, J.; Shim, H.; Kim, H. DQ: Two approaches to measure the degree of decentralization of blockchain. *ICT Express* **2021**, *7*, 278–282. [CrossRef]
45. Pervez, H.; Muneeb, M.; Irfan, M.U.; Haq, I.U. A comparative analysis of DAG-based blockchain architectures. In Proceedings of the 2018 12th International Conference on Open Source Systems and Technologies (ICOSST), Lahore, Pakistan, 19–21 December 2018; IEEE: Piscataway, NJ, USA, 2018; pp. 27–34.
46. Park, S.; Kim, H. DAG-based distributed ledger for low-latency smart grid network. *Energies* **2019**, *12*, 3570. [CrossRef]
47. Cui, L.; Yang, S.; Chen, Z.; Pan, Y.; Xu, M.; Xu, K. An efficient and compacted DAG-based blockchain protocol for industrial Internet of Things. *IEEE Trans. Ind. Inform.* **2019**, *16*, 4134–4145. [CrossRef]
48. Xie, J.; Zhang, K.; Lu, Y.; Zhang, Y. Resource-efficient DAG blockchain with sharding for 6G networks. *IEEE Netw.* **2021**, *36*, 189–196. [CrossRef]
49. Park, J.; Chitchyan, R.; Angelopoulou, A.; Murkin, J. A block-free distributed ledger for p2p energy trading: Case with iota? In *Proceedings 31, Proceedings of the Advanced Information Systems Engineering: 31st International Conference, CAiSE 2019, Rome, Italy, 3–7 June 2019*; Springer: Berlin/Heidelberg, Germany, 2019; pp. 111–125.
50. Fan, C.; Khazaei, H.; Chen, Y.; Musilek, P. Towards a scalable DAG-based distributed ledger for smart communities. In Proceedings of the 2019 IEEE 5th World Forum on Internet of Things (WF-IoT), Limerick, Ireland, 15–18 April 2019; IEEE: Piscataway, NJ, USA, 2019; pp. 177–182.
51. Xiang, F.; Huaimin, W.; Peichang, S.; Xue, O.; Xunhui, Z. Jointgraph: A DAG-based efficient consensus algorithm for consortium blockchains. *Softw. Pract. Exp.* **2021**, *51*, 1987–1999. [CrossRef]
52. Anceaume, E.; Guellier, A.; Ludinard, R.; Sericola, B. Sycomore: A permissionless distributed ledger that self-adapts to transactions demand. In Proceedings of the 2018 IEEE 17th International Symposium on Network Computing and Applications (NCA), Cambridge, MA, USA, 1–3 November 2018; IEEE: Piscataway, NJ, USA, 2018; pp. 1–8.
53. Zhang, Z.; Vasavada, V.; Ma, X.; Zhang, L. Dledger: An iot-friendly private distributed ledger system based on dag. *arXiv* **2019**, arXiv:1902.09031.
54. Cullen, A.; Ferraro, P.; King, C.; Shorten, R. On the resilience of DAG-based distributed ledgers in IoT applications. *IEEE Internet Things J.* **2020**, *7*, 7112–7122. [CrossRef]
55. Ferraro, P.; King, C.; Shorten, R. On the stability of unverified transactions in a DAG-based distributed ledger. *IEEE Trans. Autom. Control.* **2019**, *65*, 3772–3783. [CrossRef]
56. Sang, C.L.; Adams, M.; Hörmann, T.; Hesse, M.; Porrmann, M.; Rückert, U. An analytical study of time of flight error estimation in two-way ranging methods. In Proceedings of the 2018 International Conference on Indoor Positioning and Indoor Navigation (IPIN), Nantes, France, 24–27 September 2018; IEEE: Piscataway, NJ, USA, 2018; pp. 1–8.
57. Decawave DW1000 Manual. Available online: https://www.decawave.com/sites/default/files/resources/dw1000_user_manual_2.11.pdf (accessed on 14 October 2023).
58. Gavin, H.P. *The Levenberg–Marquardt Algorithm for Nonlinear Least Squares Curve-Fitting Problems;* Department of Civil and Environmental Engineering, Duke University: Durham, NC, USA, 2019; Volume 19.
59. Popov, S. The tangle. *White Pap.* **2018**, *1*, 30.

Disclaimer/Publisher's Note: The statements, opinions and data contained in all publications are solely those of the individual author(s) and contributor(s) and not of MDPI and/or the editor(s). MDPI and/or the editor(s) disclaim responsibility for any injury to people or property resulting from any ideas, methods, instructions or products referred to in the content.

electronics

MDPI

Article

Content Caching and Distribution Policies for Vehicular Ad-Hoc Networks (VANETs): Modeling and Simulation

Irene Kilanioti *, Nikolaos Astrinakis and Symeon Papavassiliou *

School of Electrical and Computer Engineering, National Technical University of Athens, 157 80 Athens, Greece; nikos.astrin@gmail.com

* Correspondence: eirinikoilanioti@mail.ntua.gr (I.K.); papavass@mail.ntua.gr (S.P.)

Abstract: The paper studies the application of various content distribution policies for vehicular ad hoc networks (VANETs) and compares their effectiveness under various simulation scenarios. Our implementation augments the existing Veins tool, an open source framework for vehicular network simulations based on the discrete event simulator OMNET++ and SUMO, a tool that simulates traffic on road networks. The proposed solution integrates various additional features into the pre-existing Veins realizations and expands them to include the modeling and implementation of proposed caching and content distribution policies and the measurement of respective metrics. Moreover, we integrate machine learning algorithms for distribution policies into the simulation framework in order to efficiently study distribution of content to the network nodes. These algorithms are pre-trained neural network models adapted for VANETs. Using these new functions, we can specify the simulation parameters, run a plethora of experiments and proceed to evaluate metrics and policies for content distribution.

Keywords: content caching; content distribution policies; VANETs; machine learning

1. Introduction

Wireless vehicular ad hoc networks (VANETs) are multi-hop networks of vehicles equipped with various sensors and technologies, that enable them to communicate with each other within their transmission bandwidth. The VANET topology is dynamic and constantly changes as the nodes of the network move. This results in the frequent creation and destruction of links between network nodes, and, thus, several routing algorithms used for mobile ad hoc networks are not applicable for road ad hoc networks. The path that the nodes take is, moreover, affected by the road network. Vehicles send and receive content and conduct wireless routing. Vehicle-to-Vehicle (V2V), Vehicle-to-Infrastructure (V2I) and Vehicle-to-Everything (V2X) communication technologies cover these functionalities [1].

The nodes that comprise VANETs are [2]:

- Vehicles. Vehicles can be either private or public transport vehicles.
- Servers/Origin server. Servers act as data repositories for content that exists in the network and they communicate directly and exclusively with Roadside Units (RSUs).
- Roadside Units (RSUs). Static or mobile RSUs function as an intermediary between servers and vehicles and may also contain stored network information that vehicles can access.

The paradigm shown in Figure 1 handles bandwidth-demanding content for real-time automotive applications with offloading for computation and content delivery. Current policies in V2X communications leverage Dedicated Short-Range Communications (DSRC), that employ the IEEE 802.11p standard, or alternatively Cellular V2X (C-V2X), based on the Long Term Evolution (LTE) and New Radio (NR) standards [3].

High mobility, lack of fixed infrastructure in the form of RSUs, scalability issues, intermittent communication due to the dynamically changing network, shadowing because of the existence of obstacles, signal path loss and restricted communication window among vehicles make IP-based communication inappropriate for handling multi-modal and time-varying content of high-definition maps, dynamic planning information, infotainment multimedia, etc.

The content exchanged among vehicles and between vehicles and fixed infrastructure is delay-constrained, and often critical for traffic management and the safety of driver and passengers. Practical applications of Intelligent Transportation Systems (ITS) with VANETs include also solutions for environment monitoring, infotainment and healthcare of passengers [4]. Ad hoc network study may lead to improved road safety, because efficient content delivery and in-time driver update can lead to avoidance of accidents. Collisions are avoided as routes dynamically change, leveraging timely updates about hazardous road conditions and on-going accidents. Furthermore, VANETs are used to monitor traffic congestion and suggest alternative routes, and various applications include information about the nearest gas station, automatic toll payment, etc.

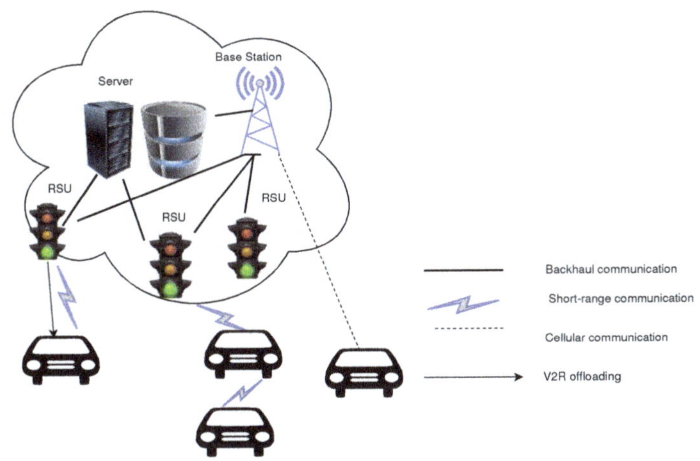

Figure 1. Diagram of an ad hoc road network.

Lack of real 5G vehicular datasets make a simulator appropriate for evaluating the performance of our scenarios for real-time automotive applications. Veins [5,6] employs IEEE 802.11p standard, which precisely captures frame timing, modulation, coding, and channel models [7]. Veins is also based on an optimized physical layer implementation that can carry out short-circuit evaluation of loss models for tasks like determining if received power surpasses a specified threshold, thus accelerating simulations by up to an order of magnitude [7]. Moreover, our implementation within Veins was not tailored to specific video codecs and packetizers. In our work, we integrate novel features into the pre-existing Veins realizations that model the proposed caching and content distribution policies. We also incorporate machine learning algorithms for distribution policies into the simulation framework, which, to the best of our knowledge, have not been integrated before, and use the respective metrics that we introduce for measurements.

Section 2 describes related research work. Section 3 models content caching and distribution policies for VANETs and describes the respective algorithms that we propose. Section 4 presents a case study of simulations. Section 5 presents and discusses the measurements associated with simulations. Section 6 concludes and mentions potential future extensions. The framework structure and instructions for installation of required programs are described in the Appendix A.

2. Related Work

Content delivery optimization refers to fixed [8] and self-organizing infrastructure [9,10]. A wide range of work addresses data collection, content caching and distribution especially for VANETs, e.g., in [11] Chaves et al. use an emulator to compare their system and the obtained results depict similar performance to Sprinkler V2V protocol and substantially better than BitTorrent. In [12], Luo et al. observe that an increase in the packet sending rate beyond a certain value results in a sharp increase in the average delay, and when the average length of vehicle buffer queues on a road segment surpasses a threshold value, they use alternate road segments to balance the network load. The decision about which content needs to be cached can be taken collectively, e.g., Kuo et al. [13] jointly optimize caching, computing and communication among vehicles and RSUs of the vehicular network to ensure quality of delivered videos.

Data homogeneity over the whole vehicular network along with increased copy costs suggest that the so-called ALWAYS caching in VANETs, where all incoming packets are cached in all nodes by default, is not the most effective strategy. Works that have proposed deviations from this strategy include the Leave Copy Everywhere (LCE) strategy [14] in V2V communication, where vehicles are treated as subscribers that consume data or nodes for caching, the PeRCeIVE approach [15], where proactive caching takes place near the consumer, and Information-Centric Network solutions that combine social collaboration mechanisms with improved media services [16]. An extensive comparison among various V2V protocols and our approach is not feasible, since the protocols were tested under different times, mobility and connectivity scenarios, and due to the presence of loading mechanism and existence of specific multimedia content codecs/ packetizers in many cases.

Lack of real 5G vehicular datasets make a simulator appropriate for evaluating the performance of our scenarios for real-time automotive applications. Veins [5,6] employs IEEE 802.11p standard, which precisely captures frame timing, modulation, coding, and channel models [7].

Implemented simulators for evaluation of content distribution over VANETs include [17–19] and are often tailored to specific video codecs and packetizers, e.g., the simulator proposed in [20] compares the quality of MPEG4 video transmissions in terms of the peak signal-to-noise ratio (PSNR), jitter, loss rate, etc.

Veins is also based on an optimized physical layer implementation that can carry out short-circuit evaluation of loss models for tasks like determining if received power surpasses a specified threshold, thus accelerating simulations by up to an order of magnitude [7]. Moreover, our implementation within Veins was not tailored to specific video codecs and packetizers.

We augmented Veins simulation framework for VANETs with functions that efficiently implement various policies for content delivery over vehicular networks, so that network resilience and efficient content management are achieved. There is no mechanism for integrating multimedia content policies into Veins. The suggested extension includes the implementation of content sharing policies and mechanisms for sending and receiving multimedia content. Metrics were implemented to detect the most central nodes. Machine learning (ML) algorithms were implemented for selecting the most efficient content distribution. Use of ML approaches is corroborated by network softwarization advances, high dimensionality of vehicular raw data, and impediments during their collection. We conducted simulations and compared the results in terms of response times for different policies and centrality metrics.

3. Modeling and Algorithms

3.1. Simple Message Handling

Our simulations include two types of messages: Content messages and simple messages, the management of which differs significantly. Initially, for each simple incoming message a MessageData structure containing the message information is created and saved.

For brevity, lists will still be referred to as message lists. There are two simple message storage lists: storedMessages and candidateMessages, the capacity of which depends on the network. Recent messages are stored in storedMessages and transferred to candidateMessages, when these messages are considered stale. The process of transferring old messages is carried out periodically. Data stored in storedMessages are sorted based on creation time (Figure 2). The Caching Policy defines which messages will be deleted, when new messages arrive and the maximum capacity is reached:

- First In First Out (FIFO) sorts the list to delete based on the receivedAt field of the stored MessageData.
- Least Recently Used (LRU). Every time we receive a message, we set that time to the last used time (field lastUsed of the MessageData structure) and if we are asked for this message again, the used time is refreshed. The list is sorted and we delete the specified number of messages with the smallest lastUsed value.
- Least Frequently Used (LFU) determines which messages will be deleted based on their usage frequency (usedFrequency field of the MessageData structure).

In case messages exist in candidateMessages and maximum capacity is reached, the size of candidateMessages is checked. If it is smaller than the selected number of messages we want to delete, we simply empty this list. Otherwise, the specified caching policy is applied. If candidateMessages has no messages and there is no more space to store, the algorithms are applied to storedMessages. So again, if the size of storedMessages is less than the number of messages to delete, we empty storedMessages list. Otherwise, we apply the selected caching policy and then, we resort storedMessages based on the timestamp field of the MessageData data stored in the list.

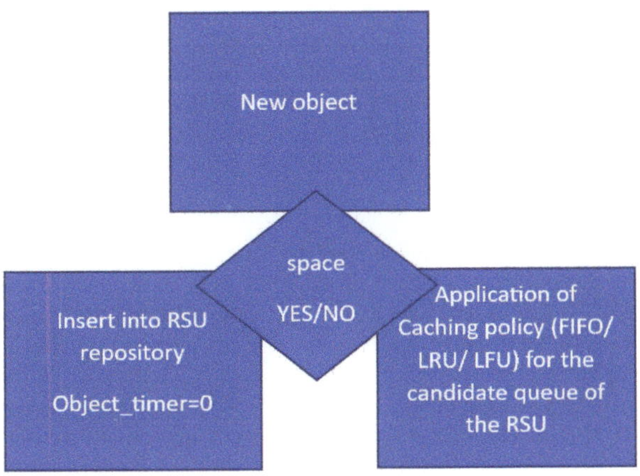

Figure 2. Simple and content message handling pipeline.

3.2. Content Message Handling

The segmentedMessages list holds all accepted and stored content messages in such a way that all segments of the same message are located one after the other, from the message with the smallest segment number to the one with the largest. Also, this list does not have a maximum capacity, so we do not need to apply a caching policy.

After inserting the message into this list, a function is called to reconstruct the original message from the individual parts present in the list. From the received content message (fields Content ID, Segments, SegmentNumber and Multimedia), we extract the number of segments the overall message consists of and whether it is multimedia content or not. We then search segmentedMessages for the first occurrence of a message with the same Content ID, Segments and Multimedia as our message. Then we keep traversing the list, and, because of its structure, all parts of the same message must be contiguous.

When we have all the segments, we create a ContentWrapper. The Segments field within the ContentWrapper has the same value as the original message. The Content field of ContentWrapper is the recomposed string we constructed. We insert this message into a dictionary. The entry key is the ContentId of the original message and the entry value is the ContentWrapper. There are two dictionaries for storing content: In multimediaData we enter multimedia data content, and in roadData all the rest data entries. Finally, we delete all the segments that are still in the segmentedContent list, since we have now reconstructed the original content.

3.3. Shortest Paths

First, we define a dictionary as the routing table of a node A with keys other nodes' addresses and values the minimum paths from node A to the node in the key. We take the path to examine and the address that exists in the last position of the path, assuming it is the address of node B. We then examine if that address exists in our routing table. If it exists, we compare the length of the path that is stored with the length of the path that exists in the path we examine, counting from the end of the path to the location where the address of node B exists. In the event that the length of the path we have stored in the routing table is shorter than the length of the path present in the message path, the function terminates. If the path we have stored in the routing table is greater than or equal to that of the message, we simply define the segment from the last position of the path to B as the minimum distance to the node B. We repeat this process until we reach the beginning of the path. For each update of the routing table we set a variable for each node with value equal to the time of the update.

Due to the nature of VANETs, the network topology changes rapidly and when a message arrives, old data in the table are emptied. However, the minimum paths between RSUs and the Origin are fixed. For this reason, there is a separate routing table for these routes. At the beginning of each simulation the Origin sends a special routing request to the RSUs and they in turn send other special routing requests. At the end of this process RSUs and Origin know all the optimal routes between them and these routes are used to quickly send messages.

3.4. Calculation of Network Metrics

VANETs can be simulated using the synthetic topology of random geometric graphs (RGGs): wireless ad hoc networks form links based on the geographical distance between nodes and in the case of VANETs this distance represents the range of the signal. In order to analyze an ad hoc network, snapshots of the network are captured at specific time points (Figure 3). These snapshots allow us to ignore the temporal variability of ad hoc networks and to apply social network analysis methods to these snapshots and investigate the VANETs, since we can construct the network graph from the snapshot, creating links according to signal range [2].

Initially, at specific time intervals, the Origin node, sends a request to the RSUs to start calculation. The RSUs forward this request to other RSUs and simultaneously send a centrality calculation request to the RSUs and vehicles in the network. They also schedule a message to be sent to themselves 10 s after receiving the request to start computation. The value of the State field of the message is Collecting. Messages with such State are used to finally compute the data we obtained from other nodes during the request and send that computation to the Origin. Additionally, the Centrality field of the launch request contains the type of centrality we want to calculate.

3.4.1. Degree Centrality

For degree centrality calculation (Algorithm 1), the requesting RSU sets the maximum number of hops of the message to 1, ensuring that the request will only reach the immediate neighbors of the node. Neighbor nodes in turn send a degree centrality response and the respective counter increases.

(**a**)

(**b**)

(**c**)

Figure 3. Topology snapshots of the VANET network at various time points. The content request for each screenshot was made at a different time, in order for us to obtain more varied measurements of the performance of each content sharing policy. The three request times with 20 RSUs were at time points: (**a**) t = 130 s, (**b**) t = 210 s and (**c**) t = 290 s.

Algorithm 1 Degree Centrality Calculation

$degree\_centrality \leftarrow 0$
RSU sends request with $maxHops = 1$
Other nodes receive request and answer
$degree \leftarrow$ number of answers received
RSU sends $degree\_centrality$ to Origin

3.4.2. Closeness Centrality

For closeness centrality calculation (Algorithm 2), the RSU that made the routing request sums the length of all paths present in its routing table and sends to the Origin the length of the routing table divided by this sum.

Algorithm 2 Closeness Centrality Calculation

$closeness\_centrality \leftarrow 0$
RSU sends request with $maxHops \geq number\_of\_nodes\_in\_the\_network$
Other nodes receive request and answer
Nodes forward the request to other nodes, updating traversed path
if traversed path is longer than path in the routing table **then**
  discard message
else
  update routing table with new shortest path
end if
$closeness\_centrality \leftarrow (sum\_of\_all\_shortest\_paths)/(size\_of\_routing\_table)$
RSU sends $closeness\_centrality$ to Origin

3.4.3. Betweenness Centrality

For betweenness centrality calculation (Algorithm 3), each node that receives a calculation request forwards the request to each node in the network and then generates a routing request. Thus, each node in the network calculates its own routing table. The receivers of a betweenness centrality calculation request call the collection. This time, however, the Dest field of the collection message is not the address of the calling node itself, but, instead, that of the node that made the request for the centrality calculation. The function of this is twofold. First, this way the node knows which RSU made the request to send it the results. Second, RSUs are nodes that can both receive betweenness calculation requests and make such requests themselves. Thus, when the collection is called, if their address is different from the dest of the collection message, the function behaves differently. The way vehicles and RSUs are collected in the event that they are not collecting for a calculation request they themselves made is described: We loop through the routing table, and for each minimal path that contains the node listed in the Dest field of the collection message, we add 1 to a counter. At the end of this process we send a message with the result to the Dest node. Whenever the Dest node receives such a response to its request, it increments the value of a counter by the value of the message and in the collection, which takes place after a few seconds, sends this counter to Origin, while simultaneously clearing its routing table.

When the Origin receives the results, the Origin notifies the RSUs about the most central RSU in the network.

Algorithm 3 Betweenness Centrality Calculation

$betweenness\_centrality \leftarrow 0$
RSU sends request with $maxHops \geq number\_of\_nodes\_in\_the\_network$
Other nodes receive request and begin calculating the shortest paths to other nodes
Nodes that receive the request also forward it to other nodes
if a node has calculated their routing table **then**
  $result \leftarrow$ number of times that requested RSU appears in a shortest path
  Send $result$ to the RSU
  $betweenness\_centrality \leftarrow betweenness\_centrality + result$
  RSU sends $betweenness\_centrality$ to Origin
end if

3.5. Machine Learning

We determine the most crucial RSUs for content dissemination introducing machine learning algorithms for VANETs. The main advantage of this method is the fact that it does not burden the network, as centrality calculation does, because there is no need to detect paths.

The algorithms detect clusters of vehicles taking as input their position on the x-axis, their position on the y-axis, their speed and direction. Each algorithm takes this data and creates 3 clusters. The algorithms used in this paper are K-Means (Algorithm 4) and Agglomerative Clustering (Algorithm 5). Agglomerative Clustering for VANETs belongs to the category of hierarchical clustering algorithms (Hierarchical Clustering). We start with n clusters, one for each point in our data set. Then, depending on the parameters set, we successively join clusters of elements with the shortest distance, gradually forming a tree from bottom to top. Its main difference from K-Means is that the distance between two points calculated by K-Means is always their Euclidean distance, whereas this is not the case in Agglomerative Clustering, where we can we choose in advance which metric will be used.

In order to apply the selected machine learning algorithm, we have to extract at certain time intervals the characteristics of the vehicles we are interested in. Therefore, a vehicle connected to TraCIMobility module of SUMO undertakes to write in the form (PosX, PosY, Speed, Direction) the coordinates, speed and direction of all vehicles located in the network (vehicles.csv file).

The machine learning algorithm writes its results to cluster_centers.csv and deletes vehicles.csv. The results written by the machine learning algorithm form a list, each node of which contains a pair of coordinates x and y corresponding to the center of each cluster created. Once the machine learning algorithm terminates, we resume the simulation. Origin periodically checks if the file cluster_centers.csv exists and if it does, it extracts the data. Then, it compares the coordinates of each cluster with the coordinates of the RSUs present in the network, and the RSUs with the shortest distance from each cluster are the most central ones. Once this calculation of the most important RSUs is done, the network is updated with the same procedure described in the centrality calculation section.

Algorithm 4 K-Means for VANETs

The algorithm reads the *vehicles.csv* file
A dataframe is created with the columns *PositionX*, *PositionY*, *Speed*, *Direction*
The data from the dataframe is fitted to the K-Means model
The model clusters the data based on the four-dimensional distance between objects
The centroids of the clusters are extracted
These centroids are inserted in the new dataframe *result*
The results are written in the new *cluster_centers.csv* file

Algorithm 5 Agglomerative Clustering for VANETs

The algorithm reads the *vehicles.csv* file
A dataframe is created with the columns *PositionX*, *PositionY*, *Speed*, *Direction*
$metric \leftarrow cosine$
$linkage \leftarrow average$
The data from the dataframe is fitted to the Agglomerate Clustering model
Each node from the original dataframe gets labeled
We calculate each centroid by finding the average coordinates of the vehicles in each cluster
The results are written in the new *cluster_centers.csv* file

4. Case Study–Simulations

4.1. Content Sharing Policies

There are two content sharing policies in our simulations:

- Pull: When a resource is requested from the RSU, the RSU checks if it has it and if it does not, it forwards the request to the Origin and copies content to its cache as well. Therefore, the object is copied to the RSU the first time it is requested.

- Push: Push policy copies content from Origin to RSUs proactively, before even the content is requested.

Case study includes simulations for various time points (Figure 3) and increasing number of RSUs. All possible combinations of content sharing policies, machine learning algorithms and centralities are taken into account for the produced simulation scenarios. For the purposes of this work, 120 simulations (240 in total) were performed for each of the 2 data sharing policies. More specifically, requests were sent at 3 different times from different vehicles, so that we obtained different topologies and, consequently, different metric results. Simulations were also performed for both Push and Pull content sharing policies and for each of the 3 centralities and 2 machine learning algorithms implemented. The caching policy FIFO was used. In our simulations, there are up to 100 vehicles that periodically enter the network and at time 83 accidents start to occur in the network. These accidents change the routes followed by the vehicles in the network, changing their topology. It is worth adding that in each message sending there is a random delay, that follows a uniform distribution and is in the field (0.01, 0.5) seconds, along with other fixed delays. Still, there is a case of rejecting a message, which is influenced by various factors that fall within the scope of this paper.

4.2. Content Message Storage

For each message segment that we obtain, we traverse the segmentedContent list. If we find a segment from the same original message, i.e., it has the same ContentId, Segments and Multimedia, we check its SegmentNumber. If we have a smaller segment number, we insert our message in that position, and return a message about successful insertion. But if we have the same segment number, the function terminates. If our message has a higher segment number, we proceed to the next position and compare again. The next message we compare may not be from the same original message, so we stop the search and place it there. If no other segment of the original message is found, we place our message at the end of segmentedMessages.

4.3. Message Acceptance

An important function, called every time a node receives a message, is a function that checks whether a message will be accepted or not. First, we check if the creator of the message we received is the sender. In this case the message is rejected directly. Then there is a check for the Recipient field, namely, if a direct recipient has been specified and its address is not the same as the address of the node that received the message, the message is ignored. We also define two variables, insertion and update with an initial value set as false. As a first step, we obtain the value of the OriginMessage field. If it is false and the routing table was last updated at least 20 s ago, the table is flushed.

Next, we check the UpdatePaths value of the message. If its value is true, we extract Route and PreviousNodes from the message. If Route is not empty and contains the address of the node that received the message, it means that the message went round, which we do not want in minimum path finding functions, so we discard the message. Next, we call the function that updates the minimal paths. It is first called for Route, if it is not empty, and if it returns false, the message is discarded. Otherwise, we make the variable update equal to true. The same procedure is followed for PreviousNodes.

We check next the capacity of simple messages. If the total number of messages in storedMessages and candidateMessages equals or exceeds the maximum capacity, the message delete function is called.

Finally, depending on whether it is a content message or not, the corresponding insertion function is called, the result of which (false or true) is inserted into the insertion variable. A message is therefore accepted provided that one of the variables update, insertion is true. This happens because two messages with different Route may have the same other elements, so they will not be accepted by the import function. But they might have a better path than what already exists, so the update variable will be true and we want the message to be accepted. Accordingly, the variable update is initially set to false, and in the event that either Route, PreviousNodes is empty, or UpdatePaths is false, it will remain so until the end of the function, but we may have a message that should be inserted into the message list.

4.4. Transmission Confirmation

For content messages, it is important to have a way for the node sending the content to know that its messages have successfully reached their destination. When a node wants to send a content message, the following procedure takes place: First, we insert into pendingAck vector the content message to be sent in the form of MessageData. We then send a self-message after 5 s, which has in the field State the value Repeating and in the fields ContentId, Segments, SegmentId and Multimedia the values of the content message we are going to send. When a node receives an auto-message with this value in the State field, it looks for pendingAck and if it finds a message there that has the same content characteristics as the auto-message, it resends it and reschedules same self-message 5 s later. Every time it resends a pendingAck message, it increments the attempts field by 1. When it exceeds the limit we have set, this message is automatically deleted.

At the same time, when a node receives a content message, it sends a confirmation message with the characteristics of this message. When the node that sent the content receives an acknowledgment message, it deletes the content message described by the acknowledgment from the pendingAck table.

5. Measurements of Simulations

The results are categorised based on the number of involved most central RSUs and the time points of the content requests. There are two broad categories of measurements, including transmission of single segment and of multiple segment multimedia content.

5.1. Single Segment Multimedia Content

We notice in Figure 4 that for a number of RSUs equal to 5, the machine learning algorithm Agglomerative Clustering has a better response for both policies. Here, for RSU Count = 10, we see that Agglomerative Clustering again outperforms both Push and Pull policies. This time, for RSU number equal to 15, K-Means performs better in Push policy, while Agglomerative Clustering performs in Pull. In the case where we have 20 RSUs, the Closeness Centrality metric has a better overall performance.

For request time point 210 (Figure 5) and 5 most central RSUs the best performance in the Push policy is given by the machine learning algorithm K-Means, while for the Pull policy Agglomerative Clustering for VANETs performs best. For 10 RSUs K-Means performs better for both policies. For 15 RSUs Agglomerative Clustering depicts a better performance in Push Policy and Betweenness Centrality in Pull. For 20 RSUs the Closeness Centrality metric performs better.

For request time point 290 (Figure 6) in the case of 5 most central RSUs for copying, we observe that K-Means for VANETs has the best performance in the case of Push policy and Agglomerative Clustering for VANETs performs best in the case of Pull policy. For 10 RSUs we notice that K-Means performs better for both content sharing policies and Degree Centrality depicts delays. For 15 RSUs we observe that Betweenness Centrality performs best in the Push policy, while Agglomerative Clustering performs best in the Pull policy. For 20 RSUs we observe that Agglomerative Clustering has the best performance for the Push policy, while Closeness Centrality performs best for the Pull policy. We notice again that Degree Centrality performs poorly when the Push policy is applied.

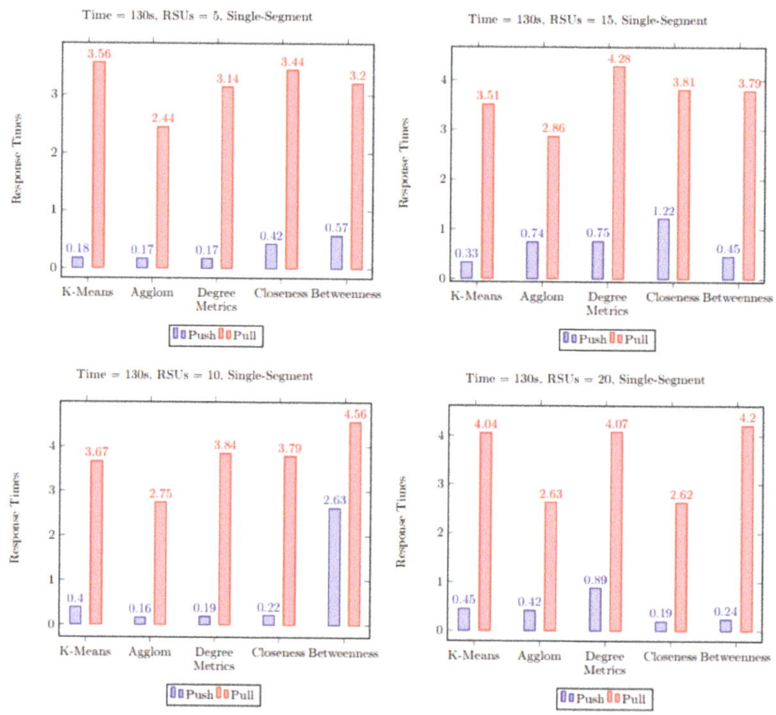

Figure 4. Response times for 5, 10, 15 and 20 most central RSUs, where copying of single segment multimedia content takes place for both content sharing policies and all ML and centrality calculation algorithms, and for request timestamp = 130.

5.2. Multiple Segment Multimedia Content

It should be noted that due to the random delays that exist when sending parts of the content, various algorithms may depict best response times for the first segment, but not for the entire message. Also, for the sake of variety of measurements, the fixed delay with which the request is sent differs from the case of the single segment multimedia content.

For time point 130 (Figure 7) and copying of content in the 5 most central RSUs we see that for the first segment, Closeness Centrality depicts best performance time for Push policy, while Betweenness Centrality performs best in the case of Pull content sharing policy. For message completion Agglomerative Clustering has better performance time for Push policy and Betweenness Centrality for Pull. For 10 RSUs we observe that for the first segment K-Means depicts the best performance time for Push policy and Closeness Centrality for Pull policy. For message completion Agglomerative Clustering performs best in terms of time for the Push policy and Closeness Centrality for Pull. In the case of 15 RSUs, for the first segment Betweenness Centrality depicts the best performance for the Push policy and K-Means for Pull. For message completion, Agglomerative Clustering shows best performance for Push policy and Betweenness Centrality for Pull policy. For 20 RSUs we observe that for the first part, the best performance time in Push policy is depicted by K-Means, while in the Pull approach by Closeness Centrality. For message completion, K-Means has the best performance for the Push policy and Closeness Centrality performs again best for Pull policy.

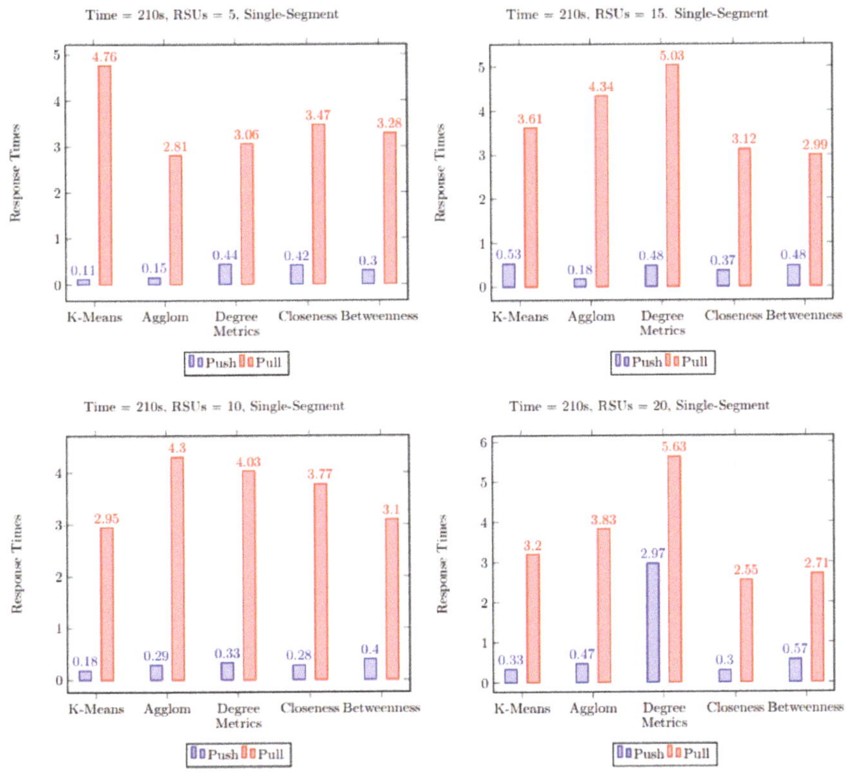

Figure 5. Response times for 5, 10, 15 and 20 most central RSUs, where copying of single segment multimedia content takes place for both content sharing policies and all ML and centrality calculation algorithms, and for request timestamp = 210.

For request time point 210 (Figure 8) and for 5 RSUs, Agglomerative Clustering has the best performance for the first segment for Push and Betweenness Centrality performs best for Pull content sharing policy. For message completion, K-Means has the best performance time for Push policy, while Betweenness Centrality performs best for Push. For 10 RSUs we observe that for the first segment Betweenness Centrality has the best performance time for both policies. For message completion Agglomerative Clustering has best performance time for Push policy and K-means performs best for Pull policy. For 15 RSUs we notice that for the first segment Betweenness Centrality depicts the best performance time for the Push policy and Closeness Centrality performs best for Pull. For message completion K-Means has the best performance time for Push policy, while Closeness Centrality has again best response time for Pull. Regarding the case of 20 RSU at time 210, for the first segment, the best performance time in the Push policy is given by Agglomerative Clustering for VANETs and by it is enDegree Centrality for Pull. To complete the message, K-Means has the best performance time for the Push policy and Degree Centrality performs best for Pull. It is worth noting, that at this point we can observe a rare case when the response times of Push policy surpass that of Pull for Degree Centrality. This can happen due to a variety of reasons, e.g., random delays, message discardings, or choosing a host RSU too far away from the node that requested the message.

For request time point 290 (Figure 9) and for 5 RSUs, the best performance time for the Push policy is depicted by Agglomerative Clustering for VANETs, while Betweenness Centrality outperforms other approaches for Pull. For message completion, K-Means has the best performance time for Push policy and Betweenness Centrality performs again best for Pull. For 10 RSUs at time 290, the best performance time in the Push policy appears in the case of Agglomerative Clustering, while Closeness Centrality gives best response time for Pull. For message completion, Agglomerative Clustering has the best performance time in the case of Push policy, while Closeness Centrality performs best again forPull. In the case of 15 RSUs, Agglomerative Clustering has the best performance time for Push policy and Betweenness Centrality for Pull. To complete the message, Betwenness Centrality has best performance time in both policies. In the last case, Betweenness Centrality has the best performance time in the Push policy scenario, while K-Means for VANETs outperforms other approaches for Pull. For message completion, K-Means has better throughput time in both policies.

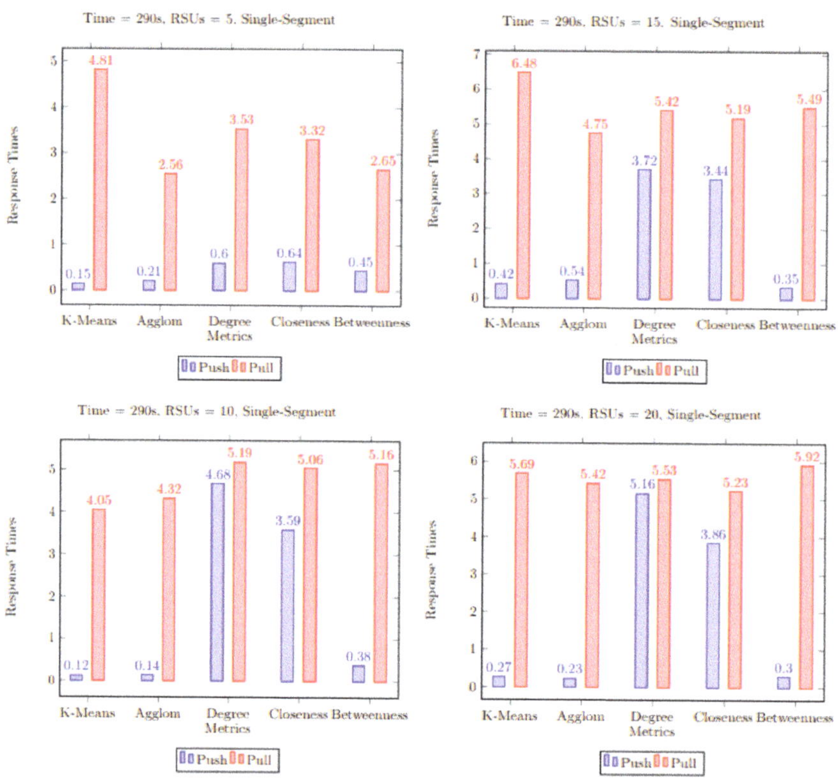

Figure 6. Response times for 5, 10, 15 and 20 most central RSUs, where copying of single segment multimedia content takes place for both content sharing policies and all ML and centrality calculation algorithms, and for request timestamp = 290.

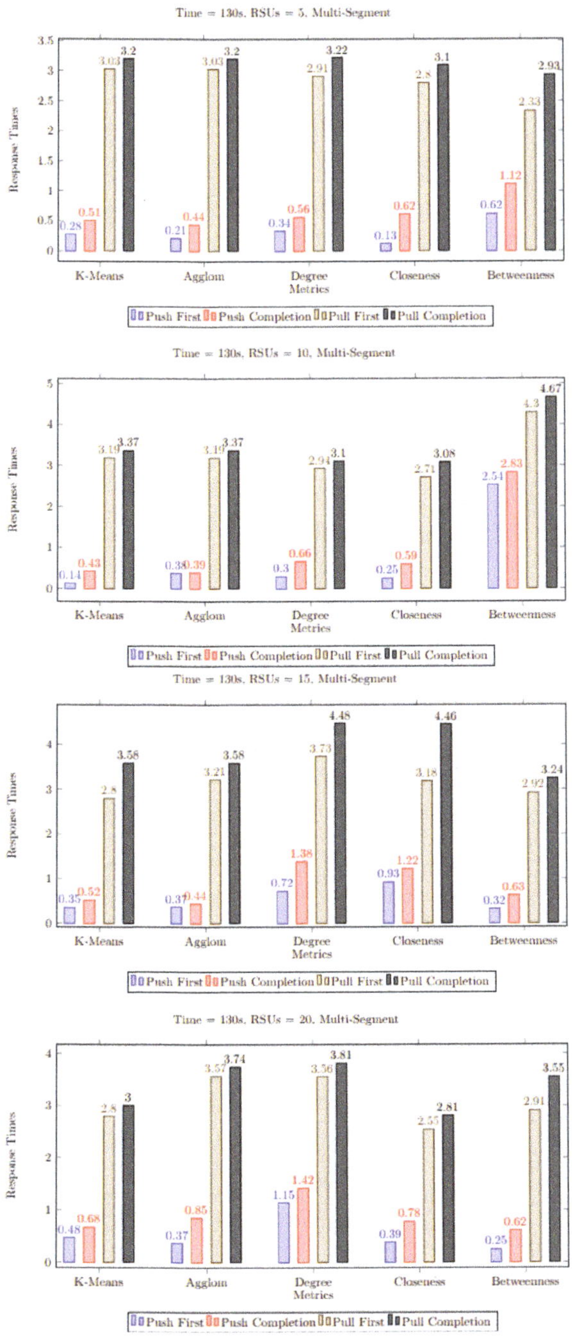

Figure 7. Response times for 5, 10, 15 and 20 most central RSUs, where copying of multiple segment multimedia content takes place for both content sharing policies and all ML and centrality calculation algorithms, and for request timestamp = 130.

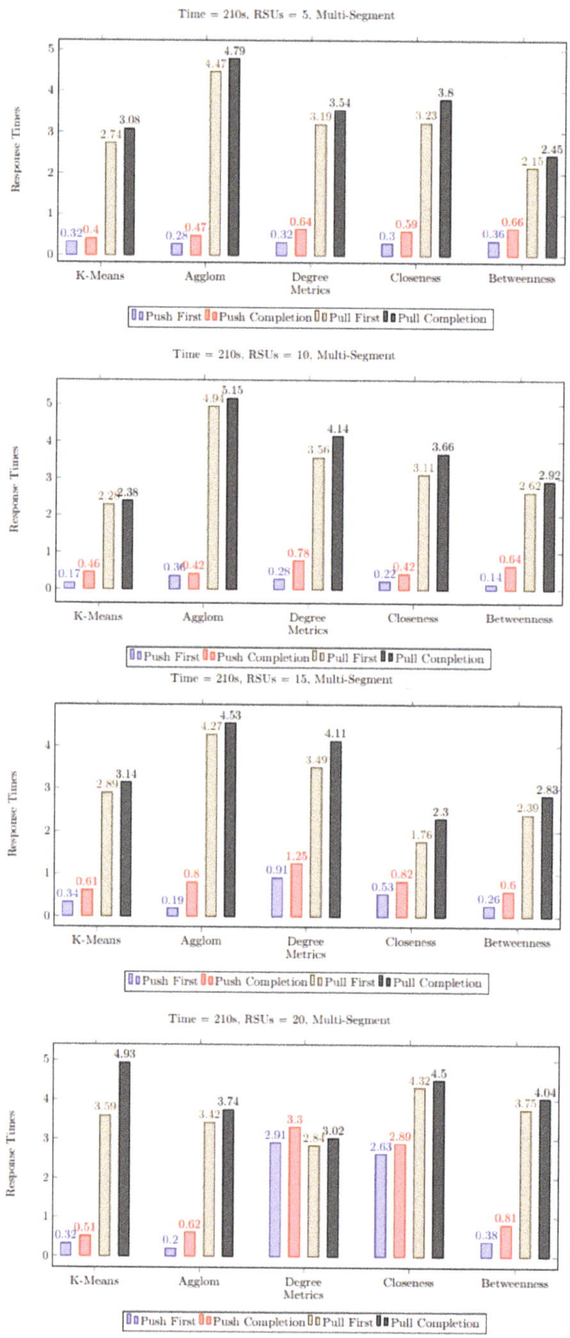

Figure 8. Response times for 5, 10, 15 and 20 most central RSUs, where copying of multiple segment multimedia content takes place for both content sharing policies and all ML and centrality calculation algorithms, and for request timestamp = 210.

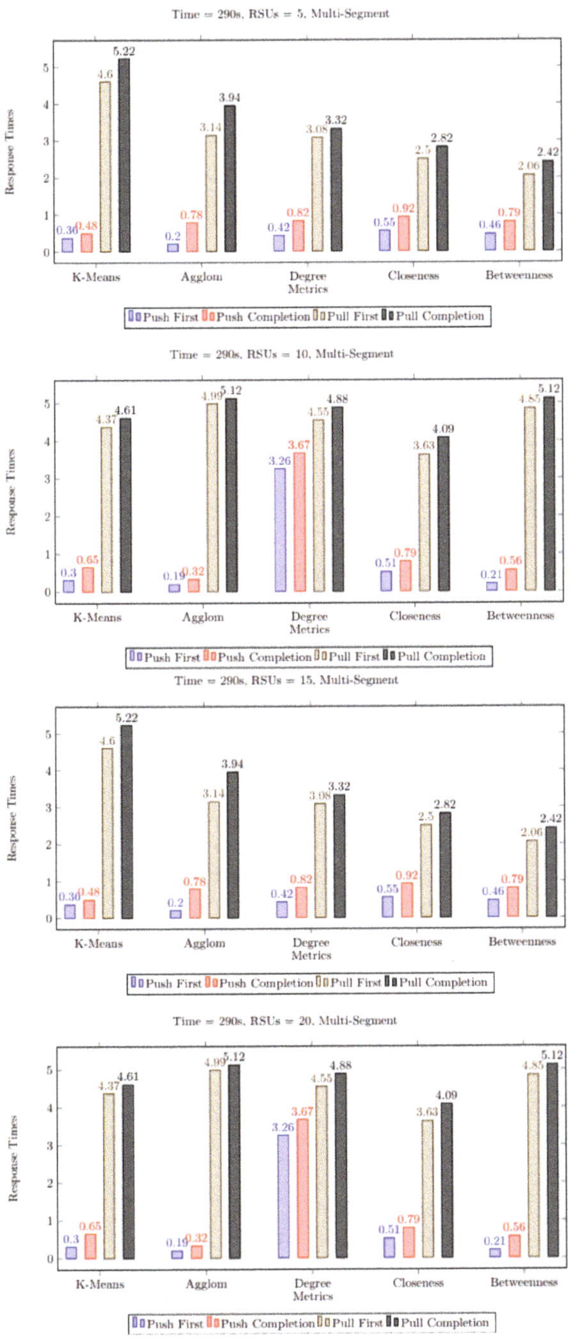

Figure 9. Response times for 5, 10, 15 and 20 most central RSUs, where copying of multiple segment multimedia content takes place for both content sharing policies and all ML and centrality calculation algorithms, and for request timestamp = 290.

5.3. Discussion of Simulation Results

The presented aforementioned measurements concern the response time, i.e., the time from the creation of the request to the receipt of the response. Simulations were based on single segment and multiple segment multimedia content requests. For multi-part data, additionally to the response time, the time when the requesting vehicle received the first part of the response is recorded.

There is no generic conclusion derived about the most feasible metric due to a number of parameters. Some of these parameters include: random delay while sending the message, the position of the vehicle requesting content and the message rejection possibility. We can say, though, that when we use the machine learning algorithms we adapted for VANETs, we obtain better response times compared to centrality measurement schemes. Moreover, the reduced times for implementation of machine learning algorithms and the reduced network congestion as there is no need to calculate shortest paths (as in closeness and betweenness centrality) make the proposed algorithms appear as a good solution for finding the most central RSUs.

We notice also that, regarding Push and Pull content delivery policies, Push is much quicker, but leads to network congestion, since content is proactively sent to the most central RSUs and may not be finally consumed.

6. Conclusions

To sum up, we focused on aforementioned measurements and their comparison. The functions of the simulator were extended, so that multimedia content could be sent and various metrics were calculated. We implemented machine learning algorithms adapted to VANETs and compared their performance.

We observe that the machine learning algorithms we adapted for VANETs perform better than centrality measurement schemes in terms of request times. Moreover, the reduced times for implementation of machine learning algorithms and the reduced network congestion make the proposed algorithms appear as a good solution for finding the most central RSUs. Concerning Push and Pull content sharing policies, Push is much quicker, but leads to network congestion, since content is proactively sent to the most central RSUs and may not be finally consumed.

The code extension of Veins allows for easy execution of a simulation scenario with the introduction of the required parameters. Future improvements include a GUI development for the efficient batch execution of simulations. Concurrently, this GUI could provide choices that modify certain simulation features (such as the stored content at the simulation start or the node that originally makes the request). Future extension includes augmentation of multimedia files with additional specific multimedia features, e.g., additionally to ID and its segments, information adapted to its specific type (video or map), fps, etc.

Author Contributions: Conceptualization, methodology, validation and formal analysis: I.K.; software and data curation, N.A.; writing—original draft, I.K.; review and editing, S.P. and I.K.; visualization, I.K. and N.A.; supervision, S.P. and I.K. All authors have read and agreed to the published version of the manuscript.

Funding: This research received no external funding.

Data Availability Statement: Code and data are available at: https://github.com/nickastrin/VANETs-Project (accessed on 25 May 2023).

Conflicts of Interest: The authors declare no conflict of interest.

Appendix A. Veins

Veins is an open-source framework for road network simulations. Veins extends OMNET++, a modular C++ based library, and Simulation of Urban MObility (SUMO), an open-source program, which allows the simulation of road network traffic with various vehicles and pedestrians and importing networks in a plethora of file formats.

In this context many simulation models are provided. The choice of models to be used in each simulation is determined by the user depending on his/ her needs [7]. To use Veins properly, the network simulator built with OMNET++ and the road traffic simulator SUMO [21] must be running. These two simulators communicate with one TCP socket and the TraCI protocol is used. In this way, OMNET++ and SUMO combination matches the movement of vehicles in the road traffic simulator with the movement of nodes in the [7] network simulation.

At this point, our implementation will be presented. It is worth noting that the code written extends the example that comes with Veins, an example that uses the map of the university of Erlangen in Nuremberg to simulate road traffic through SUMO.

The nodes of our network are vehicles (cars), static RSUs and an Origin.

Appendix A.1. Simulation Files

Starting with code written in C++, there are four main files and their corresponding headers. Note that for each of them, there is also a corresponding NED file used to declare the class in OMNET++. There follow the files:

- UnitHandler. This file is the "parent" class used by every node in the network. It contains mostly basic functions used by all child classes, such as the function that handles how to send messages. The UnitHandler class definition is a subclass from the DemoBaseApplLayer class, which is present in the example that comes with Veins.
- OriginHandler. OriginHandler is the class that defines the behavior of Origin nodes and adds some functions exclusively for them, e.g., the function that instructs the RSUs to calculate the centrality requested by the Origin.
- RsuHandler. This class, in turn, contains functions that describe how RSUs handle various scenarios. An example of this is handling requests from vehicles for some content stored in an RSU or the Origin of the simulation.
- CarHandler. Finally, CarHandler also has functions that deal with handling and creating requests. The vehicles in the network are constantly moving, so there is also a function that is called at each moment of the simulation for purposes such as accident detection.

Additionally, **KMeansML.py** and **AgglomML.py** are two files used in the simulation. They are written in Python and run a machine learning algorithm for VANETs, that writes the results of the algorithm to a .csv file. KMeansML.py runs the K-Means algorithm, while AgglmoML.py runs Agglomerate Clustering adapted for VANETs.

Appendix A.2. Message Archive

The Message.msg file is also important to describe. .msg files have a special syntax and when the code is compiled, OMNET++ constructs the corresponding files and their header in C++ (Message_m.cc and Message_m.h).

Message defines a packet type that describes the information contained in the messages sent by the network nodes during the simulation. Each node that sends a message first schedules the sending time and then the message is sent to itself (self-message). Then, depending on certain information of the message, it is determined whether it will be sent to other nodes or used for other purposes.

We will divide the information contained in a message according to the type of information and state their function. The information related to the sending details of the message includes:

- **SenderAddress**. The address of the last node that sent the message. It is used so that we can obtain an indication of the path of the message.
- **Recipient**. The address of the node that will receive the message. Any node whose address is different from the one listed in this field receives the message and ignores it.
- **Source**. The address of the node that generated the message. It is used for various purposes, mainly to know which node to respond to when a request comes to us.

- **Dest**. The address of the node to which we want the message to arrive. If a node with an address other than the one in Dest receives this message, it forwards the message to reach its destination.
- **SenderPosition**. The coordinates of the latter.

Also, the information concerning the characteristics of the message includes:

- **MaxHops**. It is the maximum number of bounces the message can make before it is deleted.
- **Hops**. It is the number of bounces the message has made so far. If it exceeds MaxHops, the message is deleted.
- **Type**. This field specifies the type of message being sent. There are 18 different message types, each for a different occasion and function.
- **State**. As mentioned at the beginning of this subsection, each message is first sent as a "self-message". State defines which operating state this message corresponds to. For example, Sending, which is the default state, sends the message to the specified nodes. In contrast, the Caching state does not send a message to other nodes, and, instead, starts the process of managing content storage. There are a total of 7 different message intents.
- **Centrality**. This field is only used for centrality calculation requests and their responses. It specifies the kind of centrality we want to calculate.

Additionally, information used to store numerical and simulation time data includes:

- **MsgInfo**. In this field we store numerical information that we want to send, e.g., centrality results.
- **AckInfo**. Information of type simtime_t, i.e., simulation time, is stored here. The field is used to send acknowledgments and ensure that received acknowledgment corresponds to content.

Another type of message fields are fields used to find an optimal path or to traverse paths that do not change as time passes (e.g., paths between RSUs).

- **Route**. A vector storing the path taken so far. Mainly used for best path finding messages. Another use is for messages between Origin and RSU. Since Origin and RSU are static, if the optimal path between them is calculated, we can use that path to send a message faster.
- **PreviousNodes**. A vector that stores the path taken by a message reply while we are traversing a path in reverse, removing nodes from Route and adding nodes to PreviousNodes.

In a message, there are also some fields which are used to change how messages should be treated:

- **OriginMessage**. This field, if set to true, indicates that this is a message between RSUs and Origin. Therefore, all vehicles that receive this message ignore it. Accordingly, if it is false then it is ignored by Origin.
- **UpdatePaths**. A message that has Route is subject to a routing check. If this variable is false, we can ignore this check.

The last category of information present in a message is the information used to convey content.

- **ContentId**. It is used to identify the content we request or send.
- **Content**. In this field we store the content we want to send.
- **Segments**. Here we define how many segments the content has. The content requester does not know how many segments the content has in the first place.
- **SegmentNumber**. In the case of sending multipart content this variable is used to define which part we are sending.

- **Multimedia**. Again for content transmission, this field specifies whether the content we request or send is multimedia content, i.e., a video or an image, or content that contains road information.

Appendix A.3. Data Structures

For the needs of this simulation framework, it was necessary to create data structures tailored to the messages received by each node.

The first of these is called **MessageData** and has similar information to that found in a message. More specifically, it includes the SenderAddress, Source, Dest, Type, ContentId, Content, Segments, SegmentNumber and Multimedia fields of a message. It also includes the creation time (timestamp) of the message, the last time the message was used (lastUsed) and the time point, when the message was received (receivedAt) by the node that wants to save it. Finally, it contains the frequency of use of the message (usedFrequency) and the retransmission attempts that have been made (attempts).

In this structure, functions have been implemented that compare two MessageData data structures depending on various parameters.

The second structure created is called **ContentWrapper**. This structure stores the data contained in a message. This means that this structure contains the Content, Segments and SegmentNumber fields of a message.

Appendix A.4. Installing and Running a Simulation

We start by installing the prerequisite programs. The installation guide followed for this purpose can be found on the Veins [22] page. The installation took place in a Windows 64-bit environment.

First, we download the SUMO version 1.8 files [23] and extract them to the location we will use for the project. We navigate to the respective folder, open the bin subfolder and check if the file sumo.exe exists.

Continuing with the installation of OMNET++, we download version 5.6.2, for the operating system used from the official page of [24]. We unzip the file again to the same folder we unzipped SUMO and open the created folder. There we run mingwenv.cmd file. Next, we type the command ./configure and then make. If a problem occurs during either of these two commands because a package is missing, we make sure that the package is installed (it will be listed along with the error message). When this process is finished, we type the command omnetpp and its IDE opens.

Continuing with the prerequisite installation, we download the Veins 5.2 files from the official [25] page and unpack them, again, into the folder where the other two programs are located. Make sure not to download the Instant Veins version. At the same time, we download the files written in paper, and place them in the Veins folder, replacing the old files with the new paper files. We open the IDE of OMNET++, if it is not already open, select File > Import > General: Existing Projects into Workspace and the folder where Veins is located.

Now that all the files are installed, we open the MinGW command line by running the file mingwenv.cmd. In this line we type (dir)/sumo-1.8.0/bin/sumo.exe-c erlangen.sumo.cfg, where (dir) is the address of the folder, in which the folder of SUMO is extracted. Then, in the graphical interface of OMNET++, we right-click on the project that we imported and click on the option Build Project. Once this process is finished, right-click again and this time select Run As > OMNET++ Simulation.

If installed correctly, the simulation window will open. There, in the first pop-up window that will appear, we press "OK". In the next window that appears, we insert a number from 1 to 24. Each number corresponds to a different simulation scenario, including all possible implemented simulations. Then, another window appears, in which the value we will put determines the ID of the multimedia information that will be requested by the RSUs during the simulation. Media content already stored in Origin has ID equal to 4 and 5. Content with ID 4 has a single part, while content with ID 5 has 3 sections. In the last window that will appear, we put the number of RSUs we want the simulation to have. There are ready-made coordinates for up to 20 RSUs, but if we want more, the coordinates of the extra RSUs will be requested in a pop-up.

Ready-made scripts define the following parameters: Request time, Metric type and Content allocation policy. With this array of parameters, we will then match the script number to the parameter values. It is noted here that for the metric ML (Machine Learning) one of the two machine learning algorithms found in the files KMeansML.py and AgglomML.py must run.

References

1. Fourati, L.; Kilanioti, I. Radio Aspects for the Internet of Vehicles (IoV) in High Mobility Environments. In Proceedings of the URSI General Assembly and Scientific Symposium, URSI GASS 2023, Sapporo, Japan, 19–26 August 2023.
2. Do, Y. *Centrality Analysis in Vehicular Ad Hoc Networks*; Technical Report; EPFL: Lausanne, Switzerland, 2008.
3. Vukadinovic, V.; Bakowski, K.; Marsch, P.; Garcia, I.D.; Xu, H.; Sybis, M.; Sroka, P.; Wesolowski, K.; Lister, D.; Thibault, I. 3GPP C-V2X and IEEE 802.11 p for Vehicle-to-Vehicle communications in highway platooning scenarios. *Ad Hoc Netw.* **2018**, *74*, 17–29. [CrossRef]
4. Kilanioti, I.; Rizzo, G.; Masini, B.M.; Bazzi, A.; Osorio, D.P.M.; Linsalata, F.; Magarini, M.; Löschenbrand, D.; Zemen, T.; Kliks, A. Intelligent Transportation Systems in the Context of 5G-Beyond and 6G Networks. In Proceedings of the IEEE Conference on Standards for Communications and Networking, CSCN 2022, Thessaloniki, Greece, 28–30 November 2022; IEEE: Piscataway, NJ, USA, 2022; pp. 82–88. [CrossRef]
5. Bronner, F.; Sommer, C. Efficient Multi-Channel Simulation of Wireless Communications. In Proceedings of the 2018 IEEE Vehicular Networking Conference (VNC), Taipei, Taiwan, 5–7 December 2018; pp. 1–8. [CrossRef]
6. Eckhoff, D.; Sommer, C. A multi-channel IEEE 1609.4 and 802.11 p EDCA model for the veins framework. In Proceedings of the 5th ACM/ICST International Conference on Simulation Tools and Techniques for Communications, Networks and Systems: 5th ACM/ICST International Workshop on OMNet++, Desenzano, Italy, 19–23 March 2012.
7. Veins Documentation. Available online: https://veins.car2x.org/documentation/ (accessed on 22 May 2023).
8. Kilanioti, I. Improving multimedia content delivery via augmentation with social information: The social prefetcher approach. *IEEE Trans. Multimed.* **2015**, *17*, 1460–1470. [CrossRef]
9. Elsayed, S.A.; Abdelhamid, S.; Hassanein, H.S. Predictive Proactive Caching in VANETs for Social Networking. *IEEE Trans. Veh. Technol.* **2022**, *71*, 5298–5313. [CrossRef]
10. Brik, B.; Lagraa, N.; Yagoubi, M.B.; Lakas, A. An efficient and robust clustered data gathering protocol (CDGP) for vehicular networks. In Proceedings of the Second ACM International Symposium on Design and Analysis of Intelligent Vehicular Networks and Applications, Paphos, Cyprus, 21–25 October 2012; pp. 69–74.
11. Chaves, R.; Senna, C.; Luis, M.; Sargento, S.; Matos, R.; Recharte, D. Content distribution in a VANET using InterPlanetary file system. *Wirel. Netw.* **2023**, *29*, 129–146. [CrossRef]
12. Luo, L.; Sheng, L.; Yu, H.; Sun, G. Intersection-based V2X routing via reinforcement learning in vehicular ad hoc networks. *IEEE Trans. Intell. Transp. Syst.* **2021**, *23*, 5446–5459. [CrossRef]
13. Kuo, T.Y.; Lee, M.C.; Lee, T.S. Quality-aware caching, computing and communication design for video delivery in vehicular networks. In Proceedings of the ICC 2022-IEEE International Conference on Communications, Seoul, Republic of Korea, 16–20 May 2022; IEEE: Piscataway, NJ, USA, 2022; pp. 261–266.
14. Tian, H.; Otsuka, Y.; Mohri, M.; Shiraishi, Y.; Morii, M. Leveraging in-network caching in vehicular network for content distribution. *Int. J. Distrib. Sens. Netw.* **2016**, *12*, 8972950. [CrossRef]
15. Grewe, D.; Wagner, M.; Frey, H. PeRCeIVE: Proactive caching in ICN-based VANETs. In Proceedings of the 2016 IEEE Vehicular Networking Conference (VNC), Columbus, OH, USA, 8–10 December 2016; IEEE: Piscataway, NJ, USA, 2016; pp. 1–8.
16. Doan Van, D.; Ai, Q. In-network caching in information-centric networks for different applications: A survey. *Cogent Eng.* **2023**, *10*, 2210000. [CrossRef]
17. Cislaghi, V.; Quadri, C.; Mancuso, V.; Marsan, M.A. Simulation of Tele-Operated Driving over 5G Using CARLA and OMNeT++. In Proceedings of the 2023 IEEE Vehicular Networking Conference (VNC), Istanbul, Türkiye, 26–28 April 2023; IEEE: Piscataway, NJ, USA, 2023; pp. 81–88.
18. Pusapati, S.; Selim, B.; Nie, Y.; Lin, H.; Peng, W. Simulation of NR-V2X in a 5G Environment using OMNeT++. In Proceedings of the 2022 IEEE Future Networks World Forum (FNWF), Montreal, QC, Canada, 10–14 October 2022; pp. 634–638. [CrossRef]
19. Garrido Abenza, P.P.; Malumbres, M.P.; Piñol, P.; López Granado, O. A simulation tool for evaluating video streaming architectures in vehicular network scenarios. *Electronics* **2020**, *9*, 1970. [CrossRef]
20. Klaue, J.; Rathke, B.; Wolisz, A. Evalvid–a framework for video transmission and quality evaluation. In Proceedings of the Computer Performance Evaluation. Modelling Techniques and Tools: 13th International Conference, TOOLS 2003, Urbana, IL, USA, 2–5 September 2003; Proceedings 13; Springer: Berlin/Heidelberg, Germany, 2003; pp. 255–272.
21. Sommer, C.; German, R.; Dressler, F. Bidirectionally Coupled Network and Road Traffic Simulation for Improved IVC Analysis. *IEEE Trans. Mob. Comput.* **2011**, *10*, 3–15. [CrossRef]

22. Veins Tutorial. Available online: https://veins.car2x.org/tutorial/ (accessed on 22 May 2023).
23. Simulation of Urban MObility Files. Available online: https://sourceforge.net/projects/sumo/files/sumo/ (accessed on 22 May 2023).
24. OMNeT++ Older Versions. Available online: https://omnetpp.org/download/old (accessed on 22 May 2023).
25. Veins Download. Available online: https://veins.car2x.org/download/ (accessed on 22 May 2023).

Disclaimer/Publisher's Note: The statements, opinions and data contained in all publications are solely those of the individual author(s) and contributor(s) and not of MDPI and/or the editor(s). MDPI and/or the editor(s) disclaim responsibility for any injury to people or property resulting from any ideas, methods, instructions or products referred to in the content.

MDPI AG
Grosspeteranlage 5
4052 Basel
Switzerland
Tel.: +41 61 683 77 34

Electronics Editorial Office
E-mail: electronics@mdpi.com
www.mdpi.com/journal/electronics

Disclaimer/Publisher's Note: The title and front matter of this reprint are at the discretion of the Guest Editor. The publisher is not responsible for their content or any associated concerns. The statements, opinions and data contained in all individual articles are solely those of the individual Editor and contributors and not of MDPI. MDPI disclaims responsibility for any injury to people or property resulting from any ideas, methods, instructions or products referred to in the content.

www.ingramcontent.com/pod-product-compliance
Lightning Source LLC
Chambersburg PA
CBHW061438210926
51850CB00008B/1541
* 9 7 8 3 7 2 5 8 7 3 8 4 5 *